AP* Achiever Advanced Placement Biology Exam Preparation Guide

to accompany

Biology
Tenth Edition
Sylvia S. Mader

Prepared by
Audra Brown Ward
Jennifer Pfannerstill

* Pre-AP, AP, and Advanced Placement program are registered trademarks of the College Entrance Examination Board, which was not involved in the production of and does not endorse these products.

 **Higher Education**

Boston   Burr Ridge, IL   Dubuque, IA   New York   San Francisco   St. Louis
Bangkok   Bogotá   Caracas   Kuala Lumpur   Lisbon   London   Madrid   Mexico City
Milan   Montreal   New Delhi   Santiago   Seoul   Singapore   Sydney   Taipei   Toronto

D1465966

The **McGraw·Hill** Companies

AP* Achiever Advanced Placement Biology Exam Preparation Guide to accompany
BIOLOGY, TENTH EDITION
SYLVIA S. MADER

Published by McGraw-Hill Higher Education, an imprint of The McGraw-Hill Companies, Inc., 1221 Avenue of the Americas,
New York, NY 10020. Copyright © 2010 by The McGraw-Hill Companies, Inc. All rights reserved.

1 2 3 4 5 6 7 8 9 0 QPD/QPD 0 9

ISBN: 978-0-07-892840-6

MHID: 0-07-892840-0

www.mhhe.com

About the Authors

Audra Brown Ward

Audra has taught AP Biology at Marist School in Atlanta, Georgia, for four years and has nine years of teaching experience in Atlanta-area private schools. She has served as an AP reader for the past six years and has led numerous workshops at College Board Teaching and Learning Conferences. Audra has also presented sessions at AP Biology Summer Institutes at Marist under the guidance of Tricia Glidewell. She received the 2007 NABT Outstanding Biology Teacher Award for the state of Georgia and is a 2009 American Physiological Society Frontiers in Physiology Fellow. Audra currently serves as chair of the Science Department at Marist, where in addition to teaching AP Biology, she also teaches Biology and Anatomy & Physiology. She is involved with several faculty committees and facilitates professional development workshops on innovative teaching strategies for new teachers. She also serves the National Association of Biology Teachers as the OBTA Director for Georgia. In her spare time, Audra enjoys reading, entertaining, and spending time with her husband and family.

Jennifer Pfannerstill

Jennifer has taught AP Biology, Anatomy & Physiology, and General Biology at Tomahawk High School in Tomahawk, Wisconsin, for the past twelve years after receiving approval to offer one of the school's first AP courses. She has participated at the College Board's Annual AP Reading as a Reader and Table Leader and is currently the Assistant to the Chief Reader. At Tomahawk High School, Jennifer participates on many committees and offers professional development in technology integration in AP Biology, SMART Board technology and the importance of writing in math and science. Jennifer also presents review sessions throughout the spring in New York City with the Rewarding Achievement, or REACH, program, focusing on improving scores on AP exams. Jennifer also is the Varsity Volleyball Coach and club director and enjoys reading (for pleasure), gardening, and playing sports of any kind with her husband and two daughters.

Contents

ABOUT THIS BOOK

The purpose of this book is to help you review for the Advanced Placement Biology exam. It assumes that AP Biology is your second course in Biology and that you are preparing to take the AP Biology exam. This book is organized to correlate with *Biology*, Tenth Edition, by Sylvia S. Mader; the charts, tables, and figures are taken from that textbook. It is not necessary, however, for you to have used that textbook in your class in order to use this book for an effective review for the exam.

There are a number of features included in this book to aid you in your review process:

- **Take Note** boxes include helpful hints about information that you must know in order to successfully prepare for the AP Biology exam. Be sure to pay close attention to these tips.

- Information that tends to be heavily tested on the exam is more extensive in the chapter summaries.

- There are multiple-choice and free-response questions at the end of each chapter to give you more practice. The answers are explained in detail to so that you can better understand those concepts with which you may struggle.

- Important vocabulary words are bolded.

- Two complete practice exams are included in the third section of this book. The exams are formatted exactly like the exam that you will take in May, so be sure to learn from your mistakes so that you can improve on any areas of weakness. The answers with complete explanations are included to help you. In addition, there is a chart that shows a correlation with the content of the exam to help you determine your areas of strength and weakness.

HOW TO USE THIS BOOK

This book should be used to supplement the textbook and materials your teacher provides. During the course of the year, your primary source of information should be your textbook in order for you to gain maximum understanding. Use this book to review and practice before the AP Biology exam.

The way that you use this book will depend largely on your strengths and weaknesses in understanding a particular topic. For those topics where you have a strong background, you may wish to skim the corresponding chapters and answer the questions at the end of the chapter. Important words are bolded and the sections are clearly labeled to help you. For other topics, you may decide to read the summary along with your textbook to reinforce the concepts you may have struggled with during the year. This book is not meant to replace your textbook, but to assist you in your review of material that you have learned throughout the school year.

CONTENT OF THE AP BIOLOGY EXAM

AP Biology is a college level course taught in high school, and is a preparation for the College Board Advanced Placement Biology Examination. This course is equivalent to a two-semester college course taken by Biology majors in the first year of college. The subject matter is more wide-ranging and in more depth than that usually covered in a high school biology course and includes: the chemical composition, structure and functions of cells, classical and molecular genetics, evolutionary biology, diversity and classification of organisms, structures and functions of plants and animals, and ecology. The concepts covered in this course fall under three main topics: Molecules and Cells, Heredity and Evolution, and Organisms and Populations. The learning expectations are centered around eight major themes:

1. Science as a Process
2. Evolution
3. Energy Transfer
4. Continuity and Change
5. Relationship to Structure and Function
6. Regulation
7. Interdependence in Nature
8. Science, Technology and Society

The AP Biology exam is organized in the same way your class is organized, as described in the Course Description provided by the College Board:

I. **Molecules and Cells (25%)**
  A. **Chemistry of Life (7%)**
    1. Water
    2. Organic molecules in organisms
    3. Free energy changes
    4. Enzymes
  B. **Cells (10%)**
    1. Prokaryotic and eukaryotic cells
    2. Membranes
    3. Subcellular organization
    4. Cell cycle and its regulation
  C. **Cellular Energetics (8%)**
    1. Coupled reactions
    2. Fermentation and cellular respiration
    3. Photosynthesis
II. **Heredity and Evolution (25%)**
  A. **Heredity (8%)**
    1. Meiosis and gametogenesis
    2. Eukaryotic chromosomes
    3. Inheritance patterns
  B. **Molecular Genetics (9%)**
    1. RNA and DNA structure and function
    2. Gene regulation
    3. Mutation
    4. Viral structure and replication
    5. Nucleic acid technology and applications
  C. **Evolutionary Biology (8%)**
    1. Early evolution of life
    2. Evidence for evolution
    3. Mechanisms of evolution
III. **Organisms and Populations (50%)**
  A. **Diversity of Organisms (8%)**
    1. Evolutionary patterns
    2. Survey of the diversity of life
    3. Phylogenetic classification
    4. Evolutionary relationships

B. **Structure and Function of Plants and Animals (32%)**
1. Reproduction, growth, and development
2. Structural, physiological, and behavioral adaptations
3. Response to the environment
C. **Ecology (10%)**
1. Population dynamics
2. Communities and ecosystems
3. Global issues

Thus, 25% of the questions on the AP Exam will focus on Molecules and Cells, 25% on Heredity and Evolution, and 50% on Organisms and Populations. Laboratory is an essential component of the exam. Laboratory questions that appear on the exam focus on skills such as experimental design, observation, proficiency in the use of laboratory equipment, gathering and interpreting data, and relating empirical information to scientific theory. Be sure to ask your teacher to provide you with materials that will help you to review the objectives, procedures, data and conclusions of the twelve required labs at a minimum, plus any additional labs that you may have performed in class.

OVERVIEW OF THE TEST

There are two sections of the AP Biology exam, which is three hours long and typically administered on the second Monday in May. The first section of the test consists of 100 multiple-choice questions. You have 80-minutes to complete this portion of the test, which comprises 60% of your overall score. The second section of the test involves a 10-minute reading period during which you will be able to read over the free response questions and begin to develop the outlines of your answers. You will then have 90 minutes to complete your responses. There will be four free-response questions, and all of them are required. This section comprises 40% of your overall score. Expect that one question will concentrate on Molecules and Cells, one on Heredity and Evolution, and two on Organisms and Populations. One or more questions may be based on one of the twelve required laboratory activities that you completed during the year.

HOW THE EXAM IS SCORED

The overall scoring of the AP Biology exam is the same as all the other AP exams:

5   Extremely well qualified
4   Well qualified
3   Qualified
2   Possibly qualified
1   No recommendation

The free-response questions are scored using a standard rubric. Each question is worth a maximum of 10 points. Students earn points based on the information that is contained in their responses. Points are lost only in the case where a student contradicts himself on a fact for which he has already earned a point. For instance, if an essay is about cellular respiration, a point is earned for stating that cellular respiration is the complete *oxidation* of glucose to form CO_2 and H_2O. If a student later writes that cellular respiration is the *reduction* of glucose, this is a contradiction and the earned point will be lost.

A complex formula is used to calculate your overall score from your raw scores in the multiple-choice and free-response sections. This formula is determined after the test is administered, and includes the consideration of historical standards as well as the performance of students on the exam in a given year. Students typically receive their overall scores in July following the May administration of the exam.

PREPARING FOR THE EXAM

Your review period before the AP Biology exam should begin no later than mid-March. Each week, you should select a particular topic to review so that by the day of the exam you have reviewed each major topic. It is a good idea to organize your review schedule so that you review those topics you feel weakest on first. There is no right or wrong way to organize your review schedule. You should feel comfortable doing what is best for you. Just be sure that your schedule is realistic, so that you are not overwhelmed with lots of cramming in the final days preceding the exam. Here is a sample schedule that you may wish to use as a reference:

Date		Topics	Chapters
March			
	Week 1	A View of Life The Cell	1–5
	Week 2	The Cell, cont'd, Genetics	6–10
	Week 3	Genetics, cont'd	11–14
	Week 4	Evolution	15–19
April			
	Week 1	Microbiology & Evolution Animal Evolution & Diversity	20–22 28–30
	Week 2	Plant Evolution and Biology	23–27
	Week 3	Comparative Animal Biology	31–36
	Week 4	Comparative Animal Biology, cont'd	37–42
May			
	Week 1	Behavior and Ecology	43–47

You do not have to wait until the weeks before the AP Biology exam to find ways to do well on the exam. Use these helpful hints to develop your own strategy for success in AP Biology throughout the school year. Some of these strategies will work for you, others will not. Find the ones that work best for you and consistently maintain them so that you can achieve maximum success not only in your class, but also on the AP Biology exam.

- **Develop a routine.** Review your notes each night instead of cramming the material one or two nights before the test. This is difficult at first, but once you have a routine, it becomes much easier to be consistent.
- **Each night, read over the sections that were discussed in class that day.** Immediate review of the material will help you remember it better. If you find you are confused, ask your teacher for clarification right away.
- **On the weekends, read over the sections that were discussed in class during the week, and go over your notes.** Fill in any information from your textbook that may help clarify information in your notes. This is particularly important if you have a heavy extracurricular load during the week.
- **Don't wait until the last minute to do your homework or assignments.** Pace yourself and do a little at a time. This will give you time to seek help from your teacher on things that you do not understand.

- **Study in groups.** The interaction with your classmates will help you remember concepts that you may forget if you always learn information by yourself. Don't depend on the group however; study on your own BEFORE reviewing with the group. It will help the study session to go more smoothly.
- **If you don't understand, ask.** Find out when your teacher is available to help you and make it a point to be consistent about seeking help as soon as you find yourself struggling. The sooner you get help, the easier it will be to understand when more difficult information is taught later in a chapter or unit.
- **Manage your time wisely.** Sometimes, it may be hard to find time to review your notes at night if you are heavily involved in extracurricular activities. Use travel time to and from school or sports events to review your notes. If you have a study hall, you can use that time to review as well.
- **Use study guides.** If your teacher provides study guides, take advantage of them. Write out the answers to the objectives so that you are sure that you understand the concepts that you need to know.
- **Attend after-school and weekend review sessions.** If your teacher offers review sessions outside of school, be sure to attend. The more you see and hear the information, the greater the chance that you will remember it.
- **If you need extra help, get a tutor.** If you decide to hire your own tutor, ask your teacher to talk to him or her so that he or she can better help you with what you need.

TIPS FOR TAKING THE EXAM

Before the Exam
- Familiarize yourself with the format and timing of the exam.
- Take as many practice tests as you can. Ask your teacher if he or she has any practice exams to share with you and ask him or her to assist you with finding practice exams that you can do on your own.
- Try to finish most of your review a few days before the exam so that you have time to ask your teacher for clarification on those concepts with which you may have trouble.
- Be sure to get at least eight hours of sleep the night before the test, and eat a nutritious breakfast that morning.
- Dress comfortably. It is a good idea to dress in layers in case the testing environment is at a temperature that is uncomfortable for you.

What to Bring:
- Several No. 2 pencils, sharpened, and an eraser
- Several ballpoint pens with navy blue or black ink (Try to avoid erasable ink that can easily smear.)
- A watch that does not make noise
- Your school code
- Your Social Security number
- A photo identification

What Not to Bring:
- Calculators or computers
- A watch with an alarm that beeps
- Cameras, cell phones, pagers, or PDAs
- Portable listening devices such as radios, iPods, MP3 players, and CD players

- Books or notes
- Scratch paper
- Clothing with subject-related information

During the Exam
Multiple Choice Questions
Sixty percent (60%) of the AP Biology Exam grade is based on 100 multiple-choice questions. Please take heed to the following recommendations for answering these questions:

- Manage your time wisely. Answering 100 questions in 80 minutes equates to 48 seconds per question. Some questions will require less time, while others will require more. Pace yourself.
- Read each question carefully. Watch for questions that read "…all of the following EXCEPT…" so that you don't get confused. If a narrative and/or figure is associated with a question, be sure to study it carefully before you select your answer.
- Plan to read through the test three times. The first time, answer all of the questions about which you are absolutely sure of the answer. Skip the questions that you are not sure about. The second time, answer the questions you were not sure about the first time, as long as you do not have to spend too much time on any one question. If you spend more than one minute thinking about a question, move on. You don't want to miss questions later in the test that you can answer quickly by spending too much time on questions at the beginning. The third time through the test, try to answer the questions that you were not sure about the first two times.
- The decision to guess on a question is an individual one. Some teachers advise their students to guess if they can narrow the choices down to two, while others advise against it. There is a ¼ point penalty for guessing, so incorrect answers result in a loss of 1.25 points, while blank answers result in a loss of 1 point. Make your decision to guess according to what you feel most comfortable doing.

Free Response Questions
Forty percent (40%) of the AP Biology Exam grade is based on four (4) required free-response (essay) questions. Please take heed to the following recommendations for writing answers to the free-response questions:

- **Read the question carefully**. Be sure to take note of the instructional words (describe, explain, compare, design, etc.) of the question, which are always bolded. Frame your answer around the instruction in order to earn as many points as possible. When the instruction is to "compare," be sure to discuss differences as well as similarities. If you are asked to "describe the effect," write about both positive and negative effects in order to earn maximum points.
- **Organize your thoughts** during the reading period. Spend approximately two minutes writing out ideas for each question. Organize your answer in the same way as the question is asked. If it has three distinct parts, then organize your answer in three distinct parts, and label each part accordingly.
- **Attempt to answer all four questions.** Write as much as you know, even if you feel you know very little. You may earn one or two points, whereas if you make no attempt to answer a question, you will not earn any.
- **Answer the question about which you know the most first.** You are not obligated to answer questions in any specific order. During the reading period, rank the questions from 1–4, with #1 being the question you know the most about and #4 being the one you are least comfortable with. Answer the questions in the order in which you feel most knowledgeable and comfortable.
- **Restating the question earns no points.** Dive right into the answer to save time. An introductory or summary paragraph is unnecessary and usually earns no points.

- **Answer each question on a separate page** to make it easy for the reader to find each of your answers. For your convenience, each question is printed on a separate page in the answer booklet so that you can write your answer immediately following the question. Answering more than one question per page can make finding all of your answers a challenge for the reader. Use the entire answer booklet if necessary.
- **Pace yourself.** Look at the clock at the beginning of the free-response section. Break the time into four 22-minute periods. Stop writing your answer to a particular question when your time is up. You can always come back to a question if you finish another one before your allotted time has expired.
- **Use dark blue or black ink.** Avoid pencil and felt-tip pen. Avoid excessive scratch-outs if possible.
- **If the question breaks naturally into parts, leave several lines between the answers** to insert more information later as it comes to you. You may even prefer to answer each part on a separate page; just mark your paper plainly so that it is obvious that there is more on the next page.
- **Develop your ideas as completely as possible, but avoid repeating the same information.** Elaboration points are often given for well-written answers, so it is advantageous to include as much information as you can without being redundant.
- **Show your work** if you are asked to do a calculation. Be sure to include units in your final answer if applicable.
- **Label the axes with units** if you are asked to draw a graph, and be sure to include a scale and title.
- **Write legibly.** If necessary, skip lines to make your essay easier to read.

CHAPTER 1
A VIEW OF LIFE

1.1   How to Define Life

Living things are called **organisms**. Organisms are often hard to define because they are so diverse; however, they share many common characteristics:

1. *Living things are organized.*
2. *Organisms obtain and use energy.*
3. *Organisms respond to the environment.*
4. *Organisms reproduce and develop.*
5. *Organisms adapt.*

1.2   Evolution, the Unifying Concept of Biology

Life on Earth is diverse, but the theory of evolution unifies life and describes how all living organisms evolved from a common ancestor. **Taxonomy** is the study of identifying and classifying organisms into groups based on the characteristics an organism shares with others of its same kind.

From the least inclusive to the most inclusive category, or **taxa**, each **species** belongs to a **genus, family, order, class, phylum, kingdom,** and **domain**. As the taxa become more inclusive, the number of organisms in each increases according to the number of characteristics they share.

The three domains of life are **Archaea, Bacteria,** and **Eukarya**. The first two domains contain prokaryotic organisms that are structurally simple but metabolically complex. They lack membrane-bound organelles, such as a nucleus. Archaea and bacteria are different in the organization of their DNA and composition of their cell walls. As well, archaea live in more extreme environments than bacteria.

Organisms in domain Eukarya have membrane-bound organelles, and include the protists, fungi, plants, and animals. Protists range from unicellular to multicellular organisms and include the protozoans and most algae. Among the fungi are the familiar molds and mushrooms. Plants are well known as the multicellular photosynthesizers of the world, while animals are multicellular and ingest their food. An evolutionary tree shows how the domains are related by way of common ancestors. Neither archaea nor bacteria have been classified into kingdoms at this time.

> **Take Note:** *It is important for the AP Biology exam that you are able to compare (i.e., describe similarities and differences among) the organisms in different kingdoms and/or domains. As you learn the characteristics and physiological processes of various organisms, be sure to make a chart or table for yourself that clearly shows these relationships.*

Charles Darwin defined evolution as "descent with modification." According to scientific evidence, the mechanism that best explains how evolution occurs is natural selection. Natural selection describes the process by which living organisms are descended from a common ancestor. Mutations occur within a population, creating new traits. Some of these new traits are adaptive advantages for organisms; that is, the new traits help organisms to survive in a changing environment. As organisms adapt to their environments, species change over time and may create new species from existing ones.

> **Take Note:** *Evolution is an overarching theme in your study of AP Biology that will be stressed on the AP Biology exam. As you proceed with your study, be sure to look for characteristics (physical adaptations, behavioral adaptations, and physiological processes) that are adaptive advantages for organisms and relate them to the survival of the species.*

1.3   How the Biosphere is Organized

The **biosphere** is the zone of air, land, and water where life exists. A **population** consists of all members of one species in a particular area. A **community** consists of all of the local interacting populations. An **ecosystem** includes all aspects of a living community and the physical environment (soil, atmosphere, etc.).

There are many interactions that occur within an ecosystem. Organisms interact with each other as well as with their environment. Ecosystems are characterized by chemical cycling and energy flow. Ecosystems stay in existence because of a constant input of solar energy and the ability of photosynthetic organisms to absorb it and convert it to chemical energy. The energy is passed through the ecosystem as the heterotrophs in a **food chain** consume the autotrophs and/or other heterotrophs, and convert it to metabolic energy by cellular respiration. Unused energy is released back into the ecosystem as heat. Interactions between various food chains make up a **food web**.

The human population modifies existing ecosystems for its own purposes. This can be dangerous not only for the modifieded ecosystem but also for humans, as we depend on healthy working ecosystems for food, medicines, and raw materials. Two biologically diverse ecosystems, **rain forests** and **coral reefs**, for instance, are severely threatened by the human population.

Biodiversity is the total number of species, their variable genes, and their ecosystems. There are many more species existing on Earth than have been identified. **Extinction** is the death of a species or larger group; perhaps 400 species become extinct every day due to human activities. The continued existence of all species and the preservation of biodiversity are both dependent on the preservation of ecosystems and the biosphere.

1.4   The Process of Science

When studying the natural world, scientists use the scientific process. The scientific method is the process used by scientists to gather information and reach conclusions about the natural world. The process includes observation, forming a hypothesis, performing an experiment, analyzing the results, and making conclusions. Conclusions may lead to new hypotheses which can be further tested to learn even more information about the world around us.

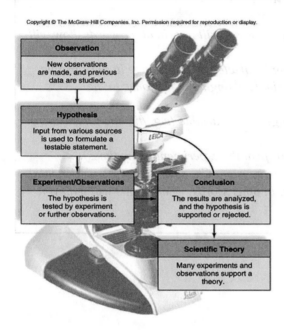

Flow diagram for the scientific method

Observations, along with previous data, are used to formulate a hypothesis. Making **observations** requires using the senses to perceive an object of study or interest. A **hypothesis** is a *testable* explanation of your observations. New observations and/or **experiments** are carried out in order to test the hypothesis. A good experimental design includes an experimental variable and a control group. The experimental and observational results are analyzed, and the scientist comes to a **conclusion** as to whether the results support the hypothesis or do not support the hypothesis. If a hypothesis is supported, a scientist may ask additional questions to further his or her understanding. If a hypothesis is rejected, the experiment is repeated to ensure it was done correctly. If the hypothesis is rejected after repeating an experiment, it is revised and retested.

Several conclusions in a particular area may allow scientists to arrive at a **theory**, such as the cell theory, the gene theory, or the theory of evolution. The theory of evolution is a unifying concept of biology.

> **Take Note:** *Experimental design is a very important aspect of the AP Biology exam. Be sure to consider the following criteria when answering a question that asks you to design an experiment:*
>
> - *Clearly state a hypothesis using the "If..., then..." format.*
> - *Set up a control group and indicate its purpose.*
> - *Use a large sample size (dozens of organisms versus one or a few).*
> - *Identify the dependent and independent variables.*
> - *Indicate which variables you will hold constant.*
> - *Indicate how you will manipulate the variable you will test. In other words, how will you conduct your experiment?*
> - *Describe how the function you are testing is affected by the change in the variable. In other words, how is the dependent variable changing in response to the independent variable?*
> - *Describe how you will measure the change in the variable (e.g., titration, measuring temperature every five minutes, etc.).*
> - *Describe how you will verify your data (repeat the experiment many times).*
> - *Indicate how you will analyze your data (e.g., run a Chi-square test, graph the data).*
> - *Relate possible results to the hypothesis. In other words, what results will lead you to confirm OR reject your hypothesis?*

Multiple Choice Questions

1. Which of the following terms includes the others?
   A. ecosystem
   B. population
   C. community
   D. biosphere
   E. organism

2. Organisms from which of the following categories can survive in harsh environments devoid of oxygen?
   A. Kingdom Plantae
   B. Domain Eukarya
   C. Domain Bacteria
   D. Kingdom Fungi
   E. Domain Archaea

3. A multicellular photosynthetic organism would be classified as a
   A. plant.
   B. animal.
   C. fungus.
   D. prokaryote.
   E. bacteria.

4. Which of the following statements is NOT evidence that all living things have a common ancestor?
   A. All life forms are composed of cells.
   B. Similar metabolic processes occur in all life forms.
   C. All life forms contain genes.
   D. Amino acids are the genetic material of all life forms.
   E. Life forms respond to their environment and maintain homeostasis.

5. Over time, one group of squirrels becomes separated from another due to erosion and the subsequent formation of a river the squirrels are unable to cross. Over time, the squirrels are unable to interbreed with one another. The inability of the squirrels to interbreed causes the squirrels to be classified as different
   A. populations.
   B. ecosystems.
   C. communities.
   D. organisms.
   E. species.

6. Organisms in which of the following groups can be unicellular, colonial, or multicellular?
   A. Domain Archaea
   B. Kingdom Fungi
   C. Kingdom Plantae
   D. Kingdom Protista
   E. Kingdom Animalia

7. Organisms of the same Family would NOT necessarily be classified in the same
   A. kingdom.
   B. phylum.
   C. class.
   D. order.
   E. genus.

8. Interacting populations in a particular area comprise a
   A. habitat.
   B. ecosystem.
   C. biosphere.
   D. community.
   E. species.

A student set up an experiment using *Drosophila* fruit flies. The student wished to determine the LC-50 dose of the exposure of caffeine to fruit flies. The lethal concentration is reached when 50% (LC-50) or more of the flies die when exposed to a toxin. The student mixed various concentrations of caffeine with the fruit fly media and then added 50 eggs to each vial. Results of the study are shown below.

Concentration of caffeine (micrograms of caffeine/10 grams of media)	Number of eggs added	Number of flies after 7 days in the vial
.001	50	45
.01	50	46
.1	50	35
1	50	6
5	50	2

9. What would be a major flaw in this experiment?
   A. lack of a measureable value
   B. too many different concentrations of caffeine
   C. lack of a control
   D. too low of an initial concentration of caffeine
   E. too many trials

10. One way to improve this experiment would be to include
    A. a second toxin with the caffeine.
    B. additional trials at each level of exposure.
    C. additional eggs in each vial.
    D. higher concentrations of caffeine.
    E. eggs of another insect to compare results.

Free Response Question

A farmer wishes to determine the minimum amount of fertilizer he needs to grow corn. You have been hired as an agricultural expert to conduct an experiment for the farmer.

    A. **Design** an experiment to determine the minimum amount of fertilizer needed (in pounds per acre) to grow corn. For your experiment, be sure to **identify** your independent and dependent variables and **discuss** a clear hypothesis.

    B. A second farmer who grows soy beans comes to you after performing the following experiment. Three separate fields were planted with 3 different varieties of soy beans. Field one was given 2 pounds of fertilizer per acre, field two was given 4 pounds of fertilizer per acre, and field three was given 6 pounds of fertilizer per acre.

        1. **Identify** the independent and dependent variables in this experiment.
        2. **Identify** at least three ways that the experiment could be improved.

Annotated Answer Key (MC)

1. **D**; While an ecosystem contains both abiotic and biotic factors, the biosphere contains all ecosystems of the world and is the most inclusive.

2. **E**; The Domain system of classification is based upon molecular differences between organisms. The Archaea include methanogens, extreme halophiles, and thermophiles. It is hypothesized that their membranes may be more stable than organisms in the other domains are more able to survive in various harsh environments.

3. **A**; While some bacteria are capable of photosynthesizing, they are not multicellular. Animals and fungi are not capable of photosynthesis.

4. **D**; Nucleic acids are the basis of the genetic code for all organisms.

5. **E**; Organisms from different communities, populations, and ecosystems are capable of interbreeding. The definition of *species* implies the ability breed with one another.

6. **D**; All organisms in Archaea are unicellular. All fungi, plants, and animals are multicellular. Organisms classified in Kingdom Protista can be unicellular, colonial, or multicellular.

7. **E**; Kingdom is the most inclusive group in the list. Therefore, if the organisms are in the same family they must be in the same kingdom, phylum, class, and order. Organisms can be in the same family, but not in the same genus. For example, dogs and wolves are in the same family (Canidae) but not the same genus.

8. **D**; Organisms of the same population must be of the same species. A habitat is not biotic and an ecosystem implies an interaction between abiotic and biotic factors, as does the biosphere.

9. **C**; The control serves as a comparison for the experimental value. In this case, a control would have been a vial with no caffeine added.

10. **B**; Science is repeatable. Without verification of repeated trials, science is meaningless.

Answer to FRQ

Part A (MAX 7 POINTS)
- Independent variable identification: amount of fertilizer
- Dependent variable identification: productivity of corn, with units, pounds of corn per acre
- Hypothesis: If more fertilizer is added, then the growth of corn will be maximized.
- <u>At least one</u> constant specified: type of corn, same soil, same amount of water on each field, same type of fertilizer on each field, equal sized patches of fields
- Repeated trials, at least two patches of field with each treatment
- Identification of a control: a field with no fertilizer added, or with the amount the farmer used last year
- Statistical treatment: repeated trials are averaged
- Range of independent variable (i.e., 1 pound per acre, 2 pounds per acre, 3 pounds per acre, 4 pounds per acre)

Part B (MAX 5 POINTS)
1. INTERNAL MAX 2
   - Independent variable: amount of fertilizer
   - Dependent variable: growth or yield of soy beans
2. INTERNAL MAX 3
   - Control: plot with no fertilizer added
   - Constants identified
   - Repeated trials: multiple fields with each amount of fertilizer
   - Specify the hypothesis to be tested

CHAPTER 2
THE CHEMISTRY OF LIFE

Take Note: *While the AP Biology exam does not test directly on this material, it is important to have a solid understanding of the concepts in this chapter in order to properly understand the biochemistry that is to follow. It is important that you have a working knowledge of the elements that are important for life, the role of isotopes in biology, and the different types of bonds that form between elements. The properties of water are extremely important, as well as the behavior of acids and bases.*

2.1   Chemical Elements

Matter is defined as anything that has mass and takes up space. Both living and nonliving matter is composed of **elements**, the simplest forms of matter that cannot be broken down to simpler substances with different chemical or physical properties. Six of the elements that occur in nature—carbon, hydrogen, nitrogen, oxygen, phosphorus, and sulfur (CHNOPS)—are important for life and make up 98% of the body weight of organisms.

Elements consist of tiny particles called **atoms**, the smallest unit of an element that displays the properties of the element. Atoms contain specific numbers of **protons, neutrons,** and **electrons**. Protons and neutrons are in the nucleus of an atom; electrons move around the nucleus. **Protons** are positively charged particles; neutrons have no charge. Both have 1 *atomic mass unit* (amu) of weight. **Electrons** are negatively charged particles located in orbitals outside the nucleus. Protons and neutrons in the nucleus determine the mass number of an atom. The atomic number indicates the number of protons and the number of electrons in electrically neutral atoms.

Isotopes are atoms of a single element that differ in their numbers of neutrons. A carbon atom with 8 rather than 6 neutrons is unstable; it releases energy and subatomic particles and is thus a **radioactive isotope**. Radioactive isotopes have many uses, including serving as tracers in biological experiments and medical procedures. An example is radioactive iodine that is used to detect tumors in the thyroid gland. High levels of radiation can destroy cells and cause cancer; careful use of radiation can sterilize products and kill cancer cells.

2.2   Compounds and Molecules

Ions form when atoms lose or gain one or more electrons. An ionic bond is an attraction between oppositely charged ions. It is formed when electrons are transferred from one atom to another atom. For example, sodium loses an electron, forming a positive charge, and chlorine gains an electron to give it a negative charge. The ionic bond that forms between them results in the formation of sodium chloride, NaCl.

Covalent bonds form when atoms share one or more pairs of electrons. There are single, double, and triple covalent bonds. In polar covalent bonds, the sharing of electrons is not equal. If the molecule is polar, the more electronegative atom carries a slightly negative charge and the other atom carries a slightly positive charge. In **nonpolar covalent bonds**, sharing of electrons is equal (i.e., the electrons are not attracted to either atom to a greater degree).

> **Take Note:** *The properties of water are important in all of Biology. Be sure you fully understand the properties of water so that you can apply the knowledge gained in this section to other topics as you proceed through your study of Biology.*

2.3   Chemistry of Water

All living things are 70–90% water. The polarity of water molecules allows hydrogen bonding to occur between them. In a water molecule, the sharing of electrons by oxygen and hydrogen is not equal; the oxygen atom is more electronegative, attracting the electrons closer to it, and thus assuming a partial negative charge. As a result, the hydrogens develop a partial positive charge.

A hydrogen bond is a weak attraction between a slightly positive hydrogen atom and a slightly negative oxygen or nitrogen atom within the same or a different molecule. Many hydrogen bonds taken together are relatively strong and help maintain the structure and function of cellular molecules such as proteins and DNA. Water's polarity and hydrogen bonding account for its unique properties. These properties, described below, allow living things to exist and carry on cellular activities. Because of hydrogen bonding, water is liquid between 0° C and 100° C, which is essential for the existence of life.

1. *High heat capacity.* The temperature of liquid water rises and falls more slowly than that of most other liquids because the hydrogen bonds between water molecules hold more heat. This protects organisms from rapid temperature changes and helps them maintain homeostatic temperature.

2. *High heat of vaporization.* Hydrogen bonds between water molecules require a relatively large amount of heat to break. This property moderates Earth's surface temperature; permitting living systems to exist. When animals sweat, evaporation of the sweat removes body heat, thus cooling the animal.

3. *Water is a solvent.* Water dissolves a great number of substances (e.g., salts, large polar molecules). Ionized or polar molecules that are attracted to water are **hydrophilic** ("water loving"); whereas non-ionized and nonpolar molecules that cannot attract water are **hydrophobic** ("water fearing"). A **solution** contains dissolved substances called **solutes.**

4. *Cohesion and adhesion.* **Cohesion** allows water to flow freely without molecules separating. **Adhesion** is ability to adhere to polar surfaces. These characteristics allow water to rise up a tree from the roots to the leaves through small tubes. Adhesion of water to walls of vessels prevents water columns from breaking apart. Cohesion allows evaporation from leaves to pull water columns from roots.

5. *Water has a high surface tension.* Water is relatively difficult to break through at its surface. This property permits a rock to be skipped across a pond surface and supports insects walking on the surface of water.

6. *Ice is less dense than water.* Below 4° C, hydrogen bonding becomes more rigid but more open, causing expansion. Because ice is less dense, it floats; therefore, bodies of water freeze from the top down. If ice was heavier than water, ice would sink and bodies of water would freeze solid. This property allows ice to act as an insulator on bodies of water, thereby protecting aquatic organisms during the winter.

> **Take Note:** *A solid understanding of acids and bases is important for your understanding of the biochemistry presented in the next few chapters. Take time now to be sure you understand these concepts, so that you can properly understand the information that is to come.*

2.4   Acids and Bases

A small fraction of water molecules dissociate to produce an equal number of hydrogen ions and hydroxide ions. The **pH scale** indicates acidity and basicity (alkalinity) of a solution. **pH** is the measurement of free hydrogen ions, expressed as a negative logarithm of the H^+ concentration (-log $[H^+]$). Solutions with equal numbers of H^+ and OH^- are neutral. In acidic solutions, there are more hydrogen ions than hydroxide ions; these solutions have a pH less than 7. In basic solutions, there are more hydroxide ions than hydrogen ions; these solutions have a pH greater than 7. Cells are sensitive to pH changes.

Biological systems often contain buffers that help keep the pH within a normal range. **Buffers** keep pH steady and within normal limits in living organisms. Buffers stabilize pH of a solution by taking up excess hydrogen (H^+) or hydroxide (OH^-) ions, keeping the pH of the blood in a narrow range so that the body can maintain homeostasis. For example, carbonic acid helps keep blood pH within normal limits: $H_2CO_3 \rightarrow H^+ + HCO_3-$.

Multiple Choice Questions

1. Which of the following statements is NOT true regarding the properties of water?
    A. Hydrogen bonds require a good deal of energy to break.
    B. Water cools more slowly than other liquids.
    C. The hydrogen bonds between water hold more heat energy.
    D. Water reaches its boiling point more slowly than substances with a lower specific heat.
    E. Water's surface tension arises due to hydrogen bonds between water molecules.

2. In a molecule of DNA, the bonds holding the side chains must be very strong, whereas the bonds holding the rungs of the ladder between the nucleic acids must break during DNA replication. What types of bonds are present between adenine and thymine and cytosine and guanine in a DNA molecule?
    A. nonpolar covalent
    B. ionic
    C. polar covalent
    D. hydrogen
    E. hydrophobic interactions

3. Van der Waals interactions and hydrogen bonds and ionic bonds are weak interactions. Therefore,
    A. their cumulative effects reinforce molecular structure, making them biologically significant.
    B. they are not found in organic molecules.
    C. they occur between molecules that are a long distance apart.
    D. they occur when two atoms that are equally electronegative are attracted to one another.
    E. they occur when cations and anions attract each other and transfer electrons.

4. Ionic bonds involve the
    A. equal sharing of electrons.
    B. transfer of electrons.
    C. uneven sharing of electrons.
    D. asymmetric distribution in molecules.
    E. hydrogen atom bonding of a hydrogen atom to one electronegative atom.

5. In the following reaction

    $C_6H_{12}O_6 + 6O_2 \rightarrow 6CO_2 + 6H_2O$

    glucose is
    A. reduced.
    B. an anion.
    C. oxidized.
    D. a cation.
    E. electronegative.

6. One of the most common biological fluids which may be outside of the neutral pH range of 6–8 is
    A. saliva.
    B. gastric juice.
    C. mucus.
    D. semen.
    E. urine.

7. A buffer works by
    A. accepting hydrogen ions from the solution when they are in excess.
    B. irreversibly combining with hydrogen ions to neutralize a base.
    C. dissociating to yield OH⁻ ions to neutralize an acid.
    D. donating hydrogen ions when they are in excess.
    E. balancing the reaction between CO_2 and H_2O.

8. Evaporative cooling occurs because
    A. molecules with the greatest kinetic energy are the most likely to leave as a gas.
    B. water's low heat of vaporization converts water to a gas.
    C. when water is heated, the kinetic energy of molecules decreases and the liquid evaporates more rapidly.
    D. molecules do not move fast enough to overcome attractions to other molecules.
    E. hydrogen bonds form as water evaporates absorbing heat energy.

9. As water travels through xylem tissue
    A. hydrogen bonding attracts water to the walls of the cells to counteract gravity.
    B. it forces water out of the root hairs as it travels from source to sink.
    C. it is actively transported by the use of ATP.
    D. it moves through aquaporins from one xylem cell to the next.
    E. hydrogen bonds are not able to form because of the speed of the water molecules moving up the xylem.

10. It is improbable that the hydrogen atoms in two water molecules would be attracted to one another and the oxygen atoms would repel one another because
    A. hydrogen atoms have a partial negative charge and will be attracted to one another.
    B. the slightly positive hydrogen of one molecule is attracted to the slightly negative oxygen of another molecule.
    C. the opposite ends of water molecules have similar charges.
    D. hydrogen is more electronegative than oxygen, and the electrons spend more time closer to the oxygen atom.
    E. its two hydrogen atoms are joined to the oxygen atom by single covalent bonds.

Free Response Question

Water is a very unique molecule and is very important to biological systems. As an example, water striders, spiders, and other arthropods are able to walk across water.

A. **Discuss** three properties of water that make this possible.

B. **Explain** why humans are not able to walk across water.

C. **Propose** a hypothesis for the structure of arthropod legs to facilitate this phenomenon.

Annotated Answer Key (MC)

1. **C**; Hydrogen bonds do not hold heat energy, rather they require heat energy to be broken. They are strong enough to maintain the integrity of the DNA molecule.

2. **D**; Hydrogen bonds are readily broken and reformed, an ideal combination for the location between nucleic acids of the DNA strand.

3. **A**; Van der Waals forces are found in many organic molecules. These interactions usually occur between molecules that are a great distance apart and are not the same as covalent or ionic bonds, as represented by choices D and E. When viewed as a whole these interactions are very important biologically, especially in protein structure.

4. **B**; Ionic bonds involve attractions between ions of opposite charge. The charges of these atoms or compounds are acquired due to the loss or gain of one or more electrons. This transfer of electrons is a relatively weak bond as compared to the sharing of electrons in a covalent bond.

5. **B**; During aerobic respiration, a molecule of glucose is oxidized to produce ~38 molecules of ATP. When a molecule is oxidized, it loses electrons. Molecules that are reduced actually gain electrons.

6. **B**; Saliva, semen, urine, and mucus are all found within a relatively neutral pH range of 6–8. The pH scale ranges from 0–14, with 7 being neutral. Gastric juice has a very acidic pH, usually around 2, due to the presence of HCl in the stomach for chemical digestion.

7. **A**; Buffers function in pairs. When the weak acid dissolves in water it forms a weak base. The two work together to counter any small shift in pH to maintain a suitable environment for many biological processes to occur.

8. **A**; Water has a very high heat of vaporization. When water is heated, the kinetic energy of molecules actually increases and those water molecules are the most likely to leave. Hydrogen bonds actually break as water evaporates.

9. **A**; As water travels from the roots to the leaves, the force of gravity works against this upward movement. Adhesion (the attraction between water and the walls of the xylem) and cohesion (the attraction of water molecules to other water molecules in the xylem) create a string of water molecules that moves upward through the xylem by capillary action. This is a passive process requiring no energy or transfer of water molecules between cells.

10. **B**; A common misconception when describing hydrogen bonding is that it is a hydrogen atom bonded to a hydrogen atom. This is an unlikely probability due to the fact that the slightly positive hydrogen atom of one molecule will be attracted to slightly negative atoms of other molecules, NOT another positive hydrogen atom.

Answer to FRQ

Part A (MAX 8 POINTS)
One point identification, one point definition.

- Polarity: unequal distribution of electrons
- Cohesion: attraction of water molecules to water molecules
- Surface tension: property of water arising from cohesion of water molecules
- Hydrogen bonding: bond between hydrogen atom and an electronegative atom

Part B (MAX 2 POINTS)
- Discussion of weight (i.e., too heavy)
- Discussion of surface area (i.e., feet too small – low SA:V ratio)
- Other plausible explanations

Part C (MAX 4 POINTS)
- High surface area to volume ratio
- Webbing
- Fringes on toes
- Pressure of step (i.e., smacking)
- Circular motion (like the oar of a boat, creation of drag)
- Height of motion (creation of air pocket under the feet)
- Speed of movement
- Shallow movement

CHAPTER 3
THE CHEMISTRY OF ORGANIC MOLECULES

3.1   Organic Molecules

The chemistry of **carbon** accounts for the diversity of organic molecules found in living things. Carbon has six electrons, four of which are valence electrons. Thus, carbon forms covalent bonds and can share electrons with as many as four other atoms, usually H, O, N, S, and P. Carbon can form a combination of single, double, or triple bonds.

Carbon can also bond with itself and hydrogen to form both chains and rings called hydrocarbons. Because the covalent bond between carbon and hydrogen is nonpolar, these carbon skeletons are hydrophobic. **Functional groups** can be added to carbon skeletons to make them more hydrophilic. Differences in the carbon skeleton and attached functional groups cause organic molecules to have different chemical properties. The chemical properties of a molecule determine how it interacts with other molecules and the role the molecule plays in the cell. Some functional groups are hydrophobic and others are hydrophilic. Take a moment to review six functional groups that are important for life.

Functional Groups			
Group	**Structure**	**Compound**	**Significance**
Hydroxyl	R—OH	Alcohol as in ethanol	Polar, forms hydrogen bond Present in sugars, some amino acids
Carbonyl	R—C$\diagup^{O}_{\diagdown H}$	Aldehyde as in formaldehyde	Polar Present in sugars
	R—C(=O)—R	Ketone as in acetone	Polar Present in sugars
Carboxyl (acidic)	R—C$\diagup^{O}_{\diagdown OH}$	Carboxylic acid as in acetic acid	Polar, acidic; Present in fatty acids, amino acids
Amino	R—N$\diagup^{H}_{\diagdown H}$	Amine as in tryptophan	Polar, basic, forms hydrogen bonds Present in amino acids
Sulfhydryl	R—SH	Thiol as in ethanethiol	Forms disulfide bonds Present in some amino acids
Phosphate	R—O—P(=O)(OH)—OH	Organic phosphate as in phosphorylated molecules	Polar, acidic; Present in nucleotides, phospholipids

R = remainder of molecule

Functional groups

There are four classes of **biomolecules** in cells: carbohydrates, lipids, proteins, and nucleic acids. Polysaccharides, the largest of the carbohydrates, are polymers of simple sugars called monosaccharides. The polypeptides of proteins are polymers of amino acids, and nucleic acids are polymers of nucleotides. Polymers are formed by the joining together of monomers. Dehydration reactions, in which a molecule of water is removed, connect monomers together.

In order to break up a polymer, a molecule of water is added in a hydrolysis reaction.

> **Take Note:** *You should understand the characteristics of carbon that make it so versatile and allow it to be the basis of organic molecules. You should also be able to identify the functional groups that can be attached to hydrocarbons.*

3.2 Carbohydrates

Carbohydrates are made of carbon, hydrogen, and oxygen in a 1:2:1 ratio. Monosaccharides, disaccharides, and polysaccharides are all carbohydrates. **Monosaccharides** are simple sugars with three to six carbon atoms in the carbon skeleton. Examples of monosaccharides are glucose, fructose, and galactose, which each have six carbons. Glucose ($C_6H_{12}O_6$) is the immediate energy source of cells. Examples of five-carbon sugars include ribose and deoxyribose, found in the nucleic acids RNA and DNA, respectively. Disaccharides are formed from the dehydration reaction between two monosaccharides, as in sucrose or lactose.

Polysaccharides such as starch, glycogen, and cellulose are polymers of glucose. Starch in plants and glycogen in animals are energy storage compounds. Starch is found in the stems and roots of plants, and glycogen is found in the liver of animals. The structural polysaccharides are cellulose in plant cell walls, and chitin in arthropods and the cell walls of fungi. Chitin's monomer is glucose with an amino group attached. The structural polysaccharide of bacteria is peptidoglycan.

> **Take Note:** *You should be able to identify the structures of glucose (and the polymers it forms), ribose, deoxyribose, and sucrose.*

3.3 Lipids

Lipids include a wide variety of compounds that are insoluble in water. Like carbohydrates, fats are also made of carbon, hydrogen, and oxygen, though not in the same fixed ratio. Fats and oils, also called **triglycerides**, allow long-term energy storage and are formed from the dehydration reaction between one **glycerol** and three **fatty acids**.

Both glycerol and fatty acids have polar groups, but fats and oils are nonpolar, and this accounts for their insolubility in water. Fats tend to contain **saturated** fatty acids, and oils tend to contain **unsaturated** fatty acids. Saturated fatty acids do not have carbon–carbon double bonds, but unsaturated fatty acids do have double bonds in their hydrocarbon chain. The double bond causes a kink in the molecule that accounts for the liquid nature of oils.

Phospholipids are similar to fats, except that one of the fatty acids attached to the glycerol is replaced by a phosphate group. In the presence of water, phospholipids form a bilayer because the head of each molecule is hydrophilic and the tails are hydrophobic. The hydrophobic fatty acid tails will orient themselves away from the water, while the phosphate heads interact with the water. The cell membrane and the membranes of the organelles are all phospholipid bilayers, thus phospholipids are crucial for organisms.

Steroids have the same four-ring structure as cholesterol, but each differs by the groups attached to these rings. Cholesterol is found in the cell membrane, where it plays a role in stabilizing the structure of the membrane. It is also the basis for the steroid hormones, estrogen and testosterone.

Waxes are composed of a fatty acid with a long hydrocarbon chain bonded to an alcohol, also with a long hydrocarbon chain. As with carbohydrates and fats, waxes are formed by dehydration reactions. Waxes are important for water retention in plants. In animals, wax found in the ear prevents foreign substances from entering the body. Bees use wax to make honeycombs.

> **Take Note:** *You should be able to identify the structures of triglycerides, phospholipids, and steroids.*

3.4   Proteins

Proteins carry out many diverse functions in cells and organisms, including support, metabolism, transport, defense, regulation, and motion. Proteins are polymers of **amino acids**. A polypeptide is a long chain of amino acids joined by **peptide bonds**. An amino acid consists of a hydrogen, a carboxyl group, an amino group, and a variable *R* group that are all attached to a central carbon atom. There are 20 different amino acids in cells, and they differ only by their *R* groups. The presence or absence of polarity is an important aspect of the *R* groups, because they determine the structure of proteins.

A polypeptide can have up to four levels of structure. The **primary structure** is the sequence of the amino acids. **Secondary structure** contains α-helices and β-pleated sheets in place by hydrogen bonding between amino acids along the backbone of the polypeptide chain. The **tertiary structure** is the final folding of the polypeptide, which is held in place by interactions between *R* groups. Proteins that contain more than one polypeptide chain have a **quaternary structure**. The structure of a protein is important to its function. Both high temperatures and a change in pH can cause proteins to **denature** and lose their shape, causing a loss of function that can be detrimental to an organism.

> **Take Note:** *You should be able to identify the structure of an amino acid. You should also be able to explain the different levels of protein structure and relate the importance of structure to function.*

3.5   Nucleic Acids

The nucleic acids **DNA** and **RNA** are polymers of **nucleotides**. Variety is possible because the nucleotides can be in any order. Each nucleotide has three components: a phosphate group, a 5-carbon sugar, and a nitrogen-containing base. DNA contains a deoxyribose sugar, while RNA contains a ribose sugar.

DNA is the genetic material that stores information for its own replication and for the order in which amino acids are to be sequenced in proteins. DNA, with the help of mRNA, specifies protein synthesis. DNA is a double-stranded helix in which A pairs with T and C pairs with G through hydrogen bonding. RNA—containing phosphate, the sugar ribose, and the bases A, U, C, and G—is single stranded.

ATP (adenosine triphosphate) is a nucleotide that has three phosphate groups attached. ATP has unstable phosphate bonds and is the energy currency of cells. Hydrolysis of ATP to ADP + P_i releases energy, which is used by the cell to make a product or do any other type of metabolic work.

Multiple Choice Questions

1. Which of the following atoms has four electrons available for covalent bonding and can form rings, long chains, and double bonds—making it the building block of the most versatile complex biological molecules?
    A.   carbon
    B.   silicon
    C.   hydrogen
    D.   oxygen
    E.   nitrogen

2. Hydrocarbons do NOT vary in
    A.   length.
    B.   ring structure.
    C.   branching.
    D.   bonding patterns.
    E.   composition.

3. Structural isomers
    A.   may differ in their spatial arrangements.
    B.   are mirror images of each other.
    C.   may differ in the location of their double bonds.
    D.   create left-handed and right-handed versions of the same molecule.
    E.   create one active and one inactive molecule.

4. The polysaccharide stored in plastids enabling plants to store glucose is
    A.   cellulose.
    B.   glycogen.
    C.   sucrose.
    D.   starch.
    E.   chitin.

5. The following molecule

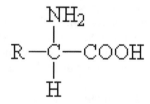

    A.   consists of a carbon atom joined to an oxygen atom by a double bond and is frequently found in carbohydrates.
    B.   can transfer energy between organic molecules.
    C.   consists of an amino group and a carboxyl group and is found in polypeptides.
    D.   is the building block of lipids.
    E.   is an alcohol and dissolves organic compounds such as sugars.

6. A dehydration reaction in the synthesis of a fat
    A.   creates glycosidic linkages between the glycerol and the fatty acids.
    B.   removes one molecule of water for each fatty acid joined to the glycerol.
    C.   bonds water to the glycerol molecule to form a triacylglycerol.
    D.   causes the glycerol molecules to separate from water and exclude the fatty acids.
    E.   requires water to form peptide bonds.

7. A change in which of the following structural components will consequently the function of each amino acid?
   A. amino group
   B. α carbon
   C. carboxyl group
   D. functional group
   E. hydrocarbon

8. The level of protein structure that involves interactions between the side chains of the amino acids rather than the interactions between molecules within the backbone is
   A. primary.
   B. helical.
   C. secondary.
   D. tertiary.
   E. quaternary.

9. Which of the following statements is true regarding chaperonins?
   A. They specify the correct structure of a polypeptide.
   B. They are carbohydrates that bond with proteins to maintain proper shape.
   C. They are organic molecules that protect against disease.
   D. They transport oxygen from the lungs to other parts of the body.
   E. They are protein molecules that assist in the proper folding of other proteins.

10. Ribonucleic acid (RNA) and deoxyribonucleic acid (DNA) differ structurally in that
    A. RNA is composed of a pentose sugar and DNA is composed of a triose sugar.
    B. DNA contains uracil.
    C. RNA bonds purines to purines and DNA bonds pyrimidines to pyrimidines.
    D. DNA contains a sugar which lacks an oxygen atom and RNA does not.
    E. DNA is single stranded and RNA is double stranded.

Free Response Question

Diabetes is a disorder in which uncontrollable blood glucose levels become extremely dangerous. After eating, blood glucose levels can become very high, resulting in hyperglycemia and tissue damage. One way to monitor the level of blood glucose is to determine the amount of glucose that becomes bonded to hemoglobin, a protein on red blood cells. This is often more accurate than routine blood sugar tests as it is not as affected by short term blood glucose level changes due to exercise, eating or medications.

   A. **Identify** and **describe** the levels of protein structure.

   B. **Explain** how the structure of hemoglobin affects its function.

   C. **Explain** how glucose may become attached to the hemoglobin and **hypothesize** as to what conditions may affect this test.

Annotated Answer Key (MC)

1. **A**; Carbon is the most versatile element in terms of its ability to form many different structures. It has four valence electrons, as does silicon. Hydrogen, oxygen, and nitrogen all have differing numbers of valence electrons. Silicon, however, is not capable of creating long chains of molecules due to its size.

2. **E**; Hydrocarbons contain hydrogen atoms and carbon atoms. While they may vary in their length, structure, branching, and bonding patterns, they must consist entirely of hydrogen and carbon.

3. **C**; Organic molecules are very specialized in the fact that they may contain the same numbers of specific atoms, but their arrangement defines their function. Enantiomers are mirror images of each other and create left-handed and right-handed versions of the same molecule. Often this creates one active and one inactive molecule. In a structural isomer, the location of the bonds may change, whereas in a geometric isomer there are the same covalent partnerships.

4. **C**; Cellulose, glycogen, starch, and chitin are all polysaccharides. Sucrose is a disaccharide. However, these polysaccharides differ in their functions. Cellulose and chitin are *structural* polysaccharides of plants and insects, respectively. Glycogen is the primary *storage* polysaccharide in animals, whereas starch is the primary *storage* polysaccharide in plants and is found in plastids.

5. **C**; This molecule contains an amino group (NH_2), a carboxyl group (COOH), a functional group (R), and a basic amino acid, the building block of polypeptides. Amino acids differ only in their R groups. It is not an alcohol and is not frequently found in lipids or carbohydrates.

6. **B**; Dehydration reactions *remove* water from molecules in order to create bonds between components. In carbohydrates, the bonds that hold the monomers of glucose together are called glycosidic linkages, but not in lipids. Water is removed, not added to the molecules and forms ester linkages, not peptide bonds.

7. **D**; Amino acids are composed of a carboxyl group, an amino group, a central carbon, and a functional group. This functional group, or side chain, can differ in their polarity and their pH properties, greatly influencing their biological function.

8. **D**; Primary structure involves the amino acid sequence of the protein structure. The bonds that hold these amino acids together are peptide bonds. In secondary structure, highly regular structures such as α helices or β pleated sheets result due to the hydrogen bonds within the backbone. However, in tertiary structure, interactions between atoms of the functional groups as well as the backbone occur (including but not limited to disulfide bridges, hydrophobic interactions, and ionic bonds).

9. **E**; Chaperonins are protein molecules that assist in the proper folding. Hemoglobin is the protein that transports oxygen from the lungs to other parts of the body, and antibodies protect against disease. Often, protein molecules end in the suffix *–in* as a hint in this question.

10. **D**; RNA and DNA both bond purines (adenine and guanine; double ring structures) to pyrimidines (cytosine and thymine; single ring structures). RNA contains uracil and bonds it to adenine in place of thymine. DNA is a double-stranded molecule, whereas RNA is a single-stranded molecule. Deoxyribose is a sugar in DNA derived from ribose, the sugar in RNA, but it lacks an oxygen atom, thus the prefix *deoxy-*.

Answer to FRQ

PART A (MAX 8 POINTS)
One point identification, one point definition.

- Primary: peptide bonds
- Secondary: hydrogen bonds
- Additional point for identification of beta pleated sheets and alpha helix structures
- Tertiary: interactions between R or functional groups
- Additional point for identification of Van der Waals, hydrophobic interactions, disulfide bridges, ionic interactions, etc.
- Quaternary: interaction between polypeptide chains

PART B (MAX 2 Function POINTS and MAX 2 Structure POINTS)

- Function: bind to oxygen in the bloodstream
- Function: carry it to outlying tissues
- Structure: can bond to between 1 and 4 molecules OR composed of four subunits
- Structure: porphyrin ring (Fe/N) for oxygen binding

PART C (MAX 4 POINTS)

- Irreversible covalent bond
- Blood transfusion
- Anemia
- Sickle cell anemia
- Pregnancy
- Kidney disease

CHAPTER 4
CELL STRUCTURE AND FUNCTION

4.1   Cellular Level of Organization

The **cell** is the smallest unit of life. In the 1800s, several scientists made important discoveries about cells. The work of those scientists is summarized as cell theory:

1. All organisms are composed of cells.
2. Cells are the basic units of structure and function in organisms.
3. Cells come only from pre-existing cells because cells are self-reproducing.

Cells have to be small in order to function efficiently. If a cell is too large, the **surface area-to-volume ratio** will be too small for nutrients and wastes to enter and leave cells. Small cells have large surface area-to-volume ratios for effective exchange between the cell and its environment. When cells become too large, they divide in order to restore the surface area-to-volume ratio that is needed in a metabolizing cell.

> **Take Note:** *You will soon be studying the processes that occur in the cell, therefore it is important that you understand the structure and function of each part of the cell before moving on to the next chapters.*

4.2   Prokaryotic Cells

Prokaryotic cells lack membrane-bound organelles such as a nucleus, and are found in a variety of environments: air, water, soil, and within or on other organisms. Structurally, they are less complex than eukaryotic cells, though they are capable of very complicated metabolism. Prokaryotic cells are divided into two domains: Bacteria and Archaea. Bacteria live in all environments and have various shapes. Archaea also have various shapes, but live in extreme aquatic environments where other organisms cannot because the environments are too salty, too hot, or too acidic.

Bacteria are surrounded by an envelope that consists of a cell membrane, a cell wall, and a **glycocalyx**. Some bacteria have a cell wall made of **peptidoglycan** that maintains the shape of the cell. The glycocalyx, which is sometimes called a capsule, is a layer of polysaccharides that surrounds the cell wall. It prevents dehydration and allows a bacterium to elude the immune system of a host that it is invading.

The cytoplasm is the fluid part of the bacteria that contains water and both inorganic and organic molecules, such as enzymes. The DNA is clustered in a region of the cell called the **nucleoid** region. Some bacteria may contain an additional chromosome of circular DNA called a **plasmid**. Surrounding the nucleoid region are thousands of ribosomes that are made of RNA and proteins. Cyanobacteria contain thylakoids, which convert light energy to chemical energy by the process of photosynthesis.

Bacteria have a variety of appendages: flagella, fimbriae, and pili. **Flagella** are appendages that are responsible for the motility of bacteria, **fimbriae** are involved in attachment to surfaces and **pili** are used for the exchange of genetic material between two bacteria.

> **Take Note:** *It is important for the AP Biology exam that you are able to describe several differences between prokaryotic cells and eukaryotic cells.*

4.3   Introducing Eukaryotic Cells

Eukaryotic cells differ from prokaryotic cells in that they have membrane-bound organelles such as a nucleus. Like prokaryotic cells, they are surrounded by a plasma membrane that consists of a phospholipid bilayer with embedded proteins that controls which substances enter and leave cells. Eukaryotic cells are thought to have evolved from prokaryotic cells by endosymbiosis.

Eukaryotic cells have compartments called **organelles** that carry out cellular functions. Eukaryotic cells derive three benefits from having organelles. First, each organelle has a specific structure and function, which allows cells to become specialized by having the organelles that are specific for the function of the cells. Furthermore, specialized cells can form tissues, which make up organs that carry out specific jobs within an organism. The shape of the cell is maintained by the **cytoskeleton**, a network of proteins that also assists in movement of the cell as well as movement within the cell.

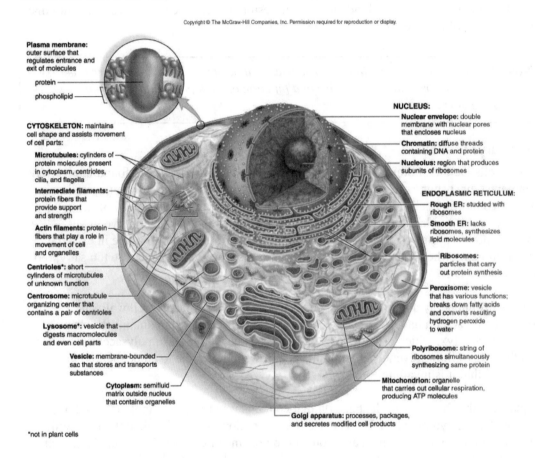

Animal cell anatomy

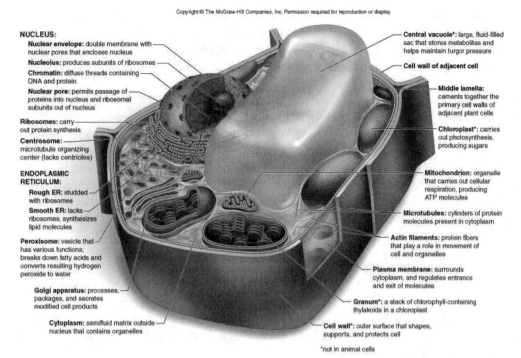

NUCLEUS:

Nuclear envelope: double membrane with nuclear pores that encloses nucleus

Nucleolus: produces subunits of ribosomes

Chromatin: diffuse threads containing DNA and protein

Nuclear pore: permits passage of proteins into nucleus and ribosomal subunits out of nucleus

Ribosomes: carry out protein synthesis

Centrosome: microtubule organizing center (lacks centrioles)

ENDOPLASMIC RETICULUM:

Rough ER: studded with ribosomes

Smooth ER: lacks ribosomes, synthesizes lipid molecules

Peroxisome: vesicle that has various functions; breaks down fatty acids and converts resulting hydrogen peroxide to water

Golgi apparatus: processes, packages, and secretes modified cell products

Cytoplasm: semifluid matrix outside nucleus that contains organelles

Central vacuole*: large, fluid-filled sac that stores metabolites and helps maintain turgor pressure

Cell wall of adjacent cell

Middle lamella: cements together the primary cell walls of adjacent plant cells

Chloroplast*: carries out photosynthesis, producing sugars

Mitochondrion: organelle that carries out cellular respiration, producing ATP molecules

Microtubules: cylinders of protein molecules present in cytoplasm

Actin filaments: protein fibers that play a role in movement of cell and organelles

Plasma membrane: surrounds cytoplasm, and regulates entrance and exit of molecules

Granum*: a stack of chlorophyll-containing thylakoids in a chloroplast

Cell wall*: outer surface that shapes, supports, and protects cell

*not in animal cells

Plant cell anatomy

4.4 The Nucleus and Ribosomes

The **nucleus** is the largest organelle, typically found in the center of the cell. Most cells have a single nucleus, however some cells, like those found in skeletal muscle, have more than one nucleus. Other cells, such as red blood cells, have no nucleus at all. Within the nucleus, there is a gel-like substance containing chromatin called **nucleoplasm**. **Chromatin** is a network of DNA and proteins that condenses into chromosomes before the cell divides. There is a concentrated area of chromatin within the nucleus called the **nucleolus**, where ribosomes are made. The nucleus is surrounded by a **nuclear envelope**, which separates it from the rest of the cell. There are protein-lined pores in the nuclear envelope that allow the nucleus to exchange ribosomes or other substances with the rest of the cell when necessary.

Ribosomes are the organelles that are responsible for protein synthesis. They have two subunits, and are made of RNA and proteins. Eukaryotic ribosomes are larger than prokaryotic ribosomes. Cells have varying numbers of ribosomes, depending on their function. Ribosomes can freely float within the cytoplasm or be attached to the endoplasmic reticulum. Free ribosomes synthesize proteins that will remain in the cell, while bound ribosomes synthesize proteins that will be secreted from a cell and used in another cell. Ribosomes are not permanently free or bound, but can adjust according to the needs of the cell.

4.5 The Endomembrane System

The **endomembrane system** is a continuous system of membranes that consists of the nuclear envelope, the endoplasmic reticulum, the Golgi apparatus, and various vesicles that transport molecules between the organelles of this system.

The **endoplasmic reticulum (ER)** is a series of flattened sacs that are made of membranes. There are two types of endoplasmic reticulum, rough and smooth, and they have different appearances and roles within the cell. Rough ER is studded with ribosomes on the cytoplasmic side and is adjoined with the nuclear envelope. The lumen of the rough ER is where secreted proteins are synthesized and modified. **Smooth ER** is continuous with the rough ER and does not have ribosomes attached. It is not flat like rough ER, but is more tubular in appearance. Smooth ER has a variety of functions, including lipid synthesis and drug detoxification.

The **Golgi apparatus** is a series of curved, flattened sacs whose appearance resembles a stack of pita bread. The cis face of the Golgi apparatus faces the ER, while the trans face faces the cell membrane. This is the place where proteins are further modified once they leave the rough ER.

Vesicles (transport organelles) bud off from the rough ER and carry proteins to the Golgi apparatus. The vesicles fuse with the Golgi at its cis face, and the protein is now in the lumen of the Golgi, where the carbohydrates attached to the proteins are modified. Another vesicle then buds off the Golgi apparatus to carry proteins to the cell membrane, where they are secreted into the extracellular space.

Lysosomes are vesicles produced by the Golgi apparatus that contain hydrolytic enzymes that work at very low pH levels. Lysosomes are responsible for digesting materials that enter the cell, such as nutrients that the cell needs as building blocks as well as bacteria that invade the cell and need to be destroyed. They also are responsible for the digestion of organelles that are no longer functional. Lysosomes fuse with vesicles that bud into the cell from the cell membrane. They can also fuse with storage vacuoles already in the cell.

4.6   Other Vesicles and Vacuoles

Peroxisomes are organelles that are similar to lysosomes in that they are membranous sacs that contain enzymes. The enzymes in peroxisomes, however, catalyze reactions that result in the formation of hydrogen peroxide, H_2O_2. Hydrogen peroxide is toxic to cells, so it is immediately broken down to water and oxygen by the enzyme catalase. Peroxisomes have a variety of functions in different cells, including lipid metabolism and the production of bile salts in animals. In plants, they are responsible for transforming fatty acids into sugars as well as carrying out reactions opposite of photosynthesis that use oxygen and produce carbon dioxide.

Vacuoles are membranous sacs that are larger than vesicles and are usually responsible for storage in a cell. In some animal cells, vacuoles store lipids. In plant cells, some vacuoles store water (e.g., the central vacuole), while others store pigments. Some protists contain contractile vacuoles which eliminate excess water as well as food vacuoles that store and digest nutrients.

4.7   The Energy-Related Organelles

Chloroplasts are the organelles in plant cells that convert light energy from the sun to chemical energy by the process of photosynthesis. They have an outer and an inner membrane. Within the chloroplast, there are stacks of **thylakoids** called **grana**. Chlorophyll is found in the thylakoid membranes. The grana are surrounded by a fluid called the **stroma**. The stroma is where sugars are produced and where the DNA and ribosomes of the chloroplast are located. Chloroplasts are a type of organelle called plastids. Other kinds of plastids include **chromoplasts**, which contain pigments other than chlorophyll, and **leucoplasts**, which synthesize and store starch.

Mitochondria are the organelles that produce energy for cells in the form of ATP. Like chloroplasts, they have a double membrane. The space between the two layers is called the **intermembrane space**. The intermembrane space is highly convoluted into shelves called **cristae**, which increase the surface area for more ATP production. Within the inner membrane there is a fluid called the **matrix**, where the DNA and ribosomes of the mitochondria are located, and where some of the reactions of cellular respiration occur.

4.8   The Cytoskeleton

The **cytoskeleton** is a network of proteins found in the cytoplasm that help the cell to maintain its structure and allow movement of organelles within the cell. The cytoskeleton is not static, but rather, it is dynamic. The proteins of the cytoskeleton are constantly assembling and disassembling in order to meet the motility needs of the cell. The three proteins that make up the cytoskeleton are actin filaments, intermediate filaments, and microtubules.

Actin filaments are long, fibrous proteins that are made of two chains of globular actin monomers twisted around each other. Actin is responsible for activities of the cell such as the formation of pseudopods in animal cells as well as cytoplasmic streaming in plants. In order to facilitate movement, actin interacts with a motor protein called myosin, using energy from ATP.

Intermediate filaments are rope-like fibrous proteins that are intermediate in size between actin filaments and microtubules. They are found just inside the nuclear envelope, just inside the plasma membrane, as well as in junctions between cells.

Microtubules are hollow cylinders that are made of globular tubulin dimers. A tubulin dimer consists of α-**tubulin** and β-**tubulin**. Microtubules are important in various cellular functions, such as cell division, the structure of cilia and flagella, and the movement of organelles within the cell. The assembly of microtubules is regulated by a **microtubule organizing center (MTOC)**, which is located in the **centrosome**, near the nucleus of the cell. Microtubules radiate throughout the cell from the MTOC and associate with the motor proteins dyenin and kinesin in order to facilitate movement using ATP as energy.

Centrioles are short cylinders that are found in the centrosomes of the cells of animals and protists. Plants and fungi also have centrosomes, but they do not contain centrioles. They are made of nine sets of triplet microtubules. They are found in pairs, are perpendicular to each other, and are found near the nucleus. Centrioles have a role in cell division in animal cells. In cells with cilia and flagella, centrioles make up the **basal bodies** at the base of the cilia or flagella.

Cilia and flagella are hair-like extensions on the surface of the cells that function either in movement of a cell or in movement of substances across the surface of the cell. Cilia are shorter than flagella, however they share a similar structure. Both cilia and flagella are made of microtubules in 9+2 arrangement. There are nine sets of doublet microtubules arranged in a circle around two single microtubules. The doublets are attached to each other by the motor protein **dyenin**, which forms side arms between the doublets. Movement occurs when the dyenin arms attach to neighboring doublets, and using energy from ATP, cause the doublets to slide past one another.

Multiple Choice Questions

1. Which of the following components gives a cell membrane its fluid characteristic?
   A. integral proteins
   B. oligosaccharides
   C. peripheral proteins
   D. glycolipids
   E. cholesterol

2. A large number of chromoplasts would be found in which of the following types of cells?
   A. xylem
   B. phloem
   C. mesophyll
   D. root
   E. cambium

3. Which of the following cells is most likely to contain projections called cilia?
   A. xylem
   B. sperm
   C. phloem
   D. oviduct
   E. cardiac muscle

4. Tay Sachs disease often affects children by progressively destroying cells in their brains due to the accumulation of a specific type of fats. This is due to the fact that
   A. the Golgi bodies do not transport the fats out of the cell in vesicles and they accumulate in the cell.
   B. the lysosomes do not contain the appropriate enzyme to metabolize the fat.
   C. the fats do not pass through the cell wall to the bloodstream and accumulate in the cell.
   D. hydrogen peroxide is not produced by the peroxisomes to destroy the fats.
   E. the smooth endoplasmic reticulum produces an excess of fats which accumulate in the cell.

5. Your biology teacher shows you an electron micrograph image of a cell that contains DNA, ribosomes, and amyloplasts and has a cell wall. Which of the following types of organisms contains this cell?
   A. red pine
   B. streptococcus
   C. amoeba
   D. mushroom
   E. jellyfish

6. Evidence supporting the fact that an increase in the amount of surface area directly increases the efficiency of cellular processes would NOT be shown by which of the following?
   A. Mitochondria contain many folds called cristae.
   B. Chloroplasts contain stacks of thylakoids.
   C. Microvilli are present on cells of the inner ear.
   D. The large central vacuole maintains pH and allows enzymes to act.
   E. Folds of the smooth endoplasmic reticulum aid in drug metabolism.

Questions 7–10 refer to the following five choices. For each question, select the choice that is most closely related. Each choice may be used once, more than once, or not at all.
   A. Golgi Apparatus
   B. mitochondria
   C. smooth endoplasmic reticulum
   D. chloroplast
   E. lysosome

7. Location of the conversion of CO_2 and H_2O to glucose

8. Produces ATP for cellular energy

9. Contains stacks called grana integral in photosynthesis

10. Found in high concentration in liver cells to detoxify drugs

Free Response Question

In the cells of individuals afflicted with a particular disorder, a protein is produced correctly but not exported from the cell. In order to diagnose the cause of this disorder:

    A. **Identify** THREE of the components of the endomembrane system and **describe** each function.

    B. **Explain** the how the structure of each of these organelles relates to its function.

    C. **Propose** two possible hypotheses for this disorder.

Annotated Answer Key (MC)

1. **E**; The unsaturated nature of the phospholipids AND cholesterol give the cell membrane optimum fluid characteristics. Cholesterol prevents interactions between hydrocarbon tails of lipids and can prevent tight packing. Integral proteins are critical in transport, peripheral proteins are necessary for attachment purposes, and glycolipids and oligosaccharides provide cell recognition properties to the cell.

2. **C**; Chromoplasts are found in the mesophyll layer of the leaves of plants and contain pigments. These photosynthetic pigments are not necessary in xylem, phloem, root, or cambium tissues.

3. **D**; The epithelial cells of the oviduct produce running currents from the internal end toward the external opening to facilitate the movement of the oocyte towards the uterus. Cilia also line air passages of the lungs and are similar, but not identical, in structure to flagella.

4. **B**; Tay Sachs is an autosomal dominant inherited disorder where progressive deterioration of the nervous system occurs. The primary cause of this disorder is the lack of the appropriate enzyme in eukaryotic lysosomes to metabolize fats. This metabolism occurs in the cell, not outside of the cell, and would not require transport through a cell *wall* of a human. Peroxisomes do metabolize fatty acids but do not produce hydrogen peroxide for this particular purpose.

5. **A**; The presence of amyloplasts, nonpigmented organelles, indicates the organism is a plant.

6. **D**; An increase in surface area does directly increase the efficiency of cellular processes. This increase can be seen in folds and stacks. While the large central vacuole is large in size it does not particularly exemplify the increase in surface area that the other examples do.

7. **D**

8. **B**

9. **D**

10. **C**

For Questions 7–10: Compartmentalization of the eukaryotic cell allows for different functions to occur in each organelle. The chloroplast of plants converts CO_2 and H_2O to glucose as plants are autotrophs and produce their own food. The chloroplast contains grana, or stacks of thylakoids, that function in photosynthesis. This glucose can be broken down in plants and animals in the mitochondria to produce ATP. While glucose is stored in the liver in the form of glycogen, liver cells also contain high concentrations of smooth endoplasmic reticulum as it does detoxify drugs. It is important to note that high numbers of specific organelles can be found in cells that require more efficient processing related to that particular organelle (i.e.. high number of mitochondria in skeletal muscle cells).

Answer to FRQ

PART A (MAX 6)
One point identification, one point description of each function.

- Nuclear envelope: protection OR control of substances moving out of the nucleus (i.e., mRNA)
- ROUGH Endoplasmic reticulum: initial modification of the polypeptide chains of proteins.
- SMOOTH Endoplasmic reticulum: lipid synthesis OR detoxification
- Golgi bodies/apparatus: final modification of proteins OR sorting and packaging
- Vesicles: transport
- Vacuoles: storage
- Cell Membrane: control of substances moving in and out of the cell

PART B (MAX 3)
- Pores of nuclear envelope assist in control of substance
- Bilayer of nuclear membrane for protection
- Folds of ER for increased surface area in processing and modification OR detoxification
- Cis/trans identification on Golgi for proper export of proteins
- Membrane bound vacuoles for efficient transport
- Bilayer of cell membrane for control of substances
- Other plausible descriptions

PART C (MAX 3)
- Improper modifications at ER
- Improper modifications at Golgi
- Inability of the cell membrane to allow export of the vesicles (i.e., recognition)
- Ineffective trafficking of proteins within the system

CHAPTER 5
MEMBRANE STRUCTURE AND FUNCTION

5.1   Plasma Membrane Structure and Function

The **fluid-mosaic model** best explains the structure of the plasma membrane because it is made mostly of lipids and proteins. The membrane is a **phospholipid bilayer**, with the phospholipids arranged with their hydrophilic heads at the surface, and the hydrophobic tails in the interior. The lipid bilayer has the consistency of oil but acts as a barrier to the entrance and exit of most biological molecules. Because cells need to be flexible, the membrane is fluid. The fluidity is determined by the degree of saturation of the fatty acid tails. More unsaturation will cause the membrane to be more fluid. The fluidity of the membrane is moderated by **cholesterol**, which is present in the hydrophobic portion of the membrane. At high temperatures, cholesterol prevents the membrane from being too fluid, whereas at low temperatures it prevents the hydrophobic tails from packing too closely together.

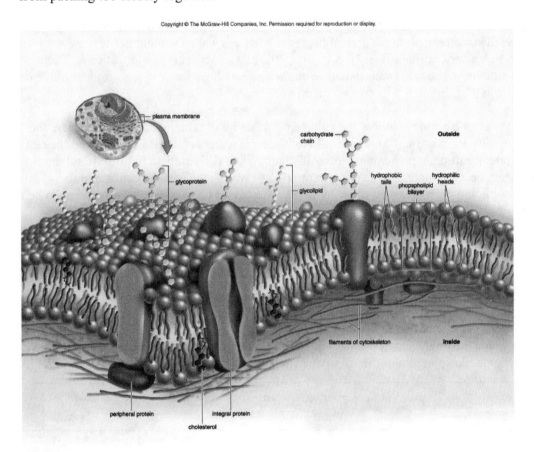

Plasma membrane of an animal cell

There are **integral proteins** embedded in the membrane. The hydrophobic portion of an integral protein lies in the lipid bilayer of the plasma membrane, and the hydrophilic portion lies at the surfaces. There are also **peripheral proteins** associated with the membrane. They are generally found on the cytoplasmic side of the membrane. These proteins act as receptors, carry on enzymatic reactions, join cells together, form channels, or act as carriers to move substances across the membrane. Some of these proteins make contact with the extracellular matrix (ECM) outside of the cell and with the cytoskeleton inside the cell. Thus, the ECM can influence the happenings inside the cell via the membrane proteins.

Carbohydrates are also important to the structure of the cell membrane. They can be attached to lipids (forming **glycolipids**) or to proteins (forming **glycoproteins**) on the extracellular surface of the membrane. Membrane glycolipids and glycoproteins are involved in marking the cell as belonging to a particular individual and tissue.

5.2   Passive Transport Across a Membrane

The plasma membrane is differentially, or selectively, permeable. This means that some molecules can pass through the membrane while others cannot. Some molecules, such as lipid-soluble compounds, water, and gases, simply diffuse across the membrane from the area of higher concentration to the area of lower concentration. No metabolic energy is required for diffusion to occur because the direction of movement is in the same direction of the concentration gradient.

The diffusion of water across a differentially permeable membrane is called **osmosis**. Water moves across the membrane into the area of higher solute (less water) content per volume. Because it is polar, water cannot freely pass through the plasma membrane of most cells. There are channel proteins in the membrane called **aquaporins** that allow water to pass from one side of the membrane to the other down its concentration gradient.

When cells are in an **isotonic** solution, the volume of the cell will remain unchanged because the concentration of water is the same on both sides of the cell. Water is entering and leaving the cell at the same rate, resulting in no *net* movement of water. When cells are in a **hypotonic** solution, they gain water, because there is a lower concentration of water inside of the cell due to the presence of solutes. Animal cells may **lyse**, or burst, when placed into a hypotonic solution. When water enters a plant cell on the other hand, **turgor pressure** increases, but the cell will not lyse due to the presence of the cell wall. When cells are placed in a **hypertonic** solution, they lose water because there is a lower concentration of water outside of the cell. Animal cells will **crenate**, or shrink. In a plant cell, **plasmolysis** will occur, and the plant may eventually wilt.

During **facilitated transport**, a substance moves down its concentration gradient either by a channel protein or by a carrier protein that spans the membrane. No energy is required. Channel and carrier proteins are specific for the ions or molecules they transport. Channel proteins have an opening for an ion to simply pass through, while carrier proteins undergo a conformational change in order to transport a molecule from one side of the membrane to the other.

> **Take Note:** *Can you compare and contrast passive transport and active transport, and give examples of each?*

5.3   Active Transport Across a Membrane

During active transport, a carrier protein acts as a pump that causes a substance to move across the membrane against its concentration gradient. Energy in the form of ATP molecules is required for active transport to occur. An example of a protein that carries out active transport is the **sodium-potassium pump**. The sodium-potassium pump carries Na^+ to the outside of the cell and K^+ to the inside of the cell, as shown in Figure 5.10 of your textbook.

Larger substances can enter and exit a membrane by bulk transport: **exocytosis** and **endocytosis**. Exocytosis involves secretion. In exocytosis, a vesicle containing cell products is formed from the

Golgi apparatus and fuses with the plasma membrane, releasing its contents outside of the cell and becoming part of the membrane. Endocytosis includes **phagocytosis**, **pinocytosis**, and **receptor-mediated endocytosis**. Phagocytosis and pinocytosis involve bringing solid and liquid material, respectively, into the cell when a vesicle pinches off from the plasma membrane. Receptor-mediated endocytosis makes use of receptor proteins in the plasma membrane. Once a specific solute binds to receptors, a coated pit becomes a coated vesicle. After losing the coat, the vesicle can join with the lysosome, or after discharging the substance, the receptor-containing vesicle can fuse with the plasma membrane.

5.4   Modification of Cell Surfaces

Animal cells have an **extracellular matrix (ECM)** that influences their shape and behavior. Tissues vary as to the amount and character of the ECM. There are several proteins in the ECM. **Collagen** and **elastin** provide support, while **fibronectins** interact with the **integrins** in the membrane to foster communication between the ECM and the inside of the cell.

Some animal cells have junction proteins that join them to other cells of the same tissue. Adhesion junctions (**desmosomes** and **tight junctions**) help hold cells together. Desmosomes are attached to the cytoskeleton and result in the formation of sheets of cells. In tight junctions, the plasma membranes of neighboring cells are in contact with each other, forming barriers that prevent substances from passing between cells. **Gap junctions** allow passage of small molecules between cells, allowing them to communicate.

Plant cells have a freely permeable cell wall, with cellulose as its main component. All cells have a primary cell wall, and some also have a secondary cell wall for added strength. The area between neighboring cell walls is called the middle lamella, which contains a polysaccharide called pectin. Plant cells are joined by narrow, membrane-lined channels called **plasmodesmata** that span the cell wall and contain strands of cytoplasm that allow materials to pass from one cell to another.

Multiple Choice Questions

1. Which of the following examples is a specific case of active transport?
    A.   A carrier protein moves sucrose across the membrane with the concentration gradient.
    B.   Cytoplasmic Na+ binds to the sodium-potassium pump, phosphorylates ATP and Na+ are released to the outside.
    C.   Water moves from an area of high free water concentration to an area of low free water concentration.
    D.   Aquaporins facilitate the diffusion of water in the red blood cells of mammals.
    E.   A neurotransmitter causes a gated channel to open and allow sodium into the cell.

2. The process by which a single pump transports a specific molecule while indirectly driving the movement of other molecules is called
    A.   phagocytosis.
    B.   ion-mediated transport.
    C.   exocytosis.
    D.   cotransport.
    E.   endocytosis.

3. Which of the following observations could lead researchers to conclude that membrane proteins move laterally within the plasma membrane?
    A. A cell is freeze fractured and splits the phospholipid bilayer into two layers, and the membrane proteins go with one of the layers.
    B. Human and mouse cells labeled with protein markers are fused, and the hybrid cell contains a mixing of proteins on the surface.
    C. Coexpression of two membrane proteins on skeletal muscle cells result in an increased amount of one of the proteins to transport calcium across the membrane.
    D. There are two types of protein receptors in plants to transport CO_2 and NH_4, however in some microorganisms proteins that transport NH_4 can also transport CO_2.
    E. Integral proteins on the surface of stomach cells can transport either acids or bases.

4. Which of the following structural characteristics of a membrane protein is NOT matched correctly with its function?
    A. Microfilaments adhere to membrane proteins to maintain cell shape.
    B. Membrane proteins from two cells stick together to form a gap junction for cell communication.
    C. Membrane proteins may be enzymes to catalyze metabolic reactions within the cell.
    D. Membrane proteins may transport substances from one side of the cell to another.
    E. Membrane proteins may contain cholesterol to serve as identification tags for other cells.

5. An artificial cell consisting of an aqueous solution with 0.5M sucrose and 0.2M glucose that is permeable only to monosaccharides is placed in a beaker of water containing 0.2 M sucrose and 0.1M glucose. Which of the following will occur?
    A. The net movement of water will be out of the cell.
    B. Sucrose will move out of the cell and glucose will move into the cell.
    C. There will be no net movement of water.
    D. Glucose will move out of the cell.
    E. Sucrose and glucose will both move out of the cell.

6. The role of the Rh protein channels on cells is NOT to
    A. increase the transport of CO_2 across the membrane in humans.
    B. increase the flexibility of the red blood cell to improve surface area.
    C. improve CO_2 transport in green algae who live in anaerobic environments.
    D. provide a site for cell recognition.
    E. decrease the pH of the blood by bonding to CO_2 in the bloodstream.

7. The rate of diffusion can be increased by
    A. increasing the temperature.
    B. decreasing pressure.
    C. increasing the size of the molecules.
    D. decreasing membrane pore size.
    E. decreasing concentration.

8. Which of the following molecules can pass through the lipid bilayer?
    A. K^+
    B. Cl^-
    C. glucose
    D. HCO_3^-
    E. CO_2

9. Plant cells placed in a hypotonic environment
    A. will lyse without special adaptations to combat water concentration fluctuations.
    B. will remain flaccid due to an ineffective central vacuole.
    C. will remain turgid due to the elastic cell wall.
    D. will plasmolysize when the cell membrane pulls away from the wall.
    E. will shrivel due to the loss of water from the large central vacuole.

10. Many fields require heavy irrigation. In desert areas much of that water evaporates, leaving behind solutes. This may cause plant cells
    A. to become turgid due to the increase in solutes in the surrounding environment.
    B. to lose water to the surrounding hypotonic environment.
    C. to grow excessively due to the increased N-P-K fertilizer in the soil.
    D. to plasmolysize due to the osmosis of water to a lower water potential in the surrounding soil.
    E. to gain water from the surrounding soil.

Free Response Question

Proteins are integral components of eukaryotic cell membranes.

    A. **Identify** and **describe** three functions of membrane proteins.

    B. **Explain** why there is a difference in the number and type of proteins on specific cell membranes. Give TWO examples.

Annotated Answer Key (MC)

1. **B**; While carrier proteins often function in active transport mechanisms, in choice A it is WITH the concentration gradient, as is water movement in choice C. Aquaporins are passive transport channels, and the opening of a gated channel that allows sodium to enter the cell WITH the concentration gradient is passive. Choice B, which describes a pump, indicates an active transport mechanism.

2. **D**; In cotransport mechanisms, two molecules are simultaneously moved across a membrane. All of the other mechanisms involve one type of molecule.

3. **B**; Proteins on the plasma membrane can move laterally only if they are found in combination with other proteins on the surface. This lateral movement can be verified by observing a mixing of those proteins, but not by simply observing that there are in fact several types of proteins on the surface of cells.

4. **E**; Membrane proteins are extremely diverse in their function. They can act as catalysts, transporters, anchors, and sites for cell communication. However, they do not contain CHOLESTEROL to promote cell recognition.

5. **D**; Since the cell membrane is only permeable to MONOSACCHARIDES, sucrose will not be able to pass through the membrane. Glucose is a monosaccharide and can pass through. The higher concentration of sucrose inside the cell will cause water to move from the beaker to the cell.

6. **E**; The Rh protein has many recently discovered functions on the surface of the red blood cell. It also may be found on other cells where CO_2 transport is necessary. This protein has been found to improve CO_2 transport and aid in cell recognition (as with the incompatibility of mothers and their fetuses depending on their Rh markers). It will NOT bond to the CO_2 molecules; removing the CO_2 from the blood would INCREASE the pH of the blood.

7. **A**; Increasing temperature increases the kinetic energy of the molecules and therefore increases the rate of molecular movement. An increase in pressure, decrease in the size of molecules or increase in pore size, and an increase in concentration would also tend to increase the rate of diffusion of some molecules.

8. **E**; The lipid bilayer is permeable to water and a few small, uncharged molecules, including O_2 and CO_2.

9. **C**; Plant cells are healthy when placed in a hypotonic environment as water will enter the plant cells from the surrounding environment. The cell wall allows the water to accumulate in the plant cell without damage to the outer layer, as with animal cells that lack cell walls. In a hypertonic environment, the water would leave the plant cell to the surrounding environment and plasmolysis might occur.

10. **D**; The surrounding soil will have a higher solute potential and lower water potential than the plant cells. This will create a tendency for water to leave the plant cells and the cell walls to separate from the cell membranes, causing plasmolysis and eventual cell death.

Answer to FRQ

PART A (MAX 6)
One point identification, one point description of each function.

- Transport Proteins: allow water soluble substances to pass through
- Receptor proteins: bind extracellular substances
- Recognition Proteins: self vs. non-self
- Enzymatic proteins
- Function: catalytic activity
- Function: identification
- Adhesion Proteins:
- Function: formation of cellular junctions
- Peripheral proteins:
- Signal transduction
- Other examples.

PART B (MAX 6)
One point for each.

- Differences in: metabolism, cell volume, pH, recognition needs, transport
- Any plausible example: includes increased enzymatic proteins for metabolic purposes
- Increased recognition proteins on red blood cells

CHAPTER 6
METABOLISM: ENERGY AND ENZYMES

6.1 Cells and the Flow of Energy

Energy is the ability to do work. There are two basic forms of energy: potential and kinetic. Potential energy is stored energy, while kinetic energy is the energy of motion. Two energy laws are basic to understanding how energy is used at all levels of biological organization. The first law of thermodynamics, the law of conservation of energy, states that energy can neither be created nor destroyed, but can only be transferred or transformed. The second law of thermodynamics states that one usable form of energy cannot be completely converted into another usable form, but rather, some energy is always lost as heat. As a result of these laws, we know that the entropy of the universe is increasing and that only a flow of energy from the sun maintains the organization of living things.

6.2 Metabolic Reactions and Energy Transformations

Metabolism is the sum of all the chemical reactions occurring in a cell. Considering individual reactions, only those that result in a negative free-energy difference can occur spontaneously. These reactions release energy and are called **exergonic** reactions. **Endergonic** reactions, which require an input of energy, occur in cells because they are coupled to exergonic reactions. The energy currency that allows this coupling to occur is adenosine triphosphate, or ATP. ATP is a nucleotide consisting of the nitrogen base adenine, a ribose sugar, and three phosphate groups. The energy in ATP is derived from the high-energy bonds between the phosphate groups. ATP goes through a cycle in which it is constantly being built up from, and then broken down to, ADP + P_i.

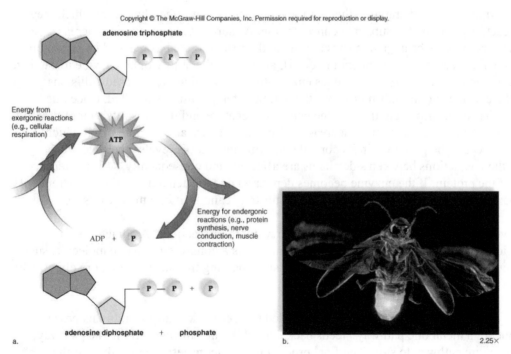

The ATP cycle

When ATP breaks down, energy is released that drives endergonic reactions. In general, ATP is used to energize or change the shape of a reactant so a reaction can occur. The sodium potassium pump discussed in Chapter 5 is one such example. ATP provides the energy for the pump to transport Na^+ and K^+ across the cell membrane against their respective concentration gradients.

Take Note: *In preparation for the AP Biology exam, be sure that you can describe the structure of ATP, explain how it provides energy, and give several examples of the role of ATP in the cell.*

6.3   Metabolic Pathways and Enzymes

Enzymes are biological catalysts that increase the rate at which chemical reactions occur in cells. Enzymes speed reactions by lowering the **energy of activation**, that is, they reduce the amount of energy that is required for a reaction to occur. Most enzymes are proteins, however, some are made of RNA and are called ribozymes. For the time being, we will focus on enzymes that are proteins. A **metabolic pathway** is a series of reactions that proceed in an orderly, step-by-step manner. Each step in the pathway—where the product of one reaction is the reactant of the next—is carried out by an enzyme.

The reactant that the enzyme binds to is called the **substrate**. Enzymes and substrates are specific for one another. The place on the enzyme where the substrate binds is called the **active site**. When a substrate binds to an active site, the active site adjusts its structure to fit the substrate better **(induced fit)**, and an enzyme-substrate complex is formed. The reaction then takes place, and the products are released from the active site, leaving the enzyme intact and unchanged. Enzymes regulate metabolism because, in general, no reaction occurs unless its enzyme is present. Which enzymes are present determine which metabolic pathways will be utilized.

Though enzymes work quite rapidly, there are several factors that affect the rate at which they catalyze reactions. The first is substrate concentration. As substrate concentration increases, so does enzyme activity; once all active sites are filled, the reaction rate has reached its maximum. Environmental factors, such as temperature or pH, also affect the ability of an enzyme to catalyze reactions. As temperature increases, so does enzymatic activity (due to increased collisions between the enzyme and the substrate) until the optimum temperature is reached. Once this temperature has been surpassed, the enzyme begins to denature and the substrate cannot fit into the active site. Thus, enzyme activity decreases. Each enzyme has an optimum pH at which it works best, because the protein is in its normal conformation. At pH levels other than the optimum, the interactions between side chains are affected, and consequently, so is the tertiary structure of the protein. If the enzyme becomes denatured due to such a disruption at extreme pH values, the rate of reaction decreases because the substrate cannot fit into the active site.

Many enzymes need cofactors or coenzymes to carry out their reactions. A cofactor is an inorganic ion, such as iron, zinc, or copper. A coenzyme is a nonprotein organic molecule, such as a vitamin. Both bind to an enzyme at the active site, enabling the substrate to more effectively bind to the enzyme.

The activity of most metabolic pathways is regulated by **feedback inhibition**. In this process, the final product of a metabolic pathway "feeds back," to inhibit an enzyme early in the pathway, causing the entire pathway to stop. The final product is a **noncompetitive inhibitor** of the enzyme that it affects. It binds to a place on the enzyme other than the active site, causing a conformational change in the enzyme that then prevents the substrate from binding. Thus, the enzyme is inhibited.

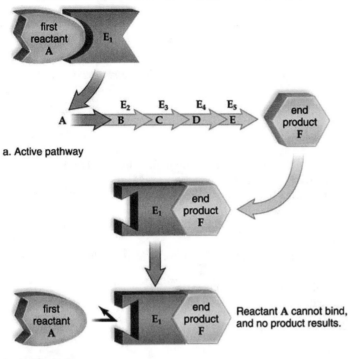

a. Active pathway

b. Inhibited pathway

Reactant A cannot bind, and no product results.

Noncompetitive inhibition of an enzyme

Take Note: *On the AP Biology exam, you may be asked to describe the structure and function of enzymes, explain the factors that affect enzyme activity, or describe how enzyme activity is regulated. It is also possible that you may be asked to design an experiment to test the effect of environmental factors on enzyme activity.*

6.4   Organelles and the Flow of Energy

There are two organelles that are responsible for the flow of energy from the sun through living things: chloroplasts and mitochondria. Energy flows through organisms because (1) photosynthesis in chloroplasts captures solar energy and produces carbohydrates, and (2) cellular respiration in mitochondria breaks down this carbohydrate to produce ATP molecules, which (3) are used to provide energy for metabolic reactions. The overall equation for photosynthesis is the opposite of that for cellular respiration. During photosynthesis, the coenzyme $NADP^+$ serves as a carrier that receives electrons from the splitting of water, while during cellular respiration, the coenzyme NAD^+ carries electrons that will eventually be used to make water.

Both photosynthesis and cellular respiration make use of an **electron transport chain**, in which electrons are transferred from one carrier to the next with the release of energy that is ultimately used to produce ATP molecules. **Chemiosmosis** explains how the electron transport chain produces ATP. The carriers of the electron transport chain have two functions: 1) to pass electrons down the chain, and 2) to use the energy lost from the passage of electrons to pump hydrogen ions (H^+) to the other side membrane. This creates an electrochemical gradient that represents potential energy. When the hydrogen ions flow down their electrochemical gradient through an ATP synthase complex, it uses the energy of the electrons flowing down their gradient to make ATP from ADP and P_i.

Take Note: *On the AP Biology Exam, you may be asked to compare and contrast chemiosmosis in chloroplasts and chemiosmosis in mitochondria. As you study the next two chapters, take note of the similarities and differences so that you will be prepared if a question of this type appears on the test.*

Multiple Choice Questions

1. Lead poisoning often causes symptoms resembling those of attention deficit disorder (ADD). Which of the following statements provides a basis for this similarity?
    A. Lead may inhibit the same enzyme in the brain that is deficient in individuals with ADD.
    B. Lead poisoning and ADD can be treated with lithium.
    C. The parts of the brain affected in ADD and lead poisoning are in different hemispheres.
    D. A deficiency of lead in the body causes ADD.
    E. Lead poisoning and ADD often occur in young children but not in older adults.

2. Substrate level phosphorylation and oxidative phosphorylation
    A. differ in the energy product formed.
    B. differ in the location of activity.
    C. occur only in animals.
    D. are evolutionary precursors of many modern metabolic reactions.
    E. occur in the cytoplasm of eukaryotic cells.

Questions 3–6 refer to the following five choices. For each question, select the choice that is most closely related. Each choice may be used once, more than once, or not at all.
    A. ATP
    B. NAD+
    C. NADP+
    D. FAD
    E. ADP

3. A coenzyme that functions as an electron carrier in photosynthesis

4. Contains adenine, ribose. and three phosphates

5. One of the products of hydrolysis of ATP

6. Carries electrons to the thylakoid membrane

7. The reduction of carbon dioxide to a carbohydrate
    A. utilizes electrons from NADPH and requires ATP.
    B. occurs in the mitochondria.
    C. is an exergonic reaction.
    D. uses FAD as a carrier molecule for the high energy electrons.
    E. occurs during aerobic respiration.

8. Alcohol induced cardiomyopathy is a disease where an increased consumption of alcohol contributes to a weakening of the heart muscle so that it cannot pump blood efficiently. Which of the following best explains the events that could lead to this disease?
    A. modification of the action of contractile proteins, actin, and myosin
    B. an increase in chemiosmotic ATP synthesis
    C. a decrease in production of NADP+
    D. an increase in the amount of smooth ER in the liver
    E. a decrease in the amount of peroxisomes in the smooth muscle cells

9. In the following enzyme cascade, an inability to produce product E could be attributed to which of the following?

    A➔ B ➔ C ➔D ➔ E ➔ F ➔G
      E1   E2   E3   E4   E5   E6

    A. an increased accumulation of product F
    B. a lack of production of product D
    C. a decreased accumulation of product G
    D. a defective enzyme E5
    E. an increase in the concentration of E3

10. Which of the following statements is true regarding enzymatic reactions?
    A. Endergonic reactions can be accelerated by coupling them to an exergonic reaction.
    B. Reactants contain less energy than products in exergonic reactions.
    C. Enzymes and their substrates possess specific complementary geometric shapes that fit exactly into one another.
    D. Enzymes increase the activation energy necessary for a reaction to proceed.
    E. Competitive inhibitors bond with an enzyme outside of the active site.

Free Response Question

Enzymes are catalytic molecules, that is they speed up the rates of reactions.

    A. **Explain** why enzymes are necessary in biological systems.

    B. **Define** THREE control mechanisms that regulate enzymatic activity. Give a biological example of each.

Annotated Answer Key (MC)

1. **A**; Lead poisoning and ADD do often have the same symptoms. However, "A" is the only choice providing a potential basis for this the identification of a cause of this disorder. Lead poisoning cannot be treated with lithium.

2. **B**; Gylcolysis and Krebs cycle use substrate-level phosphorylation. The electron transport chain uses oxidative phosphorylation. They both produce ATP and do not occur ONLY in animals.

3. **C**

4. **A**

5. **E**

6. **C**

For Questions 3–6: NADP$^+$ is produced by the light reactions of photosynthesis, is consumed in the Calvin Cycle, and is an electron carrier in photosynthesis, whereas NAD functions in glycolysis and in the citric acid cycle of cellular respiration. ATP does contain adenine, ribose, and three phosphates and is an energy carrier that can be broken down through hydrolysis to ADP and P.

7. **A**; Reduction involves the ADDITION of electrons and those are donated from NADPH in an endergonic reaction (requiring ATP). It occurs during the light-independent reactions of photosynthesis with the aid of the enzyme rubisco and releases three carbon sugars.

8. **A**; An increase in ATP would increase the pumping of blood. NADP$^+$ is an electron carrier in photosynthesis. An increase in smooth ER of the liver would not explain cardiomyopathy. The heart consists of cardiac muscle, not smooth muscle.

9. **B**; A lack of production of product D would stop the cascade before production of product E.

10. **A**; In endergonic reactions, reactants require more energy than products. Enzymes reduce the activation energy for reactions to occur; they do not increase it. Competitive inhibitors bind at the active site, not outside of it.

Answer to FRQ

PART A (MAX 5)

- Reactions could not occur at normal body temperature.
- Reactions could not produce sufficient products for efficient functioning.
- Reactants would require chemical treatment (i.e., pH change) that would destroy living tissue.
- Inability to "pre-digest" molecules to increase efficiency of metabolism.
- Increased specificity for metabolic reactions

PART B (MAX 7)

- Competitive inhibition
- Defined as: inhibitor and substrate compete for the enzyme at the active site
- Allosteric interactions
- Defined as: regulation of an enzyme or other protein by binding an effector molecule at the enzyme's allosteric site
- Noncompetitive inhibition
- Defined as: an inhibitor that binds to the enzyme at a site other than the active site
- Post translational control
- Defined as: change in protein structure after translation
- Cofactors/Coenzymes
- Defined as: nonprotein/protein components to enhance enzymatic activity
- Examples: prescription drugs, painkillers, poisons, toxins, carrier molecules (i.e., NADH), vitamins, minerals, and others

CHAPTER 7
PHOTOSYNTHESIS

7.1   Photosynthetic Organisms

Photosynthesis is the conversion of light energy from the sun to chemical energy in the form of carbohydrates, releasing oxygen as a byproduct. Sugars and oxygen are both used by the majority of living things. **Autotrophs** (cyanobacteria, algae, and land plants), which make their own food, conduct photosynthesis. **Heterotrophs**, organisms that do not make their own food, receive this chemical energy and convert it to the metabolic energy they need to do work.

In plants, photosynthesis takes place in chloroplasts. Most chloroplasts are found in the mesophyll cells of leaves. For photosynthesis to occur, plants need water and carbon dioxide. Water is obtained from the roots of the plant, while carbon dioxide enters the leaves via small openings called stomata. Recall from Chapter 4 that a chloroplast is bound by a double membrane and contains two main components: the semifluid stroma and the membranous grana made up of thylakoids. The thylakoid membranes contain chlorophyll, which absorbs light from the sun.

7.2   The Process of Photosynthesis

The overall equation for photosynthesis shows that it is a redox reaction. Carbon dioxide is reduced to form glucose, and water is oxidized, forming oxygen:

$$6CO_2 + 6H_2O \xrightarrow{\text{light}} C_6H_{12}O_6 + 6O_2$$

There are two stages of photosynthesis that take place in different parts of the chloroplast. The **light reactions** take place in the thylakoid membranes. They occur in the presence of light, and they create the ATP and NADPH that are needed for the second stage, the **Calvin cycle**. The Calvin cycle reactions take place in the stroma. They use the ATP and NADPH from the light reactions to convert carbon dioxide to carbohydrates.

7.3   Plants as Solar Energy Converters

Thylakoids contain pigments that absorb photons of light in the visible range so that photosynthesis can occur. Chlorophylls *a* and *b* absorb violet, blue, and red wavelengths best. This causes chlorophyll to appear green. The carotenoids absorb light in the violet-blue-green range and are yellow-to-orange pigments.

The **noncyclic electron pathway** of the light reactions begins when solar energy enters photosystem II (PS II). Consult your textbook for a picture of the light reactions that you can refer to as you read this explanation. In PS II, energized electrons are picked up by electron acceptors after they are excited to a higher energy level by a photon of light. The oxidation (splitting) of water replaces these electrons in the reaction-center chlorophyll *a* molecules. Oxygen is released to the atmosphere, and hydrogen ions (H^+) remain in the thylakoid space. An electron acceptor molecule passes the electrons to photosystem I (PS I) by way of an electron transport chain. When solar energy is absorbed by PS I, the energized electrons are received by another electron acceptor and are passed to NADP+, which also combines with H+ from the stroma to become NADPH. The NADPH is then available to be used to power the Calvin cycle.

The thylakoid membrane is highly organized. PS II is associated with an enzyme that oxidizes water; the cytochrome complexes transport electrons and pump H^+. PS I is associated with an enzyme that reduces NADP+, and ATP synthase produces ATP. The energy made available by passing electrons down the electron transport chain allows the proteins of the electron transport chain to pump H^+ into the thylakoid space to create an electrochemical gradient. When H^+ flows down this gradient from the thylakoid space into the stroma through ATP synthase, ATP is synthesized from ADP and P_i. This method of producing ATP by coupling its synthesis to the energy from the electrochemical gradient is called **chemiosmosis**.

7.4   Calvin Cycle Reactions

Recall that the Calvin cycle occurs in the stroma of the chloroplasts. Unlike the light reactions, the reactions of the Calvin cycle are not dependent on light to occur; rather, they need the products of the light reactions, ATP and NADPH, for energy. These molecules are used by the Calvin cycle reactions to reduce CO_2 to G3P, which is then converted to all the organic molecules a plant needs. During the first stage of the Calvin cycle, the enzyme RuBP carboxylase fixes CO_2 to RuBP, a 5-carbon molecule, producing a 6-carbon molecule that immediately breaks down to two 3-carbon molecules. During the second stage, CO_2 (incorporated into 3PG) is reduced to G3P. This step requires the NADPH and some of the ATP from the light reactions. For every three turns of the Calvin cycle, the net gain is one G3P molecule; the other five G3P molecules are used to re-form three molecules of RuBP. This step also requires ATP for energy. It takes two G3P molecules to make one glucose molecule.

Take Note: *You should be able to describe the steps of photosynthesis (both the light reactions and the Calvin cycle) for the AP Biology exam. You should also be able to trace the path of electrons and carbon dioxide.*

7.5   Other Types of Photosynthesis

The majority of land plants conduct photosynthesis as described in the previous sections. They are called **C3 plants**, because when CO_2 is fixed, it is incorporated into a 3-carbon molecule, 3PG. When the weather becomes hot and dry, a plant will close its stomata in order to prevent water loss. The downside is that this occurs at the expense of photosynthesis. When the stomata are closed, carbon dioxide cannot be absorbed and the levels of O_2 in the leaf rise. When this occurs, RuBP carboxylase binds O_2 to RuBP, resulting in **photorespiration**. Some plants have evolved to have mechanisms that avoid photorespiration so that they can retain water in hot, dry conditions without the loss of CO_2 for photosynthesis.

C4 plants have a different leaf anatomy than C3 plants in that the mesophyll cells are arranged in a concentric circle around the bundle sheath cells. Both mesophyll and bundle sheath cells contain chloroplasts, whereas in C3 plants, only the mesophyll cells have chloroplasts.

a. CO_2 fixation in a C_3 plant, blue columbine, *Aquilegia caerulea*

b. CO_2 fixation in a C_4 plant, corn, *Zea mays*

Carbon dioxide fixation in C3 and C4 plants

In C4 plants, the enzyme PEPCase fixes carbon dioxide to PEP to form a 4-carbon molecule, oxaloacetate, within the mesophyll cells. A reduced form of this molecule is pumped into bundle sheath cells, where CO_2 is released to the Calvin cycle. C4 plants avoid photorespiration by a spatial partitioning of pathways. Carbon dioxide fixation occurs utilizing PEPCase in the mesophyll cells, and the Calvin cycle occurs in the bundle sheath cells.

CO_2 fixation in a CAM plant, pineapple, *Ananas comosus*

Carbon dioxide fixation in CAM plants

In **CAM plants**, the stomata are open only at night to allow the conservation of water during the day. PEPCase fixes CO_2 to PEP only at night, and the next day CO_2 is released and enters the Calvin cycle within the same cells. This represents a temporal partitioning of pathways; carbon dioxide fixation occurs at night, and the Calvin cycle occurs during the day. CAM was first discovered in desert plants, but since then it has been documented in many different types of plants.

> **Take Note:** *You should be able to compare and contrast photosynthesis in C3, C4, and CAM plants for the AP Biology exam.*

Multiple Choice Questions

1. The light dependent reactions of photosynthesis
    A. occur in the stroma of the chloroplasts.
    B. occur when carbon dioxide and hydrogen react to form a carbohydrate.
    C. release oxygen as a by-product and split water in the process.
    D. utilize NAD+ as an electron acceptor.
    E. oxidize carbon dioxide during the formation of carbohydrates.

2. Crabgrass, corn, and sugar cane (C4 plants) are adapted to higher temperatures because
    A. CO_2 is obtained during the day and incorporated into a 3-carbon compound.
    B. they require fewer enzymes and have no specialized anatomy.
    C. CO_2 is converted to an acid and stored during the night, and later converted back into CO_2 during the daytime.
    D. CO_2 is first incorporated into a 4-carbon compound and then translocated into bundle sheath cells.
    E. photosynthesis can take place throughout the leaf and stem of the plant.

3. Which of the following organisms is an autotroph?
    A. sordaria
    B. paramecium
    C. algae
    D. *E. coli*
    E. yeast

4. Rubisco
    A. captures CO_2 from the atmosphere and converts it to two 3-carbon sugars.
    B. transports H^+ across the chloroplast membrane in chemiosmosis.
    C. combines with CO_2 to form oxaloacetate in C_4 photosynthesis.
    D. releases a molecule of CO_2 as it is converted to pyruvate.
    E. bonds to ATP and recovers PEP in the CAM pathway.

5. Oxygen is produced when
    A. electrons are accepted by water at the end of photosystem II.
    B. water is split at the initiation of photosystem II.
    C. protons from water are used in the production of carbohydrates in the stroma.
    D. CO_2 is oxidized and carbon is utilized in the production of carbohydrates.
    E. carbohydrates are oxidized in the mitochondria of plants for energy.

6. Which of the following processes will occur with or without oxygen?
    A. fermentation
    B. glycolysis
    C. oxidative phosphorylation
    D. Krebs cycle
    E. citric acid cycle

7. Chlorophyll is similar in structure to a
    A. phospholipid, due to its amphipathic nature.
    B. triglyceride, due to its three long hydrocarbon chains.
    C. protein, due to the magnesium core.
    D. disaccharide, due to its ratio of carbon:hydrogen:oxygen.
    E. cellulose molecule, due to its alternating bonds in the chain.

8. The ATP and NADPH produced in the noncyclic electron pathway
    A. facilitate the decomposition of water to H+ and O_2.
    B. remain in the thylakoid and enter photosystem I.
    C. are used in the light-dependent reactions in the grana.
    D. will be used by the light-independent reactions to reduce CO_2.
    E. combine with electrons to form $NADP^+$.

9. In autumn, the leaves of many trees change from green to red or yellow. Which of the following statements explains this observation?
    A. Leaves produce more chlorophyll as temperatures drop.
    B. Carotenes and xanthophylls are produced only in autumn, when leaves turn yellow and orange.
    C. The chlorophyll breaks down and the accessory pigments are more noticeable.
    D. Additional pigments are manufactured to protect the leaves against the cooling temperatures.
    E. Reduced amounts of sunlight reflect less green light, making it easier to see the other pigments.

10. Which of the following is correctly matched with its function?
    A. NADPH – electron donor
    B. H_2O – electron donor
    C. Thylakoid membrane – site of CO_2 reduction
    D. CO_2 – reducing agent for rubisco
    E. Cytochrome complex – reduction of CO_2

Free Response Question

A student makes the statement that plants do not have mitochondria because they produce ATP in their chloroplasts.

    A. How would you **respond** to this statement?

    B. **Explain** where and how light is absorbed in a plant cell for photosynthesis.

Annotated Answer Key (MC)

1. **C**; The light-dependent reactions of photosynthesis occur in the thylakoid membranes. During this process H_2O is split during the process of photolysis, producing hydrogen ions, electrons, and oxygen. The fixation of carbon dioxide occurs during the Calvin cycle.

2. **D**; The first intermediate formed when crabgrass, corn, and sugar cane photosynthesize is oxaloacetate, a 4-carbon molecule. This is then transferred to the bundle sheath cells where CO_2 is fixed AGAIN, allowing C4 plants to get by with smaller stomata to survive their dry conditions.

3. **C**; Autotrophs can synthesize their own organic molecules from an inorganic source of carbon. Plants and algae are common examples of photosynthetic autotrophs.

4. **A**; Rubisco captures CO_2 to convert it to two 3-carbon sugars in C3 photosynthesis.

5. **B**; The photolysis of water produces oxygen, hydrogen atoms, and electrons.

6. **B**; Glucose is broken down to pyruvate during glycolysis in aerobic (with oxygen) or anaerobic (without oxygen) respiration.

7. **A**; Chlorophyll molecules are located in and around photosystems in the thylakoid membranes of chloroplasts. At the head of the molecule is a porphyrin ring with a magnesium center and a hydrocarbon tail.

8. **D**; ATP and NADPH are used to reduce CO_2 during carbon fixation.

9. **C**; During autumn in the northern hemisphere, when temperatures drop and the daylight hours diminish, chlorophyll breaks down and the accessory pigments are more noticeable. Accessory pigments absorb energy that chlorophyll *a* does not. Carotene, an accessory pigment, is more stable than chlorophyll and persists when chlorophyll has been broken down due to the changing environmental conditions. When it disappears, the remaining carotene causes the leaf to appear yellow. Other pigments, like the anthocyanins, appear when the sugars react and the leaves may appear red.

10. **B**; Water donates the electrons during the light-dependent reactions of photosynthesis. The reduction of CO_2 does not occur in the thylakoid membrane, and rubisco is the reducing agent for CO_2.

Answer to FRQ

PART A (MAX 3)

- Incorrect statement: plants do have mitochondria.
- Mitochondria produce ATP for cellular respiration.
- ATP produced in the chloroplast are used for the production of sugars.

PART B (MAX 8)

- Thylakoid membrane of chloroplast
- Photosystems contain pigments
- Pigment molecules are embedded in a phospholipids bilayer - amphoteric
- Green light reflected, not absorbed OR absorbs most heavily in red and blue regions
- Chlorophyll *a* is at center of photosystem and uses the light energy to boost e-
- Accessory pigments trap light at other wavelengths and transfer the energy to chlorophyll *a*
- Elaboration of reactions in chloroplast (MAX 2 pts)
  - Reduces C in CO_2
  - Specific reactions
  - 9 ATP used to produce one G3P
  - others possible
- cyclic vs. noncyclic electron pathways

CHAPTER 8
CELLULAR RESPIRATION

8.1   Cellular Respiration

Cellular respiration, during which glucose is completely broken down to CO_2 and H_2O, consists of four phases: glycolysis, the prep reaction, the citric acid cycle, and the passage of electrons along the electron transport chain. Like photosynthesis, cellular respiration is also a redox process. Glucose is oxidized to CO_2 and water is formed by the reduction of oxygen:

$$C_6H_{12}O_6 + 6\ O_2 \longrightarrow 6\ CO_2 + 6\ H_2O + Energy$$

Glucose is completely oxidized in a series of step-wise reactions to transform chemical energy from food into metabolic energy in the form of ATP. Oxidation of substrates involves the removal of hydrogen ions and electrons by the coenzymes NAD^+ and FADH, which carry electrons to the electron transport chain. NAD^+ becomes NADH, and FAD becomes $FADH_2$.

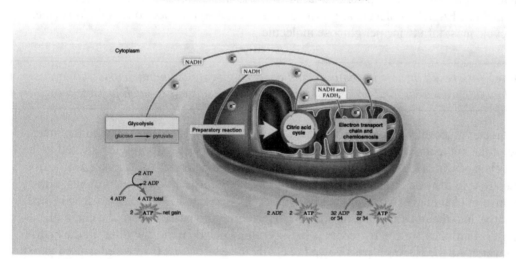

The four phases of complete glucose breakdown

8.2   Outside the Mitochondria: Glycolysis

Glycolysis is the breakdown of glucose to form two molecules of pyruvate. This anaerobic process is a series of enzymatic reactions that occur in the cytoplasm. Glycolysis has two steps: energy investment and energy harvesting. In the energy-investment step, two molecules of ATP are added to glucose and it is split into two molecules of G3P. In the energy-harvest step, there is a net gain of two molecules of ATP by substrate-level ATP synthesis and the production of 2 NADH and two pyruvates.

8.3   Fermentation

The complete oxidation of glucose requires oxygen in order for the electron transport chain to function. In the absence of oxygen (**anaerobic** conditions), however, a limited amount of oxygen can still be produced by **fermentation**. There are two types of fermentation: **lactic acid fermentation** and **alcoholic fermentation**. Fermentation involves glycolysis followed by the reduction of pyruvate by NADH either to lactate, in animals, or to alcohol and CO_2 in yeast. This reduction of pyruvate regenerates NAD^+, so that it can accept more hydrogen ions and electrons from glycolysis in order for a cell to continue to produce ATP. Although fermentation results in only two ATP molecules, it still serves a purpose. In vertebrates, it provides a quick burst of ATP energy for short-term, strenuous muscular activity in vertebrates. The accumulation of lactate as a result of fermentation results in an oxygen debt because oxygen is needed when lactate is completely metabolized to CO_2 and H_2O.

8.4   Inside the Mitochondria

In **aerobic** conditions, when oxygen is available, pyruvate from glycolysis enters the mitochondrial matrix, where **the prep reaction** takes place. This reaction takes place because pyruvate does not directly enter the citric acid cycle. During the prep reaction, CO_2 is removed from pyruvate in a redox reaction that oxidizes pyruvate and reduces NAD^+. Then, CoA is added to the 2-carbon acetyl group that remains. It is this acetyl CoA that enters the citric acid cycle. Since the reaction must take place twice per glucose molecule, two NADH and two CO_2 molecules result.

The acetyl CoA enters the citric acid cycle, a cyclical series of reactions located in the mitochondrial matrix. In the first step, the CoA is removed and the acetyl group is added to oxaloacetate for form citrate. Complete oxidation follows, and two CO_2 molecules, three NADH molecules, and one FADH2 molecule are formed. The cycle also produces one ATP molecule. The entire cycle must turn twice per glucose molecule.

> **Take Note:** *You should be able to compare and contrast glycolysis and the citric acid cycle in terms of location, reactants, and products. Can you explain the steps of each process?*

The final stage of glucose breakdown involves the electron transport chain located in the inner membrane of the mitochondria. The electron transport chain is a series of proteins located in the membrane, each of which is more electronegative than the one before. NADH and $FADH_2$ give up their electrons to the first and second proteins in the chain, respectively. The proteins of the electron transport chain have two functions: 1) to pass electrons, and 2) to pump hydrogen ions from the matrix into the inner membrane space. The electrons are passed down the chain of protein carriers until they are finally received by oxygen, which combines with H^+ in the matrix to produce water. As the electrons pass down the chain, the energy that is released is used to pump H^+ into the intermembrane space, setting up an electrochemical gradient. When H^+ flows down this gradient through the ATP synthase complex, the energy is captured and used to form ATP molecules from ADP and P_i — a process called chemiosmosis. NAD^+ and FADH then return to glycolysis and/or the citric acid cycle so that these processes can continue.

There are 36–38 ATP formed by the complete oxidation of glucose. Four are the result of substrate-level ATP synthesis in glycolysis and the citric acid cycle, and the rest are produced as a result of the electron transport chain. For most NADH molecules that donate electrons to the electron transport chain, three ATP molecules are produced. However, in some cells, each NADH formed in the cytoplasm results in only two ATP molecules because a shuttle, rather than NADH, takes electrons through the mitochondrial membrane. $FADH_2$ results in the formation of only two ATP because its electrons enter the electron transport chain at a lower energy level than those of NADH.

8.5   Metabolic Pool

Carbohydrates, proteins, and fats can be metabolized by entering glycolysis or cellular respiration at different locations. Carbohydrates are metabolized to glucose, which is the starting molecule of glycolysis. Proteins are broken down into their amino acids. The amino group is removed, and then excreted as urea. The carbon skeleton can then enter the degradative pathway after being converted to pyruvate, acetyl CoA, or an intermediate of the citric acid cycle. Fats are broken down to glycerol and fatty acids. Glycerol is converted to G3P, which is an intermediate of glycolysis. The fatty acids are broken down into 2-carbon molecules which are converted to acetyl CoA, the starting molecule of the citric acid cycle. These pathways also provide the metabolites needed for the anabolism of various important substances. Therefore, catabolism and anabolism are interconnected because both kinds of pathways use the same pools of metabolites.

Photosynthesis and cellular respiration can be compared. ATP is created in both processes by an ETC and chemiosmosis. As a result of the ETC in chloroplasts, water is split, while in mitochondria, water is formed. The enzymatic reactions in chloroplasts reduce CO_2 to a carbohydrate, while the enzymatic reactions in mitochondria oxidize carbohydrate with the release of CO_2.

Take Note: *You should be able to compare and contrast photosynthesis and cellular respiration in terms of location, reactants, products, electron carriers, and direction of H+ flow. Can you explain how nutrients other than carbohydrates are utilized for energy?*

Multiple Choice Questions

1. Glycolysis can be described as a process that
    A. is aerobic and produces 2 ATP per glucose molecule.
    B. converts glucose to pyruvate.
    C. yields 34 ATP molecules through the process of chemiosmosis.
    D. produces O2 as a byproduct.
    E. converts glycerol molecules to 36 ATP.

2. A student constructs a respirometer, measuring the changes in gas volume related to the consumption of oxygen by placing crickets in a vial with a pipette attached. Cotton is placed next to the cricket, separating it from potassium hydroxide (KOH) that combines with carbon dioxide (CO_2) to form a precipitate.

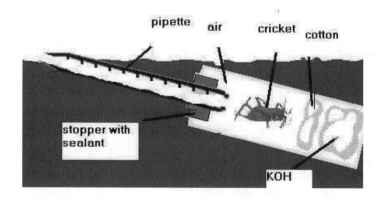

How will this apparatus provide data to calculate the rate of cellular respiration of the cricket?
- A. The student will measure the amount of CO_2 used, which will be measurable on the pipette.
- B. The student will be able to measure the amount of precipitate produced.
- C. The student will measure the amount of O_2 used, which will be measurable on the pipette.
- D. The student will measure the heat created by the cricket by monitoring the water temperature.
- E. The student will measure the total amount of CO_2 and O_2 used to calculate cellular respiration.

3. Trematol is a poison that inhibits the catabolism of lactic acid. How could strenuous exercise be confused with trematol poisoning?
- A. Both cause an increase in pH due to the accumulation of lactic acid.
- B. Both are caused by the lack of lactate dehydrogenase to break down pyruvate.
- C. Trematol inhibits the conversion of pyruvate to lactic acid and decreases the pH of the blood.
- D. Lactic acid is produced anaerobically during strenuous exercise and will accumulate in the muscles.
- E. Trematol converts glucose to lactic acid which accumulates in the mitochondria.

4. During respiration, the greatest number of ATP molecules are produced during
- A. glycolysis.
- B. fermentation.
- C. substrate level phosphorylation.
- D. electron transport system.
- E. citric acid cycle.

5. Which of the following molecules is most commonly used during cellular respiration?
- A. triglyceride
- B. kinase
- C. DNA
- D. cholesterol
- E. glucose

6. In alcoholic fermentation, each pyruvate molecule is directly converted to
- A. NADPH.
- B. NADH.
- C. acetaldehyde.
- D. lactate.
- E. ethanol.

7. You would expect to find the smallest number of mitochondria in which of the following types of cells?
- A. nerve cells
- B. fat storage cells
- C. liver cells
- D. cardiac muscle cells
- E. skeletal muscle cells

8. Which of the following events occurs first during the process of chemiosmosis?
- A. Integral proteins translocate H+ ions across the membrane, creating a pH gradient.
- B. ATP synthase molecules facilitate the movement of H+ ions across the membrane.
- C. ATP synthase phosphorylates ADP.
- D. The electrons are taken up by oxygen to make water.
- E. Oxygen moves across the membrane to bond with H+ ions to create ATP.

9. Which of the following statements is true regarding cellular respiration?
    A. Anaerobic respiration requires oxygen in order to generate ATP.
    B. The citric acid cycle produces NADPH, which releases energy to fuel chemiosmosis.
    C. Oxygen is formed as water combines with electrons at the end of the electron transport chain.
    D. The Krebs cycle occurs in the mitochondria of prokaryotes and eukaryotes.
    E. Glucose is broken down into two pyruvate molecules during glycolysis.

10. Which of the following conclusions does NOT explain the differences in oxygen consumption observed below?

Cumulative Oxygen Consumed (mL)					
Time (minutes)	0	10	20	30	40
Germinating seeds 22°C	0.0	8.8	16.0	23.7	32.0
Dry Seeds (non-germinating) 22°C	0.0	0.2	0.1	0.0	0.1
Germinating Seeds 10°C	0.0	2.9	6.2	9.4	12.5
Dry Seeds (non-germinating) 10°C	0.0	0.0	0.2	0.1	0.2

    A. Temperature increases cause increased activity and respiration in germinating seeds.
    B. The dry seeds may be dormant and metabolizing stored energy reserves very slowly.
    C. The type of seed used has an optimum germinating temperature of 10°C.
    D. To fulfill the high energy needs of a germinating seedling, cellular respiration increases as a seed emerges from dormancy and begins germinating.
    E. After emerging from dormancy, a seed is able to germinate and will respond to more familiar growth stimulating factors such as moisture, light, and soil nutrients.

Free Response Question

When making wine, yeast is added to grape juice. As time passes, the yeast uses up all of the oxygen in the flask, but continues to thrive and produce alcohol. This process is not fast, but takes considerable time.

    A. **Explain** how the yeast can function in an anaerobic environment.

    B. **Identify** what processes are occurring and **explain** why the processes take a long time.

    C. **Compare** the processes of aerobic and anaerobic respiration.

Annotated Answer Key (MC)

1. **B**; During glycolysis, glucose is converted to pyruvate, as the start of aerobic or anaerobic cellular respiration. Two ATP per glucose molecule are produced under anaerobic conditions. 34 ATP molecules are produced in the electron transport chain, O_2 is produced as a byproduct of carbon fixation, and glycerol molecules are converted to 36 ATP during lipogenesis.

2. **C**; This is a common set-up to measure the rate of cellular respiration in lab. The student can measure the amount of O_2 used by observing the pipette. Water temperature changes will not be measurable or entirely attributable to cellular respiration. CO_2 will not be measurable on the pipette, nor will the amount of precipitate produced provide data to calculate an accurate rate of cellular respiration.

3. **D**; During anaerobic respiration, lactate accumulates as a byproduct from the breakdown of pyruvate. Trematol inhibits the breakdown of lactate. Both of these scenarios would actually LOWER blood pH.

4. **D**; The greatest number of ATP (~34) are produced in the electron transport system. Glycolysis produces 2 ATP (net), as does fermentation.

5. **E**; Glucose is the substrate of glycolysis as the first stage of cellular respiration.

6. **C**; Each pyruvate molecule is converted to acetaldehyde. The NADH transfers electrons and hydrogen from NADH to convert it to ethanol, an alcoholic end product.

7. **B**; Mitochondria function in cellular respiration to produce ATP. The least amount of ATP production is required in fat storage cells, as compared to nerve cells, liver cells, and muscle cells.

8. **A**; The basis for chemiosmosis is the production of a pH gradient. Once this is established, ATP formation can occur. ATP synthases span the membrane, and ADP and inorganic phosphate are able to combine to form ATP due to the combined force of the H^+ concentration and electrical gradient.

9. **E**; Aerobic respiration requires oxygen. NADPH is produced during photosynthesis. Mitochondria are unique to eukaryotic cells.

10. **C**; There is a definite optimum germinating temperature of 40°C based upon the data, not 10°C. Temperature does increase activity and respiration in germinating seeds, and the data could be explained by the fact that the dry seeds may be dormant.

Answer to FRQ

PART A (MAX 2)

- Yeast is a facultative anaerobe.
- Facultative anaerobes can utilize oxygen but then switch to anaerobic respiration when no O2 is present.

PART B (MAX 9)

- Fermentation produces alcohol.
- Purpose is to recycle NADH
- Elaboration of NADH as electron acceptor
- MAX 2 pts for specific reactions converting pyruvate to ethanol
  - Pyruvate → acetalaldehyde
  - Acetalaldehyde → ethanol
- Fermentation only yields 2 ATP rather than 36 ATP, making it a slow process
- CO_2 released as by product
- MAX 2 pts for comparison of aerobic and anaerobic respiration
  - Location – cytoplasm vs. mitochondria
  - ETC in aerobic respiration
  - Citric acid cycle

CHAPTER 9
THE CELL CYCLE AND CELLULAR REPRODUCTION

9.1   The Cell Cycle

From the time they are formed by cell division until they divide into two new daughter cells, eukaryotic cells go through a cell cycle that includes (1) interphase and (2) a mitotic stage. **Interphase**, a portion of the cell cycle between nuclear divisions, is composed of three stages: G1, S, and G2. During the **G1 stage**, the cell is growing and performing its normal cellular functions. In the **S stage**, DNA is synthesized. In the **G2 stage**, the cell continues to grow as it prepares to divide during mitosis. The mitotic stage includes **mitosis**, division of the nucleus, and **cytokinesis**, division of the cytoplasm. During the mitotic stage (M), the chromosomes are sorted into two daughters so that each receives a full complement of chromosomes. Most cells of the body are no longer dividing and are arrested in **G0**, a nondividing state. Cells must receive signals to return to G1 from G0 and complete the cell cycle.

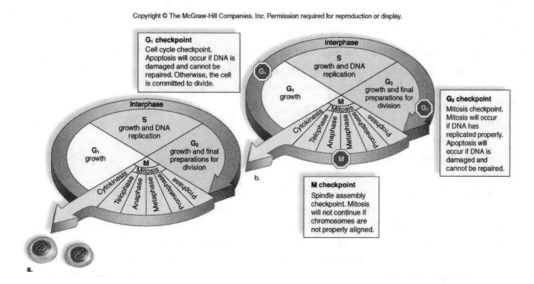

The cell cycle

The cell cycle is regulated by three well-known checkpoints—the **restriction point** (G1 checkpoint), the G2 checkpoint prior to the M stage, and the M stage checkpoint (spindle assembly checkpoint) immediately before anaphase. In order for a cell to proceed past any of these checkpoints, signaling proteins inside the cell called **cyclins** must be present. The G1 checkpoint is the most important of the three checkpoints. It ensures that conditions are favorable and that the proper signals are present, and also checks the DNA for damage. If the DNA is damaged beyond repair, the **p53** protein may initiate **apoptosis**, or programmed cell death. During apoptosis, enzymes called **caspases** bring about destruction of the nucleus and the rest of the cell. Cell division and apoptosis are two opposing processes that keep the number of healthy somatic cells in balance.

> **Take Note:** *In preparation for the AP Biology exam, you should be able to describe the steps of the cell cycle and explain how the cell cycle is regulated as well as the consequences of errors in the regulation of the cell cycle.*

9.2   Mitosis and Cytokinesis

Recall that chromatin—a mixture of DNA and proteins—is found in the nucleus during interphase. The DNA is wrapped around proteins called **histones**, and it is replicated during the S stage of interphase. During mitosis, chromatin condenses into chromosomes. A **duplicated chromosome** consists of two **sister chromatids** attached to each other at a region called the **centromere**. **Kinetochore** proteins are attached to each sister chromatid at the centromere region. The chromosome number stays constant during mitosis because when the cell divides, each daughter cell receives half of each chromosome. Each eukaryotic species has a characteristic number of chromosomes. The total number is called the diploid number, and half this number is the haploid number. The diploid number for humans is 46.

Mitosis occurs in multicellular eukaryotes, and is primarily for the purpose of development, growth, and repair of tissues. Mitosis consists of five phases summarized below.

1. **Prophase**
    - The nucleolus disappears, the nuclear envelope fragments, and the spindle forms between centrosomes.
    - The chromosomes condense and become visible under a light microscope.
    - In animal cells, asters radiate from the centrioles within the centrosomes. Plant cells lack centrioles and, therefore, asters. Even so, the mitotic spindle forms.
2. **Prometaphase** (late prophase)
    - The kinetochores of sister chromatids attach to kinetochore spindle fibers extending from opposite poles.
    - The chromosomes move back and forth until they are aligned at the metaphase plate.
3. **Metaphase**
    - The spindle is fully formed, and the duplicated chromosomes are aligned at the metaphase plate and imaginary plane at the equator of the cell.
    - The spindle consists of polar spindle fibers that overlap at the metaphase plate and kinetochore spindle fibers that are attached to chromosomes.
    - The M stage checkpoint, or spindle assembly checkpoint, must be satisfied before progressing to the next phase. If the chromosomes are properly attached to the spindle, mitosis will proceed.
4. **Anaphase**
    - Sister chromatids separate, becoming daughter chromosomes that move toward the poles. The polar spindle fibers slide past one another, and the kinetochore spindle fibers disassemble.
    - Cytokinesis begins when a cleavage furrow begins to form.
5. **Telophase**
    - Nuclear envelopes re-form, chromosomes begin changing back to chromatin, the nucleoli reappear, and the spindle disappears.
    - Cytokinesis continues and is complete by the end of telophase.

Cytokinesis in animal cells is a furrowing process that divides the cytoplasm. A ring of actin filaments forms at the cleavage furrow, and they contract in order to divide the cell in two. Cytokinesis in plant cells involves the formation of a cell plate from which the plasma membrane and cell wall are completed by vesicles from the Golgi apparatus.

Take Note: *On the AP Biology exam, you may be asked to describe the similarities and differences between mitosis in plant cells and mitosis in animal cells.*

9.3   The Cell Cycle and Cancer

Cancer is uncontrolled cell growth. The development of cancer is primarily due to the mutation of genes involved in control of the cell cycle. Most cancers begin as noncancerous, or **benign** tumors; however, they can become **malignant** as more mutations accumulate and tumor cells fail to respond to cell cycle controls. Cancer cells lack differentiation, have abnormal nuclei, do not undergo apoptosis, form tumors, and undergo metastasis and angiogenesis.

There is a genetic component to cancer, in that two kinds of genes are affected by mutations that lead to uncontrolled cell growth. Proto-oncogenes stimulate the cell cycle after they are turned on by environmental signals such as growth factors. Oncogenes are mutated proto-oncogenes that stimulate the cell cycle without need of environmental signals. They stimulate cyclins, resulting in increased division of cells. Tumor suppressor genes inhibit the cell cycle. Mutated tumor suppressor genes, such as p53, no longer inhibit the cell cycle, allowing unchecked cell division and inhibition of **apoptosis**.

9.4   Prokaryotic Cell Division

Prokaryotes reproduce by **binary fission**, which is a form of asexual reproduction. As a result, the daughter cells that are produced are genetically identical to the parent cell. The prokaryotic chromosome is long loop of DNA that is associated with a few proteins and is found in the nucleoid region of the cell. When binary fission occurs, the chromosome attaches to the inside of the plasma membrane and replicates in order to form an identical chromosome. The cell elongates, and then the chromosomes are pulled apart as the cytoplasm is evenly distributed between the two forming cells. Inward growth of the plasma membrane and formation of new cell wall material divide the cell in two.

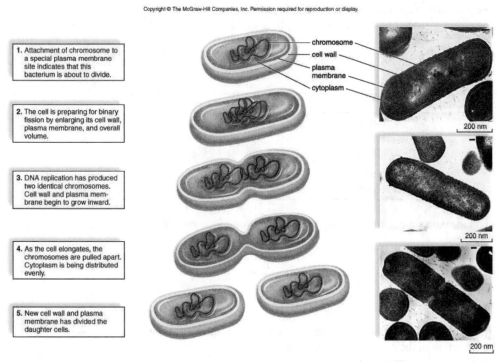

Binary fission

Multiple Choice Questions

Questions 1–4 refer to the following five choices. For each question, select the choice that is most closely related. Each choice may be used once, more than once, or not at all.

    A. centromere
    B. sister chromatids
    C. homologous chromosome
    D. kinetochore
    E. chromatin

1.   DNA being expressed and associated proteins within a nucleus

2.   Identical copies of a chromosome connected by a centromere

3.   Spindle fibers bind to this small region of disc-shaped proteins as an anchor point.

4.   The two daughter strands of a chromosome after it has duplicated

5.   During metaphase of the cell cycle
    A. a diploid set of chromosomes moves towards each pole.
    B. the nucleolus disappears.
    C. spindle formation is complete.
    D. the centromeres divide.
    E. centrosomes migrate away from one another.

6.   A major difference during cell division between plants and animals is
    A. the production of a cell plate in animals and a cleavage furrow in plants.
    B. centriole formation in plant cells.
    C. the fusing of vesicles to form a cell wall in plants.
    D. the presence of asters during spindle formation in animal cells.
    E. the disappearance of a nucleolus in telophase of plant cells.

The data in the table below shows the number of cells counted in different phases of the cell cycle. Use this data to answer questions 7-9.

Phase	Number of Cells
Interphase	350
Prophase	100
Metaphase	40
Anaphase	25
Telophase	75

7.   Based upon the data shown, the majority of cells in mitosis in normal skin cells are in
    A. anaphase.
    B. telophase.
    C. prophase.
    D. interphase.
    E. metaphase.

8.   Which of the following events occurs in the phase that takes the shortest amount of time?
    A. reformation of the nuclear envelopes
    B. division of the centromeres
    C. assembling of the spindle fibers
    D. duplication of the chromosomes
    E. shortening of the kinetochore microtubules

9. The majority of the stomach cells are engaged in what processes of the cell cycle?
   A. S phase of mitosis
   B. expressing their DNA
   C. dividing by mitosis
   D. condensing their chromatin into compact chromosomes
   E. proceeding through the G2 checkpoint into the M phase

10. Which of the following observations would NOT be seen in cancerous stomach cells?
    A. an decrease in the length of the cell cycle
    B. density dependent inhibition
    C. apoptosis of cells without appropriate cyclins
    D. a decreased number of cells in G_0
    E. cells passing through the M1 checkpoint

Free Response Question

The cell cycle consists of interphase, mitosis, and cytokinesis providing each daughter cell with DNA and appropriate cellular framework.

   A. **Identify** the phases of interphase and mitosis. If you had the means to measure the relative amount of DNA during each phase of interphase, each phase of mitosis, and during cytokinesis, what would you **predict**?

   B. **Explain** the regulatory mechanisms that maintain this balance.

   C. **Discuss** how cancer cells disregard this regulation.

Annotated Answer Key (MC)

1. **E**

2. **B**

3. **D**

4. **B**

   For Questions 1–4: Chromatin is a collection of DNA in a cell with all the proteins associated with it. DNA is often observed in chromosomes, composed of identical sister chromatids joined at the centromere. Each sister chromatid has a disk-shaped attachment site for the kinetochore microtubules of the spindle.

5. **C;** Movement towards the poles and division of the centromeres occurs in anaphase. Disappearance of the nucleolus and migration of the centrosomes away from one another occurs during prophase.

6. **D;** Plant cells and animal cells differ in the process of mitosis, but not necessarily in the end products. Plant cells do have a cell plate, whereas animal cells have a cleavage furrow. Centrioles are in animal cells, as are asters during spindle formation. The nucleolus disappears in prophase of both types of cells. However, the cell plate is synthesized by the fusion of vesicles to form the cell membrane for each of the two daughter cells, not the cell wall. This then occurs at the cell plate.

7. **C;** Interphase is part of the cell cycle, but not part of mitosis. Therefore, the 350 cells in interphase cannot be the correct answer.

8. **E**; During anaphase, shortening of the kinetochore microtubules occurs in order to pull the sister chromatids towards the poles.

9. **B**; The majority of the cells are in interphase according to the data.

10. **C**; Apoptosis will NOT occur in cancer cells without appropriate cyclin levels. They will simply keep reproducing (reducing the time spent in the cell cycle and reducing the number of cells in interphase or G_0), and the passing of checkpoints without appropriate protein levels.

Answer to FRQ

PART A (MAX 2)
- Phases of interphase: G1, S, and G2
- Phases of mitosis: prophase, prometaphase, metaphase, anaphase, telophase

PART B (MAX 6)
- Density-dependent inhibition
- Anchorage dependency
- Cell Cycle checkpoints
- Specific DNA damage checkpoints before and after S phase
- Specific spindle checkpoint at M
- Trigger of apoptosis
- Rise and fall of cyclins
- Stable levels of CdKs
- Presence of the APC

PART C (MAX 4)
- Cancer cells are NOT density dependent.
- Cancer cells are NOT anchorage dependent.
- Failure of the spindle checkpoint
- Cancer cells cannot enter G_0

CHAPTER 10
MEIOSIS AND SEXUAL REPRODUCTION

10.1   Halving the Chromosome Number

Meiosis is a type of cell division that ensures that the chromosome number in offspring stays constant generation after generation. A **diploid** somatic cell reduces its chromosome number while dividing to form **gametes**, which are **haploid**. In humans, the diploid number of chromosomes is 46, while the haploid number is 23. The formation of gametes (sperm and egg) by meiosis is important for sexual reproduction.

In the nucleus of a diploid cell, the chromosomes are arranged in pairs called **homologous chromosomes** (homologues). Homologous chromosomes have the same size, shape, and centromere position. Each chromosome of the pair contains similar genetic information because each has an **allele** for the genes located on that chromosome.

Meiosis requires two cell divisions and results in four daughter cells. As in mitosis, replication of DNA takes place before meiosis begins. During meiosis I, the homologues undergo synapsis (resulting in a bivalent or tetrad) and crossing over before they align independently at the metaphase plate. The daughter cells receive one member of each pair of homologous chromosomes. There is no replication of DNA during interkinesis. During meiosis II, the sister chromatids separate, becoming daughter chromosomes that move to opposite poles as they do in mitosis. The four daughter cells contain the haploid number of chromosomes and only one of each kind. The fate of the daughter cells produced by meiosis varies according to species. While the haploid cells of animals will become the gametes, in plants they will become spores that will produce gametes by mitosis.

10.2   Genetic Variation

Sexual reproduction is an adaptive advantage that ensures that offspring have a different genetic makeup than their parents. Meiosis contributes to genetic variability in two ways: crossing-over and independent assortment of the homologous chromosomes.

When homologous chromosomes lie side by side during **synapsis**, nonsister chromatids may exchange genetic material in a process called **crossing-over**. Due to crossing-over, the chromatids that separate during meiosis II have a different combination of genes, and the resulting chromosomes are called **recombinant chromosomes**. When the homologous chromosomes align at the metaphase plate during metaphase I, the maternal or the paternal chromosome can be facing either pole. This is called **independent assortment**. The way that one pair of chromosomes aligns at the metaphase plate is completely independent of how the other 22 pairs align themselves. Since there are 23 pairs of chromosomes in a human cell, there are 2^{23} possible combinations of chromosomes in the gametes. **Fertilization**—the fusion of gametes— enhances the genetic variation from meiosis because it is random. There are many ways that sperm and egg can come together to form a zygote.

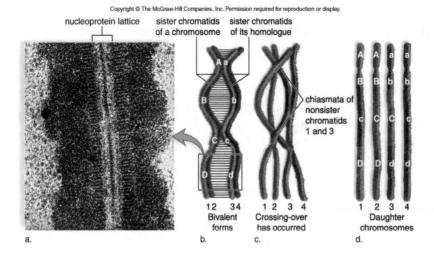

Crossing over occurs during meiosis I.

Genetic variation is important for the long term survival of a species because it confers an advantage for populations of organisms in a changing environment. The organisms that can easily adapt to the new environment can survive and reproduce.

Take Note: *For the AP Biology exam, you should have a clear understanding of how genetic variation occurs and why it is important. Consider this when you begin to study prokaryotes, as you should be able to compare and contrast the ways that prokaryotes obtain genetic variation with the ways that eukaryotic cells obtain genetic variation.*

10.3   The Phases of Meiosis

Meiosis I is divided into four phases, summarized below. Review the pictures in your textbook as you read over this section.

1. **Prophase I**
    *   Bivalents form and crossing-over occurs as chromosomes condense.
    *   The nuclear envelope fragments.
2. **Metaphase I**
    *   Bivalents independently align at the metaphase plate.
3. **Anaphase I**
    *   Homologous chromosomes separate, and duplicated chromosomes move to poles.
4. **Telophase I**
    *   Nuclei become haploid, having received one duplicated chromosome from each homologous pair.

Meiosis II is also divided into four phases, as summarized below.

1. **Prophrase II**
    *   Chromosomes condense, and the nuclear envelope fragments.
2. **Metaphase II**
    *   The duplicated chromosomes align at the metaphase plate.

3. **Anaphase II**
   - Sister chromatids separate, becoming daughter chromosomes that move to the poles.
4. **Telophase II**
   - Four haploid daughter cells are genetically different from the parent cell.

Take Note: *For the AP Biology exam, you should be able to describe the process of meiosis in detail. Can you draw diagrams of how the chromosomes change during meiosis given a diploid number of 4 or 6?*

10.4   Meiosis Compared to Mitosis

Mitosis and meiosis can be compared as outlined in the following table.

Mitosis	Meiosis
Somatic cells are produced.	Gametes are produced.
One division.	Two divisions.
Does not contribute to genetic variation.	Contributes to genetic variation by crossing-over and independent assortment.
Daughter cells are identical to parent cell.	Daughter cells are genetically different from the parent cell.
Prophase No pairing of chromosomes. No crossing-over.	*Prophase* Pairing of homologous chromosomes. Crossing-over during prophase I.
Metaphase Duplicated chromosomes plate at metaphase plate.	*Metaphase* Bivalents at metaphase plate.
Anaphase Sister chromatids separate, becoming daughter chromosomes that move to the poles.	*Anaphase* Homologous chromosomes separate and move to the poles in anaphase I. In anaphase II, sister chromatids separate.
Telophase Daughter nuclei have the parent cell number of chromosomes; chromosome number is maintained. (2n ➔ 2n)	*Telophase* Daughter nuclei have the haploid number of chromosomes; chromosome number is halved. (2n ➔ 1n)

Take Note: *In preparation for the AP Biology Exam, you should be able to describe the similarities and differences between mitosis and meiosis.*

10.5   The Human Life Cycle

Any organism that reproduces sexually undergoes meiosis. In the animal life cycle, only the gametes are haploid; the individual is always diploid. In plants, there is an **alternation of generations**, where a diploid phase alternates with a haploid phase. Meiosis produces haploid spores that develop into a multicellular haploid adult (the sporophyte) that produces the gametes. The gametes fuse during fertilization to form a diploid gametophyte. In unicellular protists and fungi, the zygote undergoes meiosis, and spores become a haploid adult that gives rise to gametes.

During the life cycle of humans and other animals, meiosis is involved in **spermatogenesis** and **oogenesis**. Whereas spermatogenesis occurs in males and produces four **sperm**, oogenesis occurs in females and produces one **egg** and two to three nonfunctional **polar bodies**. When a sperm fertilizes an egg, the **zygote** has the diploid number of chromosomes. Locate a picture of spermatogenesis and oogenesis in your textbook. Review it carefully before moving on.

Take Note: *In preparation for the AP Biology exam, you should be able to describe the similarities and differences between spermatogenesis and oogenesis.*

10.6   Changes in Chromosome Number and Structure

Nondisjunction is failure of the chromosomes to properly separate during meiosis I or meiosis II. Nondisjunction results in **aneuploidy** (extra or missing copies of chromosomes). **Monosomy** occurs when an individual has only one of a particular type of chromosome ($2n - 1$); **trisomy** occurs when an individual has three of a particular type of chromosome ($2n + 1$).

Down syndrome is a well-known trisomy in human beings resulting from an extra copy of chromosome 21. The chances of Down syndrome increase with the age of the mother. Aneuploidy of the sex chromosomes is tolerated more easily than aneuploidy of the autosomes, which is often lethal. Turner syndrome (XO), Klinefelter syndrome (XXY), poly-X females, and Jacobs syndrome (XYY) are examples of aneuploidy in the sex chromosomes.

Changes in chromosome structure are caused by abnormalities in crossing-over, resulting in deletions, duplications, inversions, and translocations within chromosomes. The accompanying figure explains each of these mutations. Many human syndromes, including Williams syndrome, cri du chat syndrome, and Alagille syndrome, result from changes in chromosome structure.

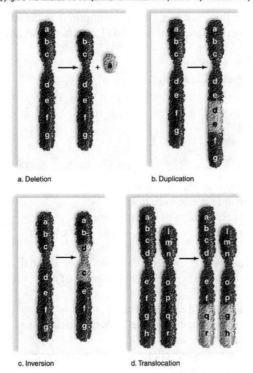

a. Deletion

b. Duplication

c. Inversion

d. Translocation

Types of chromosomal mutations.

Take Note: *For the AP Biology exam, you should be able to explain how nondisjunction occurs and give specific examples of the consequences of changes in chromosome number and structure in humans.*

Multiple Choice Questions

1. Genetic recombination occurs due to
   A. crossing-over between homologous chromosomes yielding a different combination of genes in one gamete than in another.
   B. proper alignment of paternal chromosomes and maternal chromosomes on the proper side of the metaphase plate.
   C. a differing number of homologous chromosomes in each of the daughter cells after meiosis I.
   D. the production of genetically identical daughter cells after mitosis.
   E. the lack of pairing of chromosomes during prophase II of meiosis.

2. Mitosis and meiosis are similar in that
   A. they both produce haploid daughter cells.
   B. sister chromatids move towards the poles in both.
   C. homologous chromosomes pair and undergo crossing over in both.
   D. they both contain two nuclear divisions.
   E. the cells produced in both are genetically identical to the parent cell.

3. Meiosis reduces the number of chromosomes. However, this may occur at different points in the life cycles of animals, plants, and fungi. Which of the following statements does NOT accurately describe these life cycles?
   A. In animals, the haploid stage consists only of the gametes.
   B. In plants, there is both a diploid and a haploid muticellular form.
   C. In fungi, the dominant form is usually haploid.
   D. In plants and fungi, the zygote is diploid.
   E. In plants, the spore is a multicellular diploid form.

4. Sexually reproducing organisms
   A. depend on mutations to generate variation.
   B. produce great numbers of offspring within a limited amount of time.
   C. depend on crossing-over during mitosis to generate genetic variation.
   D. increase genetic variation due to an independent assortment of chromosomes.
   E. are less likely to deal with environmental disturbances due to a lack of genetic variation.

5. Aneuploidy, defined as an abnormal number of chromosomes, is the most common chromosomal abnormality. Which of the following events will result in Down syndrome, an example of a disorder caused by the inheritance of an extra chromosome 21?
   A. the incorrect separation of this pair of sister chromatids during meiosis II
   B. the failure of two sets of homologous chromosomes to line up properly during metaphase I
   C. the nondisjunction of this set of homologues during meiosis II
   D. a crossing over of genetic information between homologous chromosomes during prophase I
   E. the incomplete formation of the cleavage furrow in telophase.

6. During pregnancy, a cell undergoes what appears to be a normal mitotic nuclear division. However, two daughter cells with 2n − 1 and 2n + 1 complement of chromosomes are produced, causing a population of cells to be abnormal. Which of the following events is most likely responsible for this disorder?
   A. Two sister chromatids did not separate into the proper daughter cells during anaphase.
   B. Homologous chromosomes were sorted incorrectly during prophase.
   C. The protein structure of nonkinetochore microtubules was defective and incorrectly shortened during anaphase.
   D. Duplication of DNA during cytokinesis did not occur properly.
   E. The cleavage furrow did not completely separate the two daughter cells.

The diagram below shows two homologous chromosomes in various possible phases of either mitosis or meiosis. Use the diagram to answer Questions 7–10.

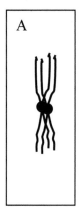   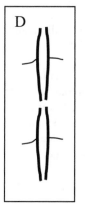 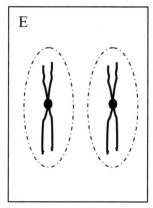

A B C D E

7. Completion of meiosis I

8. Point at which independent assortment occurs

9. Point at which crossing-over occurs

10. Which diagrams show a possible arrangement of chromosomes during either mitosis or meiosis?
    A. A and B
    B. B, C, and D
    C. A, B, and C
    D. C, D, and E
    E. B, C, D, and E

Free Response Question

Meiosis is often called reduction division.

> A. **Identify** the primary purpose of meiosis and **explain** the term "reduction division."

> B. **Compare and contrast** the processes of mitosis and meiosis.

Annotated Answer Key (MC)

1. **A**; Genetic recombination is definitely due to crossing-over during prometaphase I of meiosis between homologous chromosomes. The possible outcomes of random alignment of these homologous chromosomes during metaphase also contributes to genetic diversity of gametes.

2. **B**; Mitosis does NOT produce haploid daughter cells, contain homologous chromosomes, or have two divisions. Meiosis does NOT produce genetically identical daughter cells due to crossing-over and random alignment during the first division. However, in mitosis and in the second division of meiosis, sister chromatids do move towards the poles.

3. **E**; In plants, the sporophyte produces spores by meiosis. These develop into a gametophyte. Both the spores and the resulting gametophyte are haploid, meaning they only have one set of homologous chromosomes.

4. **D**; Sexually reproducing organisms use the process of meiosis (not mitosis as choice C asserts) to increase genetic variation. This occurs due to crossing-over and the independent assortment of chromosomes. It also occurs during fertilization based upon random joining of sperm and eggs. Sexually reproducing organisms, therefore, are more likely to have genetic variation; however, they do produce a smaller number of offspring in a given time due to reproductive complexity.

5. **A**; If the sister chromatids fail to separate during meiosis II there will be two copies of chromosome #21 instead of one in the gamete, and if fertilized will most likely combine with an additional copy of chromosome #21 to create an individual with trisomy 21 or Down syndrome. Incomplete formation of the cleavage furrow would result in a much more serious outcome than trisomy. Homologous chromosomes do not separate in meiosis II (choice C) and the failure of two sets of homologous chromosomes to align during metaphase I would not necessarily result in trisomy of one of the sets.

6. **A**; Similar to Question 5, aneuploidy may occur if the daughter cells do not have the correct complement of chromosomes. While the other choices may cause abnormalities, none would correctly produce $2n - 1$ and $2n + 1$ daughter cells.

7. **E**; At the completion of meiosis I, two haploid cells will be produced.

8. **E**; This is also where independent assortment occurs.

9. **A**; Crossing-over occurs during prometaphase I of meiosis, where they form tetrads.

10. **B**; E is only in mitosis (haploid daughter cells) and A is only in meiosis (homologous chromosomes).

Answer to FRQ

PART A (MAX 4)
- Produce haploid gametes
- Definition of diploid
- Definition of haploid
- Reduction indicates a decrease from diploid parent cell to haploid daughter cells.
- Division indicates that a separation of chromosomes takes places (actually twice).

PART B (MAX 8)

Characteristic	Mitosis	Meiosis
Number of divisions	One	Two
Separation of homologues	No	Yes
Crossing-over	No	Yes
Type of gametes produced	Diploid	Haploid
Separation of chromatids	Yes	Yes
Type of cells	Somatic	Gametic
Resulting number of cells	2	4
Number of S phases	1	1
Centromere division	Yes, at anaphase	No, at anaphase I; yes, at anaphase II
Type of process	Conservative; identical cells produced	Variation among the products of meiosis

CHAPTER 11
MENDELIAN PATTERNS OF INHERITANCE

> **Take Note:** *After completing this chapter, you should be able to determine the probability of genetic crosses using Punnett squares and interpret and analyze pedigree charts.*

11.1 Gregor Mendel

Until the late nineteenth century, it was believed that both parents contributed equally to the genetic makeup of their offspring and that an organism was intermediate to that of its parents. This concept was called the blending concept of inheritance; however, it did not explain all observed patterns of inheritance.

Gregor Mendel used the garden pea as the subject in his genetic studies. In contrast to previous plant breeders, his study involved the nonblending traits of the garden pea. Mendel studied seven different traits of the pea plants, all of which were distinct; there were no intermediate characteristics. He cross-pollinated true-breeding plants and observed thousands of plants. Mendel applied mathematics, followed the scientific method very closely, and kept careful records. His work led him to develop a **particulate theory of inheritance**, effectively disproving the blending theory of inheritance.

11.2 Mendel's Laws

In his initial experiments, Mendel conducted **monohybrid crosses**, where he cross-pollinated pea plants that were **true breeding** (producing offspring exactly like themselves) for a single trait. For instance, he cross pollinated a true breeding tall plant with a true breeding short plant. He defined these **homozygous** parents as the **P generation**. The offspring of this cross, which were all tall plants, is the F_1 **generation**. When Mendel crossed these **heterozygous** plants with other heterozygous plants from the F_1 generation, he found that the recessive phenotype reappeared in about ¼ of the F_2 plants; there was a 3:1 phenotypic ratio. From these observations, Mendel proposed his first law, the **Law of Segregation**, which states that the individual has two factors for each trait, and these factors separate from each other during gamete formation. These factors come back together during fertilization, giving each offspring two factors for each trait.

The factors that Mendel discovered were later named **genes**. Different versions of genes are called **alleles**. The place on a chromosome where an allele is located is called the gene **locus**. The **dominant** allele (always written with a capital letter) masks the expression of the **recessive** allele (written with a lowercase letter) whenever they appear together on homologous chromosomes. In Mendel's monohybrid cross, for example, the dominant allele is represented by T and the recessive allele by t. The true breeding plants in the P generation were homozygous, TT and tt, and the F_1 generation were heterozygous, Tt. These plants displayed the dominant **phenotype** (physical appearance) even though they had a different **genotype** than the tall plants in the P generation. The genotype indicates the alleles for a particular trait that are present in an individual.

Mendel also conducted **dihybrid**, or two-trait crosses, in which he studied the offspring of plants that were true-breeding tall with green pods and true-breeding short with yellow pods. Individuals of the F_1 generation showed both dominant characteristics, but there were four phenotypes among the F_2 offspring in a ratio of 9:3:3:1. This observation allowed Mendel to deduce the **Law of Independent Assortment**, which states that during gamete formation, one pair of alleles separates from one another independently of those of another pair. Therefore, all possible combinations of parental alleles can occur in the gametes.

The laws of probability can be used to calculate the expected phenotypic ratio of a cross. Punnett squares allow us to use the product of the law of probability to calculate all of the possible genotypes among the offspring, and then the sum law can be used to arrive at the phenotypic ratio. A large number of offspring must be counted in order to observe the expected results and ensure that all possible types of sperm have fertilized all possible types of eggs, as is done in a Punnett square.

> **Take Note**: *For the AP Biology exam, you should have a clear understanding of the Law of Segregation and the Law of Independent Assortment. How are these two laws related to meiosis?*

A **testcross** is used to test whether an individual showing the dominant phenotype is homozygous dominant or heterozygous. In order to determine the genotype of the dominant individual, it is crossed with another organism (of the same species) that shows the recessive phenotype, because its genotype is known. By observing whether any offspring are recessive, the genotype of the dominant individual can be determined. The two-trait testcross allows an investigator to test whether an individual showing two dominant characteristics is homozygous dominant for both traits or for one trait only, or is heterozygous for both traits. If all of the offspring from the testcross are dominant for both traits, the dominant parent is homozygous dominant for these traits. On the other hand, if the offspring of the testcross show four different phenotypes in a 1:1:1:1 ratio, the parent is heterozygous for both traits.

Studies have shown that many human traits and genetic disorders can be explained on the basis of simple Mendelian inheritance. These disorders are controlled by one pair of alleles on an **autosome** (any chromosome that is not a sex chromosome). When studying human genetic disorders, biologists often construct **pedigree charts** to show the pattern of inheritance of a characteristic within a family. The pattern shown in the pedigree indicates the manner in which a characteristic is inherited. When a disorder is **autosomal recessive**, the normal allele is dominant and the disorder is shown in individuals who are homozygous recessive because they inherited the trait from two **carrier** parents. Carriers are individuals who have the normal phenotype but "carry" the recessive allele because they are heterozygous. Examples of autosomal recessive disorders are methemoglobinemia, cystic fibrosis, and Niemann-Pick disease.

> **Take Note:** *In preparation for the AP Biology exam, you should be able to interpret and analyze a pedigree chart. Ask your teacher for some practice problems to assist you with learning this skill.*

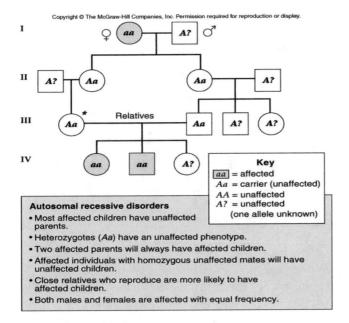

Autosomal recessive pedigree

Autosomal dominant disorders are seen in individuals who are either homozygous dominant or heterozygous, as the recessive allele is normal. Diseases that are autosomal dominant include Osteogenesis imperfecta and hereditary spherocytosis.

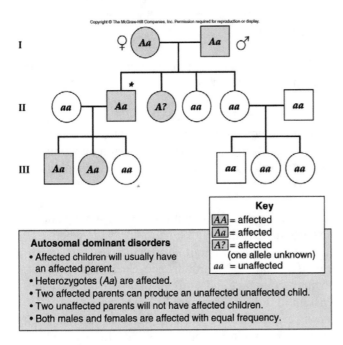

Autosomal dominant pedigree

11.3 Extending the Range of Mendelian Genetics

Other patterns of inheritance have been discovered since Mendel's original contribution. For example, some genes have more than two or **multiple alleles**, although each individual organism has only two alleles. The inheritance of blood type in human beings is an example of this. There are three alleles that determine blood type: I^A, I^B, and i. I^A and I^B are **codominant** to each other, but each is dominant to i; thus, there are multiple genotypes that result in four different phenotypes for blood type:

Genotype	Phenotype
$I^A I^A$, $I^A i$	Type A
$I^B I^B$, $I^B i$	Type B
$I^A I^B$	Type AB
Ii	Type O

With **incomplete dominance**, the phenotype of heterozygote F_1 individuals is intermediate between the parent phenotypes who are both homozygous for either allele of a trait. This does not support the blending theory, because the parent phenotypes reappear in F_2. Let's look at incomplete dominance in four-o'clocks as an example, where the alleles for flower color are red ($R_1 R_1$) and white ($R_2 R_2$). When true-breeding red flowers are crossed with true-breeding white flowers, the F_1 generation consists of plants with pink flowers ($R_1 R_2$). In the F_2 generation, however, the offspring are red, pink, and white in a 1:2:1 ratio.

An example of incomplete dominance in humans is familial hypercholesterolemia. Related to incomplete dominance is **incomplete penetrance**, in which some traits that are dominant may not be expressed due to unknown reasons. For instance, the allele for polydactyly (one or more extra digits on the hands or feet) is inherited in an autosomal dominant manner; however, not all individuals who inherit the dominant allele express the trait.

In **pleiotropy**, one gene has multiple phenotypic effects. Marfan syndrome, porphyria, and sickle-cell disease are examples of human disorders with pleiotropic effects.

Polygenic traits are controlled by several genes on different chromosomes that have an additive effect on the phenotype. The results in the offspring are quantitative variations. Usually a bell-shaped curve is seen because environmental influences bring about many intervening phenotypes, as in the inheritance of height in human beings. **Multifactorial traits** are controlled by polygenes in addition to the environment, as in the case of skin color and eye color.

Sex chromosomes in mammals determine the sex of the individual, with XX being female and XY being male. Experimental support for the chromosome theory of inheritance came when Morgan and his group were able to determine that the gene for a trait unrelated to sex determination, the white-eyed allele in *Drosophila*, is on the X chromosome. Such traits are called **X-linked traits** because the alleles are not carried on the Y chromosome. Therefore, when doing X-linked genetics problems, it is the custom to indicate the sexes by using sex chromosomes and to indicate the alleles by superscripts attached to the X. The Y is blank because it does not carry these genes. Color blindness, Menkes syndrome, adrenoleukodystrophy, and hemophilia are examples of X-linked recessive disorders in humans.

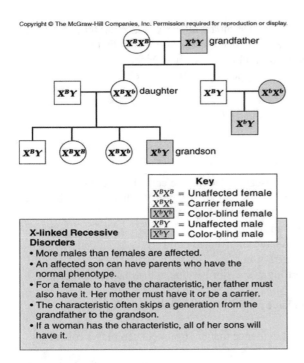

Key
X^BX^B = Unaffected female
X^BX^b = Carrier female
X^bX^b = Color-blind female
X^BY = Unaffected male
X^bY = Color-blind male

X-linked Recessive Disorders
• More males than females are affected.
• An affected son can have parents who have the normal phenotype.
• For a female to have the characteristic, her father must also have it. Her mother must have it or be a carrier.
• The characteristic often skips a generation from the grandfather to the grandson.
• If a woman has the characteristic, all of her sons will have it.

X-linked recessive pedigree

Take Note: *For the AP Biology exam, you should be able to compare and contrast the different patterns of inheritance and describe at least one example of each.*

Multiple Choice Questions

1. Phenylketonuria (PKU) is an autosomal recessive disorder where afflicted individuals are incapable of metabolizing the amino acid phenylalanine and it accumulates in nerve cells of the brain. What is the probability that a man and woman who are both phenotypically normal, but are carriers, will have a child with PKU?
   A. 1/4
   B. 1/2
   C. 1/3
   D. 1/6
   E. 1/8

2. A family has four children, whose blood types are A, B, AB, and O. What are the possible blood types of the parents?
   A. A and O
   B. O and B
   C. B and AB
   D. A and B
   E. A and AB

3. Skin color is polygenic and dependent upon many genes. Dominant alleles produce darker pigmented skin, whereas recessive alleles produce lighter colored skin. What is the probability that a fair-skinned invidual (aabbccDdEeff) and a relatively dark-skinned invidiual (AABbCCDdeeFF) will have a child who is also relatively dark skinned with the specific genotype of (AABBccDdEeFf)?

    A. 1/64
    B. 1/8
    C. 1/2
    D. 0
    E. 1/4

4. Cystic fibrosis is inherited as an autosomal recessive disease affecting the cells that produce saliva, sweat, digestive juices, and mucus. Evidence to support this particular mode of inheritance includes the fact that

    A. cystic fibrosis is more common in males than females.
    B. parents who are phenotypically normal can have an afflicted child.
    C. individuals with cystic fibrosis are all heterozygous.
    D. females have a decreased chance of inheriting cystic fibrosis due to the inactivation of the second X chromosome.
    E. if both parents are afflicted the child will have a 50% chance of inheriting the disease.

5. Which of the following statements provides the best evidence that environment can play a role in determining the phenotype of an organism?

    A. A primrose develops white flowers when grown above 32°C and red flowers when grown at 24°C.
    B. The more genes involved, the more continuous is the variation in phenotypes.
    C. True-breeding red- and white-flowered four-o'clocks produce pink-flowered offspring at any temperature.
    D. A curly-haired Caucasian and a straight-haired Caucasian will have wavy-haired offspring regardless of the time of year.
    E. Some mutations are caused by a gene that moves from another location in the genome.

6. Paula and Bernie have a child named Liz. Paula is blood type A−, and Liz is type B+. What are the possible blood types of Bernie?

    A. AB+ or B+
    B. AB− or O+
    C. A+ or B+
    D. AB− or AB+
    E. A− or O−

7. A man with type A blood and a woman with type B blood have a child. If this child has type O blood, which of the following statements must be true?

    A. The father could be homozygous A.
    B. The mother has homozygous B blood type.
    C. The maternal grandmother is homozygous A.
    D. The paternal grandfather must be blood type A.
    E. The paternal grandmother could have type O blood.

8. Coat color in Labrador retrievers is controlled by the inheritance and interaction of two genes. Black coat color is dominant to chocolate, but yellow Labrador retrievers will be produced if a second dominant gene allowing the ability to express the pigment is not inherited. Two black Labrador retrievers, heterozygous at both loci, are mated. What is the chance that they will produce yellow offspring?
    A. 1/64
    B. 9/16
    C. 3/16
    D. 1/4
    E. 1/8

9. In peas, the tall allele is dominant over the short allele. Which of the following statements is true about the cross between a homozygous dominant tall plant and a short plant and the F_1 offspring that is allowed to self-fertilize?
    A. The F_2 offspring are all tall.
    B. Half of the F_2 offspring are short.
    C. All of the F_1 offspring are short.
    D. Only ¼ of the F_2 offspring are short
    E. Half of the F_1 offspring are short.

10. Paula and Bernie are both carriers of hemophilia. What are the chances that at least two of their three children will inherit the disease?
    A. 1/4
    B. 1/8
    C. 3/64
    D. 1/16
    E. 10/64

Free Response Question

The particulate theory of inheritance as proposed by Mendel bases itself on two basic laws: the law of independent assortment and the law of segregation.

    A. **Explain** how these two laws contribute to Mendel's model of inheritance.

    B. **Discuss** how two of the following concepts go beyond Mendel's theory.
        • Pleiotropy
        • Polygenic inheritance
        • Epistasis
        • Multiple alleles
        • Incomplete dominance

Annotated Answer Key (MC)

1. **A**; As carriers, both individuals have genotypes Pp. Therefore, they have a 25% chance of producing a child who is PP, a 50% chance of producing a carrier (Pp) child, and a 25% chance of having an afflicted (pp) child.

2. **D**; Blood type is a codominant trait in humans. While A and B are both dominant over O, they are codominant to each other. A and O parents could only produce A, and possibly O, offspring. B and O parents could only produce B, and possibly O, offspring. An AB parent is incapable of having an O offspring. Both parents must be heterozygous in this scenario to produce children of all four blood types.

3. **D**; AA x aa will produce 100% Aa offspring. The offspring described in the question is AA, which is impossible given the genotype of the fair-haired parent. Therefore, the answer is 0.

4. **B**; Autosomal recessive diseases often "skip" a generation, where both parents can be carriers (Ff x Ff) and produce an afflicted child (ff). It is not sex linked (as answers A and D suggest) and only has a 25% chance of being passed on with carrier parents.

5. **A**; Primroses can develop different colored flowers at different temperatures, displaying the importance of the environment on phenotype.

6. **A**; Bernie cannot be blood type O. A parent who is blood type A can only pass on an A or an O allele. Thus, the B allele in the daughter must have come from the father. So the father must be either blood type B or AB. Likewise the (+) Rh factor must have come from the dad as well, so dad must be + for the Rh factor.

7. **E**; The father cannot be homozygous A and have an O child. Likewise, the mother cannot be homozygous B and have an O child. If the maternal grandmother is homozygous A, she could not have a daughter who is B. If the paternal grandfather is A, he does not necessarily have to have a son who is A; depending on the paternal grandmother, he could be many blood types. However, the paternal grandmother COULD have O blood type if the paternal grandfather is homozygous or heterozygous A or AB.

8. **D**; 9/16 of the puppies will be black (BBEE , BbEE , BBEe , BbEe) 3/16 will be chocolate (bbEE, bbEe) and 4/16 or ¼ will be yellow (BBee, Bbee, bbee) as they do not have the gene to deposit the color, regardless of the inheritance.

9. **D**; In this cross (TT × tt) produce an F1 offspring (Tt). This offspring is allowed to self fertilize and produces 1 TT: 2 Tt: 1tt, with only ¼ of the offspring being short.

10. **E**; As carriers, Paula and Bernie have a ¼ chance of having a child with hemophilia. In order to fulfill the question, the first and second child could have the disease, the second and third child could have the disease, the first and third child could have the disease, or all three children could have the disease. There is a 1/4 chance for each child to get the disease and a 3/4 chance each will not be afflicted. So the probabilities are as follows: 27/64 for three unaffected children (3/4 × 3/4 × 3/4), 1/64 for three with the disease (1/4 × 1/4 × 1/4), 3 × 9/64 for one with the disease and 3 × 3/64 for two with the disease.

Answer to FRQ

PART A (MAX 6)

Law of Segregation
- Diploid cells have pairs of genes.
- The genes are on pairs of homologous chromosomes.
- During meiosis, the genes are separated from each other.
- In each generation, an organism inherits two genes for a trait.
- One gene is from each parent.

Law of Independent Assortment
- During meiosis, genes on pairs of homologous chromosomes have been sorted.
- The sorting of one gene pair occurs independently of other gene pairs.
- The inheritance of one gene does not influence the sorting of another gene.

PART B (MAX 6)

Pleitropy
- The expression of alleles of one gene have many effects.
- The effects may be positive or negative.
- The effects may not be present in the same time in the individual.

Polygenic Inheritance
- A phenotype can be attributed to the inheritance of alleles of one or more genes.
- The phenotype can vary in degrees.
- Generally observable in a continuous fashion (i.e., "bell curve")

Epistasis
- An interaction between genes
- Inheritance of one gene affects the expression of another.

Multiple Alleles
- Three or more forms of the same gene
- Two forms can be present in a diploid individual.
- Interactions between genes may be dominant/recessive or codominant.

Incomplete Dominance
- One allele is not fully dominant over another.
- Both alleles are expressed.

CHAPTER 12
MOLECULAR BIOLOGY OF THE GENE

> **Take Note:** *This chapter focuses on the structure of DNA and gene expression, the synthesis of proteins from the code in DNA. Your understanding of the material in this chapter is vital as you proceed through the next two chapters.*

12.1   The Genetic Material

DNA, **deoxyribonucleic acid**, was discovered as the hereditary material in the middle of the twentieth century. There were a number of scientists who contributed to the understanding of the structure and function of DNA. Griffith injected strains of pneumococcus (*Streptococcus pneumoniae*) into mice and observed that healthy mice died when injected with live S strain bacteria, while healthy mice injected with either the heat-killed S strain or the live R strain of the pneumococcus did not. However, when heat-killed S strain was injected into healthy mice along with live R strain bacteria, virulent S strain bacteria were recovered from the dead mice. Griffith concluded that the R strain had been transformed by some substance passing from the dead S strain to the live R strain. Twenty years later, Avery and his colleagues reported that the transforming substance is DNA.

DNA is made of four different nucleotides: the **purines**, adenine (A) and guanine (G), and the **pyrimidines**, cytosine (C) and thymine (T). Recall from Chapter 3 that each nucleotide in DNA has three components, a nitrogen-containing base, a deoxyribose sugar, and a phosphate group.

To discover the structure of DNA, Chargaff performed a chemical analysis of DNA and found that the amounts of A and T and the amounts of G and C were the same, and that the amount of purine equals the amount of pyrimidine. Franklin prepared an X-ray photograph of DNA that showed that it is helical, has repeating structural features, and has certain dimensions. Watson and Crick built a model of DNA in which the sugar-phosphate molecules make up the sides of a twisted ladder, and the complementary-paired bases are the rungs of the ladder. The A-T base pair is held together by two hydrogen bonds, while the G-C base pair is held together by three hydrogen bonds. The strands of DNA are **antiparallel** to each other, meaning the 5′ phosphate end of one strand is opposite the 3′ sugar end of the other strand.

> **Take Note:** *For the AP Biology exam, you should be able to explain how DNA was discovered to be the genetic material. How is the structure of DNA suited to its function?*

12.2   Replication of DNA

During **DNA replication**, the two strands of a single DNA molecule are copied to produce two new molecules, each of which is an exact copy of the parent molecule. DNA replication is **semiconservative** because each new molecule of DNA consists of one parental strand and one new strand.

DNA replication begins when a **helicase** enzyme unwinds and unzips the two strands, forming a **replication fork**. Binding proteins attach to each strand to keep them from coming back together before the complementary bases are added by the enzyme **DNA polymerase**, which joins the nucleotides together and proofreads them to make sure the bases have been paired correctly. In order to bind to the DNA and add the complementary bases, DNA polymerase must bind to an RNA primer that is complementary to a short segment of the DNA. Because DNA is synthesized in a 5′ to 3′ direction, replication of the leading strand occurs continuously, while replication of the lagging strand produces short segments called **Okazaki fragments** that are later joined together by an enzyme called **DNA ligase**.

Prokaryotes differ from eukaryotes in that they have a single circular chromosome, whereas eukaryotes have many linear chromosomes. Replication in prokaryotes typically proceeds in both directions from one **origin of replication** to a termination region until there are two copies of the circular chromosome. Replication in eukaryotes is slower and occurs at multiple origins of replication, resulting in many replication bubbles (places where the DNA strands are separating and replication is occurring). Replication occurs at the ends of the bubbles—the replication forks. Since eukaryotes have linear chromosomes, they cannot replicate the **telomeres** (the very ends), and the chromosomes become shorter with each successive round of replication. The telomeres are added to the chromosome by an enzyme called **telomerase**, preventing the loss of important sequences with each replication.

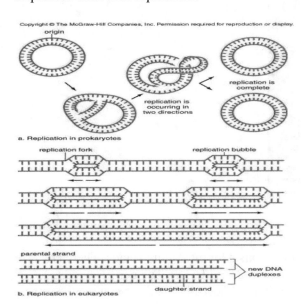

Prokaryotic versus eukaryotic replication

Take Note: *For the AP Biology exam, you should be able to describe in detail the steps of DNA replication. How does replication in prokaryotes differ from replication in eukaryotes?*

12.3 The Genetic Code of Life

RNA (**ribonucleic acid**), like DNA is a polymer of nucleotides. It differs from DNA in that it contains ribose sugars instead of the deoxyribose sugars in DNA. Whereas DNA is a double-stranded helix, RNA is single stranded and is not a helix. Like DNA, RNA contains adenine (A), guanine (G), and cytosine (C), but unlike DNA it contains the pyrimidine uracil (U) instead of thymine. The central dogma of molecular biology says that (1) DNA is a template for its own replication and also for RNA formation during transcription, and (2) the sequence of nucleotides in mRNA directs the correct sequence of amino acids of a polypeptide during translation.

In order to synthesize proteins based on the information in DNA, two steps must occur: transcription and translation. For these processes to occur, the information in DNA must be read and interpreted. The genetic code is a triplet code, and each **codon** (code word) consists of three bases. These codons are used to translate the information in DNA to the amino acid sequences that make up proteins, as shown in the accompanying chart.

First Base	Second Base				Third Base
	U	C	A	G	
U	UUU phenylalanine	UCU serine	UAU tryosine	UGU cysteine	U
	UUC phenylalanine	UCC serine	UAC tryosine	UGC cysteine	C
	UUA leucine	UCA serine	UAA stop	UGA stop	A
	UUG leucine	UCG serine	UAG stop	UGG tryptophan	G
C	CUU leucine	CCU proline	CAU histidine	CGU arginine	U
	CUC leucine	CCC proline	CAC histidine	CGC arginine	C
	CUA leucine	CCA proline	CAA glutamine	CGA arginine	A
	CUG leucine	CCG proline	CAA glutamine	CGG arginine	G
A	AUU isoleucine	ACU threonine	AAU asparagine	AGU serine	U
	AUC isoleucine	ACC threonine	AAC asparagine	AGC serine	C
	AUA isoleucine	ACA threonine	AAA lysine	AGA arginine	A
	AUG (start) methionine	ACG threonine	AAG lysine	AGG arginine	G
G	GUU valine	GCU alanine	GAU aspartate	GGU glycine	U
	GUC valine	GCC alanine	GAC aspartate	GGC glycine	C
	GUA valine	GCA alanine	GAA glutamate	GGA glycine	A
	GUG valine	GCG alanine	GAG glutamate	GGG glycine	G

Messenger RNA codons

The code is degenerate—that is, more than one codon exists for most amino acids. There are also one start codon (AUG) and three stop codons (UAA, UAG, and UGA). The genetic code is considered universal because with only a few exceptions, almost all organisms contain the same nucleotides and amino acids.

12.4   First Step: Transcription

Transcription of DNA to form messenger RNA (mRNA) occurs in the nucleus and begins when **RNA polymerase** attaches to the **promoter** of a gene on the template strand with the assistance of **transcription factors**. This first step of transcription is called **initiation**. RNA polymerase elongates the RNA strand by joining the complimentary RNA nucleotides together in a 5′ to 3′ direction. **Elongation** continues until RNA polymerase reaches a stop sequence on the DNA. Multiple molecules of RNA polymerase work to concurrently synthesize many copies of an mRNA transcript, which will ultimately result in many copies of a protein in a cell.

In eukaryotic cells, the mRNA transcript is processed once its synthesis is complete. A guanine cap is put onto the 5′ end, and a poly-A tail is put onto the 3′ end. These modifications are signals for the attachment of the ribosome for translation and transport out of the nucleus, respectively.

Finally, the **introns** (noncoding sections) are removed and the **exons** (the sequences that are expressed) are linked together by ribozyme-containing **spliceosomes**.

12.5   Second Step: Translation

Translation takes place in the cytoplasm and requires mRNA, transfer RNA (tRNA), and ribosomal RNA (rRNA). Transfer RNA is responsible for bringing the amino acids to the ribosomes to be added to a protein. Each tRNA has an **anticodon** at one end and an amino acid at the other; amino acid-activating enzymes ensure that the correct amino acid is attached to the correct tRNA. When tRNAs bind with their codon at a ribosome, the amino acids are correctly sequenced in a polypeptide according to the order predetermined by DNA.

Ribosomes are the organelles that are responsible for protein synthesis. They are made of protein and rRNA, and are assembled in the nucleus then move to the cytoplasm. Ribosomes have two subunits, a large 60S subunit and a small 30S subunit. The large subunit has three tRNA binding sites: the A site, the P site, and the E site. As the ribosome moves along the mRNA, the tRNA molecules are transferred from one binding site to the next as amino acids are added to the growing polypeptide chain. **Polysomes**, several ribosomes simultaneously translating a single mRNA molecule, are often seen in a cell and increase the efficiency of translation by synthesizing many copies of a single polypeptide.

Translation requires three steps: initiation, elongation, and termination. During initiation, mRNA, the first (initiator) tRNA, and the two subunits of a ribosome all come together in the proper orientation at a start codon. Proteins called **initiation factors** assist this assembly, resulting in the initiator tRNA, with its attached methionine, located in the P site of the large ribosomal subunit.

During elongation, the incoming tRNA comes into the ribosome at the A site and the anticodon binds to its complementary codon on the mRNA. The polypeptide is transferred from the tRNA in the P site to the tRNA in the A site, and a peptide bond is formed. When the ribosome moves to the next codon, the tRNA molecules are relocated to the next site. The tRNA in the E site exits the ribosome, the P site contains the tRNA with the polypeptide chain, and the A site is free to receive the next tRNA.

During termination the ribosome arrives at a stop codon. A protein called a release factor binds to the A site, the polypeptide is cleaved from the last tRNA, and the ribosome now dissociates from the mRNA.

Take Note: *For the AP Biology exam, you should be able to describe the similarities and differences between transcription and translation.*

12.6   Structure of Eukaryotic Chromosomes and Genes

Eukaryotic cells contain nearly 2 m of DNA, yet must pack it all into a nucleus no more than 5 um in diameter. Thus, the DNA must have a very compact structure. This is achieved by the winding of DNA around **histone** proteins to make **nucleosomes**. The nucleosomes are further compacted into a zigzag structure, which is then folded upon itself many times to form radial loops, which is the usual compaction state of **euchromatin**. **Heterochromatin** is further compacted by scaffold proteins, and further compaction can be achieved prior to mitosis and meiosis when the chromatin condenses into chromosomes.

Genes only comprise 1.5% of the human genome. The rest of this DNA is surprisingly more active than once thought. About half of this DNA consists of repetitive DNA elements, which may be in tandem or interspersed throughout several chromosomes. Some of this DNA is made up of mobile DNA sequences called **transposons**, which are a driving evolutionary force within the genome. The remaining half of the genome remains unclassified, but even these unknown DNA sequences may play an important role in regulation of gene expression.

Take Note: *For the AP Biology exam, you should be able to describe the similarities and differences between the organization of prokaryotic and eukaryotic chromosomes, and the expression of prokaryotic and eukaryotic genes.*

Multiple Choice Questions

1. Which of the following pieces of evidence aided Watson and Crick in their ultimate discovery of DNA as the hereditary material?
   A. In the DNA of several species, the percentage of adenine nucleotides approximately equaled the number of thymine nucleotides.
   B. The bonds between purines and pyrimidines must be covalent in nature to hold the double helix together.
   C. The X-ray diffraction photos showed that the DNA molecule was composed of three strands.
   D. The sugar-phosphate backbone was located on the inside of the molecule due to its hydrophobic nature.
   E. The pairing of purines with pyrimidines yielded too narrow a diameter of the double helix.

2. Which of the following statements is true about the leading strand during DNA synthesis?
   A. Primase reads the DNA and adds DNA in short segments.
   B. DNA replication proceeds in the 3′—5′ direction when laying down RNA nucleotides.
   C. The short segments of DNA are lengthened by DNA polymerase III to form Okazaki fragments.
   D. DNA polymerase III reads the DNA and adds nucleotides to the leading strand continuously.
   E. DNA polymerase unites separate Okazaki fragments on the leading strand.

3. Which of the following statements provides evidence that DNA is the material of genes?
   A. Proteins are not very diverse in size and shape.
   B. A eukaryotic cell doubles its DNA content and distributes it equally prior to and during mitosis.
   C. Injected protein molecules cause cells to produce additional DNA and proteins.
   D. The relative amount of two pyrimidine bases, adenine and guanine, were equal in many species.
   E. Proteins are not very diverse in their function, which was essential for the function of genes.

4. Which of the following is NOT true about eukaryotic DNA replication?
   A. DNA replication is semiconservative.
   B. Replication begins at multiple origins or sequences.
   C. DNA replication occurs continuously on both strands.
   D. DNA replication proceeds in both directions from each origin in a replication bubble.
   E. DNA replication is slow due to the fact that histone proteins must be replaced after replication is complete

5. Telomerase is an enzyme that adds specific DNA repeats to the 3′ end of the DNA strand. One consequence of this activity is increased stability of the molecule. Which statement below is true about telomeres and telomerase?
    A. Telomerase prevents the shortening of chromosomes.
    B. Telomerase is active throughout the lifetime of an individual.
    C. Telomeres postpone the ends of chromosomes being eroded by replication.
    D. Telomeres are specific genes which are not needed as an individual ages.
    E. Telomeres are added in both the 5–3 prime and in the 3–5 prime direction by telomerase.

6. Gene expression results in an accurate template for a protein product due to the fact that
    A. mRNA is processed at the ribosome to remove introns.
    B. tRNA codons pair complimentarily to mRNA.
    C. the mRNA molecules are able to translate the tRNA strand at the ribosome.
    D. one strand of DNA serves as a template for the formation of mRNA.
    E. the sequence of the mRNA anticodons determines the order of amino acids.

7. During translation
    A. the mRNA moves from the P site to the A site.
    B. a large ribosomal subunit translates the start codon (AUG), and then the small subunit attaches.
    C. a tRNA begins at the P site with a complimentary codon to the mRNA sequence.
    D. each tRNA leaves the A site after transferring its amino acid to the mRNA.
    E. the polypeptide is transferred and attached to the newly arriving amino acid in the A site.

8. Which of the following characteristics of DNA affects its ability to replicate continuously?
    A. The hydrogen bonds cannot be broken in a continuous fashion.
    B. DNA polymerase can only add nucleotides to the 5′ end of the growing DNA strand.
    C. The phosphate groups are attached to the nucleotides in an alternating fashion.
    D. The untwisting of the DNA molecule causes tighter twisting and strains the molecule.
    E. The thymine molecules must be replaced by uracil on the new strand.

9. What is the appropriate mRNA strand sequence for the following noncoding strand of DNA?

    5′-ATTACAGGCGGA-3′

    A. 5′-AUUACAGGCGGA-3′
    B. 5′-AUUUCUGGCGGU-3′
    C. 5′-AGGCGGACAUUA-3′
    D. 5′-ATTACAGGCGGA-3′
    E. 5′-UCCGCCUGUAAU-3′

10. If the percentage of adenine is 20% in a given sample of DNA, then what would be the percentage of guanine?
    A. 20%
    B. 30%
    C. 40%
    D. 80%
    E. 60%

Free Response Question

The discovery of DNA's structure and how it replicates was very much a collaborative effort based on the efforts of many scientists.

      A. **Identify** a researcher's work and **explain** how his or her discoveries helped unlock the secrets of DNA.

      B. **Identify** a modern application of DNA and how DNA technologies are being used to help humankind.

Annotated Answer Key (MC)

1. **A**; It was Chargaff who gave Watson and Crick the pertinent ratios of adenine:thymine and guanine: cytosine. This purine + pyrimidine equation helped them to determine the width of the molecule. The X-ray diffraction photos from Franklin showed that it was two strands, not three, as Pauling had suggested, the sugar phosphate backbone was on the outside, and the bonds between the purine and pyrimidines are hydrogen bonds.

2. **D**; Primase does not add DNA nucleotides and it proceeds in the 5′–3′ direction. The short segments of RNA are lengthened in an Okazaki fragment, and ligase is the enzyme that unites the fragments.

3. **B**; Proteins are very diverse in structure and function, size and shape. However, it was DNA that was replicated during the cell cycle with purines complimentary to pyrimidines and was the basis for Griffith's experiments.

4. **E**; DNA replication is semiconservative (one parent strand with one daughter strand), does begin at multiple origins, and can occur on both strands in both directions.

5. **C**; Telomeres are regions of repetitive DNA at the ends of strands that prevents erosion. While age may have some effect on the function and behavior of telomerase and telomeres, the genes are still needed during old age.

6. **D**; During transcription, one strand of DNA serves as the template for mRNA. mRNA contains codons, not anticodons as in tRNA, and is processed prior to leaving the nucleus.

7. **E**; mRNA moves from the A site to the P site to the E site. The small subunit translates AUG and then the large subunit joins at the A site. Each tRNA leaves the P site after transferring its amino acid to the mRNA.

8. **B**; DNA polymerase is capable of laying down nucleotides in the 5′–3′ direction. Therefore, one strand must be leading and one strand must be lagging, waiting to be able to lay down a 5′ nucleotide.

9. **A**; If this is the noncoding strand, then the coding strand is 3′-TAATGTCCGCCT-5′ and therefore the mRNA sequence would be 5′-AUUACAGGCGGA-3′, replacing any thymine (T) nucleotides with uracil (U) because it is RNA.

10. **B**; If the percentage of adenine is 20%, then the percentage of thymine must also be 20%. This accounts for 40% of the DNA sample. The remaining 60% of the sample is composed of guanine and cytosine at 30% each.

Answer to FRQ

PART A (MAX 8 pts)
One point for research, one point for finding; no points for name only.

- Griffith: no pts
- Avery: transformation of rough and smooth strains of streptococcus → DNA transforms
- Hershey and Chase: radioactive tagging of P and S → DNA is genetic material
- Chargaff: measurements of nucleotides I organisms → A and T, C and G
- R Franklin: x-ray crystallography → width of strands and DNA is a helix
- Watson and Crick: used others research → initial model of DNA with base pairing on inside of helix
- Messelsohn and Stahl: radioactive tagged viruses (P and S) → semiconservative replication of DNA

PART B (MAX 2 pts)

- One point for ID of a modern DNA technology; could range from transforming bacteria to produce insulin to GMOs (genetically modified organisms) to new emerging technologies yet to be discovered when this question was written
- One point for linking the technology to helping humankind

CHAPTER 13
REGULATION OF GENE ACTIVITY

13.1 Prokaryotic Regulation

Prokaryotes often organize genes that are involved in a common process or metabolic pathway into **operons**, which coordinately regulate the expression of the genes that code for the enzymes involved in the pathway. The operon model of gene regulation was developed by Jacob and Monod. Operons typically include a **promoter**, an **operator**, and **structural genes**. Elsewhere on the chromosome, a **regulator gene** codes for a **repressor** which can bind to the operator. When it does, RNA polymerase is unable to bind to the promoter, and transcription of the structural genes of the operon cannot take place. Alternately, operons may also be regulated by activators called **inducers**.

The *trp* **operon** is an example of a **repressible operon** because it normally exists in the "on" condition, meaning that RNA polymerase is actively transcribing the genes for the enzymes that are responsible for tryptophan synthesis. When tryptophan, the **corepressor**, is present, it binds to the repressor. The repressor is then able to bind to the operator, and transcription of structural genes does not take place because the RNA polymerase binding site is blocked.

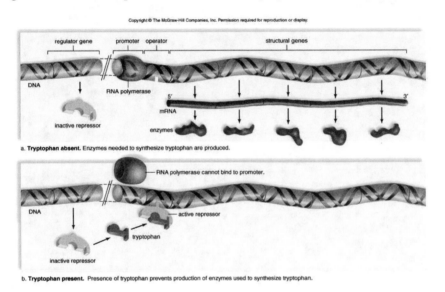

a. **Tryptophan absent.** Enzymes needed to synthesize tryptophan are produced.

b. **Tryptophan present.** Presence of tryptophan prevents production of enzymes used to synthesize tryptophan.

The trp operon

The *lac* **operon** is an example of an **inducible operon** because it normally exists in the "off" position. This means that the binding site for RNA polymerase on the promoter is blocked so that transcription cannot occur. When lactose, the inducer, is present, it binds to the repressor. The repressor is unable to bind to the operator, and transcription of structural genes takes place if glucose is absent and the cell needs to break down lactose as an alternate source of carbohydrate for the cell.

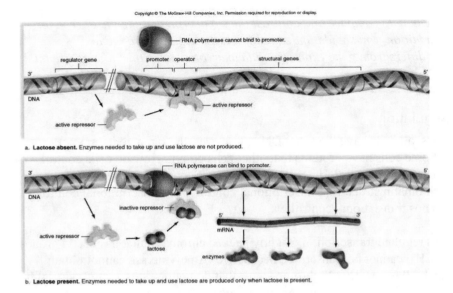

RNA polymerase cannot bind to promoter.

regulator gene   promoter operator   structural genes

3'

DNA

active repressor

active repressor

a. **Lactose absent.** Enzymes needed to take up and use lactose are not produced.

RNA polymerase can bind to promoter.

3'   5'

DNA

inactive repressor

active repressor   lactose

mRNA

enzymes

b. **Lactose present.** Enzymes needed to take up and use lactose are produced only when lactose is present.

The lac operon

Both the *lac* and *trp* operons exhibit negative control, because a repressor is involved. The lac operon also exhibits positive control. Glucose is the preferred source of carbohydrate for bacteria, therefore the structural genes in the *lac* operon are not maximally expressed unless glucose is absent and lactose is present. When this condition occurs in the cell, the levels of **cyclic AMP** (cAMP) are high. At that time, cAMP attaches to a molecule called **CAP** (catabolite activator protein), and this cAMP-CAP complex binds to a site next to the promoter. Now RNA polymerase is better able to bind to the promoter, and transcription occurs. This is an example of positive control because it amplifies the basal level of transcription by allowing RNA polymerase to bind more efficiently to the promoter.

CAP binding site   promoter operator

DNA

RNA polymerase binds fully with promoter.

cAMP   active CAP

inactive CAP

a. **Lactose present, glucose absent (cAMP level high)**

CAP binding site   promoter operator

DNA

RNA polymerase does not bind fully with promoter.

inactive CAP

b. **Lactose present, glucose present (cAMP level low)**

Action of CAP

Take Note: *You should be able to compare and contrast the trp operon and the lac operon for the AP Biology exam. In addition, you should understand how the regulation of gene expression in prokaryotes differs from gene expression in eukaryotes. Keep this in mind as you review the next section.*

13.2   Eukaryotic Regulation

In multicellular eukaryotes, all cells have the same DNA; however, the same genes are not expressed in every cell. Some cells have proteins that other cells do not. How do cells control which genes are expressed in which cells? The following levels of control of gene expression are possible in eukaryotes: chromatin structure, transcriptional control, posttranscriptional control, translational control, and posttranslational control.

Chromatin structure helps regulate transcription. Tightly packed chromatin that has acetyl groups on the histone tails ($-COCH_3$) cannot be transcribed because RNA polymerase cannot obtain access to the DNA. **Barr bodies** (inactive X chromosomes that do not produce gene products) in mammalian females are examples of highly condensed chromatin that is genetically inactive. On the other hand, loosely packed chromatin that has methyl groups ($-CH_3$) on the histone tails is easily transcribed. Examples of less-condensed euchromatin that is genetically active are the lampbrush chromosomes in vertebrates, in which the decondensed loops of chromatin radiate from the central axis of the chromosome.

Regulatory proteins called **transcription factors**, as well as DNA sequences called **enhancers** and **silencers**, play a role in controlling transcription in eukaryotes. Transcription factors bind to the promoter, then assist RNA polymerase in attaching to its binding site. Transcription **activators** bind to enhancers that are located upstream from the transcription start point. They increase the rate of transcription by enhancing the interactions of the **mediator proteins** and transcription factors, allowing RNA polymerase to begin transcription. In the same manner, the binding of repressor proteins to silencers in the DNA sequence prevents transcription by denying RNA polymerase access to the promoter.

Posttranscriptional control is achieved by creating variations in messenger RNA (mRNA) by **alternative RNA splicing**. Differential splicing of exons during intron removal may yield multiple mRNA messages from the same gene, resulting in the synthesis of a variety of different proteins in the cell. Altering the speed with which a particular mRNA molecule leaves the nucleus is another form of posttranscriptional control of gene expression. The amount of time it takes for the mRNA to leave the nucleus directly affects the amount of protein that is synthesized per unit time after transcription.

Translational control affects mRNA translation and the length of time it is translated, primarily by altering the stability of an mRNA. Whether translation takes place and how long the mRNA is active is determined by the presence of the 5′ cap and the length of the poly-A tail. Any influence that leads to the removal of the 5′ cap or shortens the poly-A tail leads to the destruction of the mRNA and reduces the amount of protein synthesized from that mRNA. **MicroRNAs** (miRNAs) do not code for proteins and are a unique example of translational control because they can either inhibit translation or trigger degradation of the mRNA, as shown in the accompanying figure.

Posttranslational control begins once a protein is synthesized and active, and affects whether or not an enzyme is active and how long it is active. Some proteins, such as trypsin produced by the stomach, are not immediately active once they are synthesized because they are made in an inactive form. Once they are modified by cleavage, they are active. Protein complexes called **proteosomes** degrade proteins, allowing them to be present only for as long as necessary to carry out their function.

13.3 Regulation Through Gene Mutations

In molecular terms, a gene is a sequence of DNA nucleotide bases, and a genetic **mutation** is a permanent change in this sequence. Mutations can happen spontaneously as a result of cellular processes such as the movement of transposons. Rarely is a mutation the result of a mistake in replication, as DNA polymerase has a proofreading function in addition to its role in adding nucleotides to the new DNA strand. Mutations are also due to environmental mutagens such as radiation and organic chemicals. Ultraviolet (UV) radiation can cause thymine dimers to form wherever two thymines are found next to each other in a DNA sequence. This results in kinks in the DNA that are removed and repaired by DNA repair enzymes that constantly monitor and mend DNA. Some mutagens are called carcinogens because they cause cancer.

Point mutations involve a change in a single nucleotide in the DNA and can have a range of effects, depending on the particular codon change. Point mutations can result in an abnormal or incomplete protein or have no effect on a protein at all. Sickle cell disease is an example of a point mutation that greatly changes the activity of the affected gene.

Frameshift mutations happen when a nucleotide is added or deleted, causing the reading frame of the codons to change. The result is usually a nonfunctional protein. Most cases of cystic fibrosis are due to a frameshift mutation that changes the structure of a chloride channel, causing it to be inactive. Nonfunctional proteins can affect the phenotype drastically, as in albinism, which is due to a single faulty enzyme, and androgen insensitivity, which is due to a faulty receptor for testosterone.

Cancer is often due to an accumulation of genetic mutations among genes that code for regulatory proteins. The cell cycle occurs inappropriately when **proto-oncogenes** become oncogenes and overstimulate the cell cycle.

Cancer is also caused when **tumor suppressor genes** are no longer effective, leading to uncontrolled cell division. Mutations that affect transcription factors and other regulators of gene expression are frequent causes of cancer.

Take Note: *It is important for the AP Biology exam that you can describe and give examples of how a mutation in the DNA affects the structure and function of a protein.*

Multiple Choice Questions

1. Barr bodies would be observable in all but which of the following conditions?
   A. XXY
   B. XXXY
   C. XXYY
   D. XYY
   E. XX

2. Based on the table shown, which of the following point mutations will cause the most drastic effect?
   A. a mutation in the third position of the codon AUU
   B. a substitution of UAC to UAG early in the gene
   C. a base change in the transcription CAU to UAU
   D. a substitution of UAA to UAG at the end of the gene
   E. a substitution from UUA to CUC

3. Eukaryotic protein synthesis differs from that of prokaryotes because
   A. transcription and translation occur simultaneously in eukaryotes.
   B. ribosomal subunits are larger in prokaryotes.
   C. operons regulate gene expression in eukaryotes.
   D. pokaryotes do not use enzymes during protein synthesis.
   E. DNA binding proteins determine the rate of transcription in eukaryotes.

4. Which of the following is true about prokaryotic gene regulation?
   A. A regulator gene codes for a repressor protein that controls the operon.
   B. An operator is the sequence of DNA where RNA polymerase attaches.
   C. Promoters code for enzymes and are transcribed as a unit.
   D. A terminator determines when the operon becomes inactive.
   E. Structural genes are located outside the operon and code for enzymes.

5. When a *trp* operon is in the "off" condition,
   A. the repressor will code for an operator and change shape.
   B. tryptophan will bind to the repressor.
   C. the binding site for tryptophan will change shape.
   D. RNA polymerase will attach to the promoter and tryptophan is produced.
   E. tryptophan codes for a regulator gene and the operon is repressed.

Questions 6–9 refer to the following five choices. For each question, select the choice that is most closely related. Each choice may be used once, more than once, or not at all.
   A. transcriptional control
   B. posttranscriptional control
   C. translational control
   D. posttranslational control
   E. ribosomal control

6. Determines which structural genes are expressed

7. Occurs before the mRNA leaves the nucleus but before there is a protein product

8. Includes transposons that move between chromosomes

9. Occurs after protein synthesis

10. Which of the following mutations occurs when nonhomologous chromosomes exchange segments?
    A. deletion
    B. duplication
    C. inversion
    D. reciprocal translocation
    E. transformation

Free Response Question

Gene expression must be regulated in both prokaryotes and eukaryotes.

   A. **Describe** the functional components of an operon.

   B. **Explain** how the regulation of gene expression in prokaryotes differs from gene expression in eukaryotes.

Annotated Answer Key (MC)

1. **D**; Barr bodies are observable in cases where there are at least 2 X chromosomes.

2. **B**; This mutation would cause a stop codon to be recognized early in the gene, prohibiting most of the translation of the protein. Often, due to the degeneracy of the code, several codons can code for the same amino acid (as in AUU to AUC or AUA). UAU and AUG are both stop codons. UUA and CUC both code for leucine.

3. **E**; Transcription and translation occur simultaneously in prokaryotes, not eukaryotes. Ribosomal subunits are larger in eukaryotes than in prokaryotes and operons regulate prokaryotes. Transcription factors (DNA binding factors) are proteins that bind to specific seuqnences of DNA. Without transcription factors, the creation of RNA from DNA cannot occur.

4. **A**; Regulators are located outside the operon and code for repressor proteins. Promoters are where RNA polymerases attach, operators are where the repressors bind, and structural genes are located within the operon and code for enzymes.

5. **B**; Some operons in *E. coli* usually exist in the "on" rather than the "off" condition. *E. coli* produces five enzymes as part of the anabolic pathway to synthesize the amino acid tryptophan. If tryptophan is already present in medium, these enzymes are not needed and the operon is turned off. The regulator codes for a repressor that usually is unable to attach to the operator. The repressor has a binding site for tryptophan (if tryptophan is present, it binds to the repressor). This changes the shape of the repressor that now binds to the operator.

6. **A**

7. **C**

8. **A**

9. **D**

For Questions 6–9: There are several levels of control that can modify the amount of gene product. Transcriptional control in the nucleus determines which structural genes are transcribed and the rate of transcription; it includes transcription factors initiating transcription and transposons (DNA sequences that move between chromosomes and shut-down genes). Posttranscriptional control occurs in the nucleus after DNA is transcribed and preliminary mRNA forms. Translational control occurs in cytoplasm after mRNA leaves the nucleus but before there is a protein product. Posttranslational control occurs in the cytoplasm after protein synthesis.

10. **D**; During a translocation mutation, a segment of DNA is transferred from one chromosome to another, nonhomologous chromosome.

Answer to FRQ

PART A MAX 6
One point for identification of the component, one point for function.

- Promoter: DNA sequence that enables a gene to be transcribed
- Operator: segment of DNA that a regulatory protein binds to
- Structural genes: DNA sequences for particular traits

PART B (MAX 6)

- Prokaryotes: usually regulated at the level of transcription
- Eukaryotes: there are several points in the process of gene expression where regulation can occur
- Prokaryotes (MAX 2 – must identify and describe)
- Regulator gene: gene that controls expression of other genes
- Repressor: inhibits the expression of structural genes
- Inducers: disable respressors
- Eukaryotes (MAX 2 – must identify and describe)
- chromatin structure (histones, RNA polymerase, Barr bodies)
- transcriptional control (bind to the promoter, activators, mediator proteins, the binding of repressor proteins to silencers)
- posttranscriptional control (alternative RNA splicing, altering the speed with which a particular mRNA molecule leaves the nucleus)
- posttranslational control (active/inactive forms, proteosomes, degradation of proteins, allowing them to be present only for as long as necessary to carry out their function)

CHAPTER 14
BIOTECHNOLOGY AND GENOMICS

14.1   DNA Cloning

DNA cloning can isolate and produce many identical copies of a gene so that it can be studied in the laboratory or inserted into a bacterium, plant, or animal. A gene may be transcribed and translated to produce a protein, which can become a commercial product or used as a medicine. Two methods are currently available for making copies of DNA: recombinant DNA technology and the polymerase chain reaction (PCR).

Recombinant DNA contains DNA from two different sources, such as a human gene and a bacterial plasmid. A **plasmid** is a small circular piece of DNA that is found in some bacteria separate from their chromosome. It serves as a "vector" that carries the human gene during the cloning process. A **restriction enzyme**, such as EcoRI, is used to cleave both the plasmid DNA and the foreign DNA. The resulting "sticky ends" facilitate the insertion of foreign (human) DNA into vector DNA. The foreign gene is sealed into the vector DNA by DNA ligase, and a recombinant DNA (rDNA) molecule is formed. Viruses can also be used as vectors to carry foreign genes into bacterial host cells. The bacteria then express the human gene as long as the plasmid also contains their unique regulatory regions. Also, bacterial DNA that is complementary to human DNA, called **cDNA**, can be made using the enzyme **reverse transcriptase**. Bacteria then use this cDNA to express human genes.

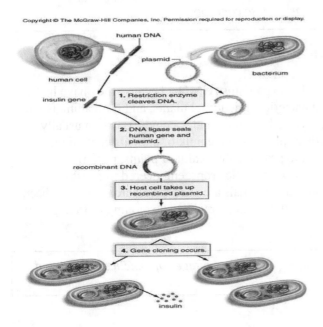

Cloning a human gene

The **polymerase chain reaction (PCR)** uses DNA polymerase and nucleotides to quickly make multiple copies of a specific piece (target) of DNA. PCR is a chain reaction because the targeted DNA is replicated over and over again. Analysis of DNA segments following PCR has many types of uses, from assisting genomic research to doing DNA fingerprinting for the purpose of identifying individuals and their paternity. Typically, the products of PCR are analyzed using **gel electrophoresis**, in which DNA fragments separate according to their size, forming distinct patterns in the gel.

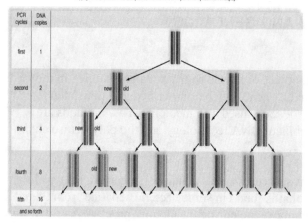

PCR cycles	DNA copies
first	1
second	2
third	4
fourth	8
fifth	16
and so forth	

Polymerase chain reaction (PCR)

Take Note: *You should be able to describe the methods and applications of DNA cloning on the AP Biology exam. You should be able to analyze a figure of a DNA molecule cleaved with restriction enzymes and the resulting electrophoresis gel. If asked, you should also be prepared to draw a gel resulting from the cleavage of DNA by restriction enzymes, as well as a restriction map of a plasmid. Ask your teacher for practice problems that will reinforce these concepts.*

14.2 Biotechnology Products

Transgenic organisms are known as **genetically modified organisms (GMOs)** because they have had a foreign gene inserted into them. Genetically modified bacteria, agricultural plants, and farm animals now produce commercial products of interest to humans, such as hormones and vaccines. Bacteria usually secrete the product into the growth medium. The seeds of plants and the milk of animals contain the product. Transgenic bacteria have also been engineered to promote the health of plants, perform bioremediation, extract minerals, and produce chemicals. Transgenic crops, engineered to resist herbicides and pests, are commercially available. Transgenic animals are used in research and have been given various genes, in particular the one for bovine growth hormone (bGH). The cloning of transgenic animals to generate a product is now possible.

Take Note: *You should be able to explain the advantages and disadvantages of genetically modified organisms.*

14.3 Gene Therapy

Gene therapy, by either *ex vivo* or *in vivo* methods, is used to correct the genotype of humans and to cure various human ills. Viruses can be genetically engineered to serve as vectors to carry a normal gene into the body to replace a dysfunctional gene. *Ex vivo* gene therapy (in which cells are removed from the body, inserted with a gene, and returned to the body) has apparently helped children with SCID lead normal lives. *In vivo* treatment (where the gene is directly inserted into the body) for cystic fibrosis has been less successful. A number of *in vivo* therapies are being employed in the war against cancer and other human illnesses, such as cardiovascular disease.

14.4 Genomics

Genomics is the study of the genomes of organisms. Researchers now know the sequence of all the base pairs along the length of the human chromosomes thanks to the 13-year Human Genome Project. So far, researchers have found only 23,000 genes that code for proteins; the rest of our DNA consists of regions that do not code for a protein. Currently, researchers are placing an emphasis on functional and comparative genomics in order to learn the functions of known genes and to compare them with the genomes of other organisms. This information will help in the discovery of possible cures for certain diseases and will serve as a tool to understand evolutionary relationships between organisms. Comparative genomics has revealed that there is little difference between the DNA sequence of our bases and those of many other organisms. Genome comparisons have revolutionized our understanding of evolutionary relations by revealing previously unknown relationships between organisms.

Genes only comprise 1.5% of the human genome; there are 20,500 genes in the genome. The rest of this DNA is surprisingly more active than once thought. About half of these intergenic sequences consist of repetitive DNA elements—sequences of two or more nucleotides—which may be repeated in tandem or interspersed throughout several chromosomes. Other **intergenic sequences** are made up of mobile DNA sequences called **transposons**, which are a driving evolutionary force within the genome. Transposons are DNA sequences that can move within or between chromosomes. They are somewhat like regulator genes in that their movements sometimes alter the expression of neighboring genes. The remaining half of the genome remains unclassified, but even these unknown DNA sequences may play an important role in regulation of gene expression.

Functional genomics aims to understand the function of protein coding regions and noncoding regions of our genome. To that end, researchers are utilizing new tools such as DNA microarrays to simultaneously monitor the expression of thousands of genes. Microarrays can also be used to create an individual's genetic profile, which can be helpful in predicting illnesses and how a person will react to particular medications.

Multiple Choice Questions

1. A restriction map of a circular DNA plasmid is constructed based upon electrophoretic separation of the DNA. From this map, which of the following statements could NOT be concluded?
    EcoRI

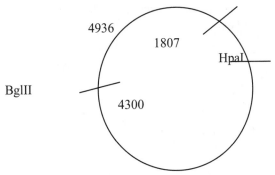

    A. The uncut DNA plasmid is 11,043 bp long.
    B. Treatment with EcoRI and HpaI will result in fragments of 1807 bp and 9236 bp.
    C. DNA cut with EcoRI will result in fragments of 4936 bp, 1807 bp and 4300 bp.
    D. Treatment with BglII and EcoRI will result in fragments of 4936 bp and 6107 bp.
    E. DNA cut with all three restriction enzymes will result in fragments of 4936 bp, 1807 bp, and 4300 bp.

2. Recombinant DNA can
    A. cleave plasmid DNA and foreign DNA.
    B. be composed of DNA and RNA.
    C. contain DNA from a human gene and a bacterial plasmid.
    D. carry foreign DNA into host cells.
    E. can be found in many common foods.

3. The enzyme that synthesizes cDNA is
    A. primase.
    B. ligase.
    C. DNA polymerase III.
    D. synthase.
    E. reverse transcriptase.

4. After completing PCR in the lab, the anticipated DNA fragment was not as expected. A possible cause for these unexpected results could have been
    A. lack of a buffered environment for optimum activity of DNA polymerase.
    B. use of two primers complimentary to the DNA.
    C. subjecting the DNA to repeated temperature changes.
    D. use of a thermal cycler.
    E. use of bacterial polymerase.

5. Transposons are NOT able to
    A. carry genes for antibiotic resistance.
    B. jump from chromosomal DNA to plasmid DNA.
    C. damage the genome of their host.
    D. be disrupted by RNA interference mechanisms.
    E. act as mutators in bacteria.

6. The direction of migration in gel electrophoresis is due to
    A. hydrogen bonds between nucleotides.
    B. number of carbon molecules in the deoxyribose sugar.
    C. anti-parallel nature of the DNA.
    D. covalent bonds between the phosphates and nucleotides.
    E. negative charge of the sugar-phosphate backbone.

Questions 7–10 refer to the following five choices. For each question, select the choice that is most closely related. Each choice may be used once, more than once, or not at all.
    A. cDNA
    B. rDNA
    C. mtDNA
    D. rRNA
    E. nDNA

7. Contain most of the genes that provide instructions for making enzymes involved in oxidative phosphorylation

8. Used by bacteria to express human genes

9. Molecular component of a ribosome

10. Inherited only maternally

Free Response Question

Gel electrophoresis is a mechanism used widely in biotechnology labs.

    A.   **Describe** how the process of gel electrophoresis is based upon the molecular properties of DNA.

    B.   **Explain** how enzymes can regulate the process to ease its use in eukaryotes.

    C.   **Discuss** one application of gel electrophoresis today.

Annotated Answer Key (MC)

1. **C**; The uncut DNA plasmid is 11,043 bp long (1807 + 4300 + 4936). Treatment with EcoRI and HpaI will results in fragments of 1807 and 9236 (4936 + 4300), and treatment with BglII and EcoRI will result in 4937 and 6107 (1807 + 4300). Finally, DNA cut with all three enzymes would yield cuts of 4936, 1807, and 4300 bp.

2. **C**; Recombinant DNA can contain DNA from a human gene and a bacterial plasmid, thus the term "recombinant." Restriction enzymes are used to cleave plasmid and foreign DNA. Vectors carry foreign DNA into host cells.

3. **E**; Reverse transcriptase synthesizes cDNA. Ligase anneals nicks in the DNA, and DNA polymerase places DNA nucleotides in place.

4. **A**; An environment that is properly buffered at the correct pH is necessary for optimum activity of enzymes. Two primers complimentary to the DNA are used in PCR. Repeated temperature changes and the use of a thermal cycler are the basis for PCR.

5. **C**; Transposons can carry genes for antibiotic resistance, can be mutagens by inserting itself into a functional gene, and certainly do jump from chromosomal DNA to plasmid DNA. Eukaryotic RNAi mechanisms can disrupt their function by reducing their activity levels.

6. **E**; The direction is due to the negative charge of the sugar phosphate backbone. The number of carbon molecules and the bonds between nucleotides do not directly impact this movement.

7. **C**

8. **A**

9. **D**

10. **C**

For Questions 7–10: Mitochondrial DNA (mtDNA or mDNA) provides instructions for making enzymes involved in oxidative phosphorylation of cellular respiration, which occurs in the mitochondria. It is inherited only maternally due to decreased number of mitochondria in the sperm, failure of the mtDNA molecules of the sperm to enter the egg, or destruction of sperm mtDNA in the fertilized egg. Ribosomal RNA (rRNA) is the chief component of ribosomes; complementary DNA (cDNA) is used by bacteria to express human genes and is actually a double-stranded copy of a single-stranded mRNA.

Answer to FRQ

PART A (MAX 5)

- Strands of DNA are separated.
- Separation is based upon charge and size (TWO POINTS).
- DNA molecules have a negative charge (at neutral pH).
- DNA strands move towards the positive field.
- Larger DNA molecules move slower than small ones or vice versa.
- Different sized strands stop at different points to leave a pattern.

PART B (MAX 5)

- Restriction enzymes cut DNA molecules in particular places.
- Production of palindromes (manageable sequences)
- Nuclear DNA consists of many large chromosomes.
- Smearing would result from the separation of large fragments; distinct banding is only possible with fragments.
- Number of fragments depends on the number of times the appropriate sequence occurs.
- Creation of sticky ends

PART C (MAX 2)

- One point – correct ID of an application (compared with other samples to match DNA – DNA fingerprinting - i.e., crime scene, paternity testing, analyze results of PCR)
- Second point – elaboration of how the process is accomplished

CHAPTER 15
DARWIN AND EVOLUTION

Take Note: *Evolution is the unifying principle in Biology and is the glue that connects all the aspects of Biology together because it explains the unity within the diversity that exists in the biosphere and has been supported by so many different lines of evidence. As you study the next five chapters, be sure you have a solid understanding of the concepts so that as you begin your study of organisms, you can make clear connections. Expect that at least one free response question on the AP Biology exam will focus on directly on some aspect of evolution.*

15.1   History of Evolutionary Thought

A century before Darwin's trip on the HMS *Beagle* in 1831, taxonomy—the classification of organisms—had been a main concern of biologists. Carolus Linnaeus, who developed the system of binomial nomenclature, thought that each species had a place in the *scala naturae* and that classification should describe the fixed features of species. He did not believe that species could change over time. Some naturalists, however, such as Count Buffon and Erasmus Darwin, put forth tentative suggestions that species do change over time, but neither recommended a mechanism by which such change might occur.

Georges Cuvier and Jean-Baptiste de Lamarck, contemporaries of Darwin in the late-eighteenth century, agreed on **evolution**—change in species over time—but differed sharply on their explanations of a possible mechanism. To explain the fossil record of a region, Cuvier proposed that a series of catastrophes (extinctions) and repopulations from other regions had occurred, and that the result was that changes in species appeared over time. Lamarck said that descent with modification does occur and that organisms do become adapted to their environments; however, he suggested the inheritance of acquired characteristics as a mechanism for evolutionary change. In other words, any characteristic that a parent acquired during its lifetime was automatically passed on to its offspring. Lamarck's explanation, however, is not consistent with the molecular mechanism of inheritance developed by Gregor Mendel.

15.2   Darwin's Theory of Evolution

Charles Darwin formulated hypotheses concerning evolution after taking a trip around the world as a naturalist aboard the HMS *Beagle* (1831–36). His hypotheses were that common descent does occur and that natural selection results in adaptation to the environment.

Darwin's trip involved two primary types of observations:

1. *Geology and fossils.* Darwin's observations in various places such as Chile and Argentina caused him to concur with Charles Lyell that the observed massive geological changes on Earth were caused by slow, continuous changes. Therefore, he concluded that the earth is old enough for descent with modification to occur.

2. *Biogeography.* Darwin's study of organisms around the world, including the animals of the Galápagos Islands, allowed him to conclude that adaptation to the environment can cause diversification, including the origin of new species.

Natural selection is the mechanism Darwin proposed for how adaptation comes about. The elements of Darwin's explanation are as follows:

1. *Members of a population exhibit random, but inherited, variations.* These variations do not arise purposefully and can be equally harmful or helpful when they appear. The variations that enable an organism to adapt to the environment are passed from one generation to the next.

2. *There is a struggle for existence.* The resources on Earth are insufficient for the survival of all members of a population; therefore, there is competition for the available resources.

3. *Organisms differ in reproductive success.* The reproductive success of an individual relative to the other members of a population is called **fitness**. The individuals that are most fit for an environment are those that obtain the most resources and produce the greatest number of viable offspring. The characteristics that determine fitness vary among populations.

4. *Natural selection can result in adaptation to a local environment.* An **adaptation** is a characteristic—physical or behavioral—that allows an organism to better survive in its environment. Adaptations result from natural selection because they lead to differential reproductive success and the accumulation of certain traits in a population.

> **Take Note:** *You should have a thorough understanding of natural selection as the mechanism of evolution and be able to explain it on the AP Biology exam. Remember that organisms do not "decide" to adapt and that natural selection is not predetermined. Natural selection occurs because organisms that just happen to have the traits that allow them to survive in a particular environment will survive and reproduce.*

15.3 Evidence for Evolution

The hypothesis that organisms share a common descent is supported by many lines of evidence. The fossil record, biogeography, anatomical evidence, and biochemical evidence all support the hypothesis.

The fossil record gives us the history of life in general and allows us to trace the descent of a particular group. **Fossils** are the remains and traces of past life. By studying the fossil record, particularly **transitional fossils** (common ancestors for two different groups of organisms), we can observe the changes that have appeared in an organism as it has descended over time from its ancestors.

Biogeography is the study of the range and distribution of plants and animals throughout the world. It shows that the distribution of organisms on Earth is explainable by assuming that related organisms evolved in one locale then spread to other regions. This explains why some continents have a different mix of plants and animals than other continents.

Comparing the anatomy and the development of organisms reveals a unity of plan among those that are closely related. **Homologous structures**, structures that are anatomically similar, show evidence of a common ancestor shared by different kinds of organisms, such as the forelimbs of vertebrates. **Analogous structures** on the other hand, such as the wings of a bird and the wings of a butterfly, perform the same function but do not have the same structure. **Vestigial structures**, which are fully developed and have a function in one group of organisms but are reduced in structure and/or function in another group, are also evidence of a common ancestor. The vestigial legs of some snakes and whales are an example. In addition to having homologous structures, vertebrates also share similarities in embryonic development, such as a postanal tail and pharyngeal pouches. Though each group of vertebrates now has its own pattern of development, the basis of each pattern is based on that obtained from the common ancestor.

All organisms have certain biochemical molecules, including proteins and nucleic acids, in common. The DNA of all organisms contains the same nucleotides, and the proteins contain the same 20 amino acids. Any differences in DNA or amino acid sequences of the same protein in different organisms indicate the degree of relatedness. The more similarities in an amino acid sequence, the more closely related are the organisms. Likewise, two organisms are not as closely related when there is a greater the number of differences in an amino acid sequence of a shared protein.

Multiple Choice Questions

1. Homologous structures provide evidence of evolution because they
    A.   have been modified for a new purpose.
    B.   allow biologists to trace organisms to a common ancestry.
    C.   perform the same function.
    D.   are features within the same organism that have similar structure.
    E.   disappear during embryonic development.

2. Tapeworms do not have digestive systems. This can be attributed to the fact that
    A.   the digestive system was not used and therefore became absent.
    B.   a mutation occurred to cause the loss of this system.
    C.   smaller digestive systems enabled them to dedicate more internal space to reproduction.
    D.   their ancestors did not have digestive systems.
    E.   digestive systems are not found in worms.

3. Upon study of several populations of butterflies you notice that there is a prevalence of orange butterflies to blue butterflies. This can be attributed to
    A.   stabilizing selection.
    B.   disruptive selection.
    C.   balancing selection.
    D.   directional selection.
    E.   frequency dependent selection.

4. Which of the following is NOT a component of the theory of natural selection?
    A.   Members of a population have inheritable variations.
    B.   A population produces more offspring in each generation than the environment can support.
    C.   The individuals with favorable traits acquire more resources than the individuals with less favorable traits.
    D.   An increasing proportion of individuals in each succeeding generation will have the favorable characteristics.
    E.   Increasing complexity is the result of a natural force is inherent in all organisms.

5. Which of the following explains the theory of inheritance of acquired characteristics?
    A.   Phenotypic changes acquired during an organism's lifetime result in genetic changes that can be passed on to subsequent generations.
    B.   Selection of a phenotypic form is based on the environmental pressures.
    C.   Mutations in the gene pool accumulate over time, altering the fitness of certain forms,
    D.   Organisms with certain inherited characteristics are more likely to survive and reproduce than others with different characteristics.
    E.   Populations are subject to immigration and emigration that can alter the genetic makeup of the gene pool.

6. Which of the following is an example of a homologous structure?
   A. wing of an insect and the wing of a bird
   B. leg of a horse and the wing of a bat
   C. wing of a bat and the shape of a fish
   D. jointed leg of an insect and the leg of a human
   E. human eye and the eye of a squid

Questions 7–10 refer to the following five choices. For each question, select the choice that is most closely related. Each choice may be used once, more than once, or not at all.
   A. Charles Darwin
   B. Charles Lyell
   C. Georges Cuvier
   D. Carolus Linnaeus
   E. Jean Baptiste de Lamarck

7. Hypothesized that the environment can bring about inherited change via the mechanism of acquired characteristics

8. Hypothesized that a series of mass extinctions accounted for the appearance of new fossils

9. Hypothesized that some individuals have favorable traits that enable them to better compete for resources

10. Hypothesized that erosion and volcanism acting gradually over a long period of time best explained geological features of Earth

Free Response Question

The theory of evolution explains the world of living things, how it came to be, and why it is so diverse.

   A. **Discuss** the contributions of TWO of the following scientists towards the discovery of evolutionary theory.
   - Charles Lyell
   - Carolus Linnaeus
   - Thomas Malthus
   - Jean Baptiste de Lamarck

   B. **Explain** the expressions "descent with modification" and "survival of the fittest" in terms of evolutionary theory.

Annotated Answer Key (MC)

1. **B**; Some vestigial organs disappear during embryonic development. They do not necessarily have the same function as analogous structures.

2. **C**; Tape worms are a parasite absorbing food from their host. Therefore, the worms with smaller digestive systems could dedicate more space and energy to reproduction.

3. **D**; Directional selection shifts the population towards one extreme phenotype. Disruptive selection favors both ends and would show equal numbers of blue and orange, while stabilizing selection would favor an intermediate phenotype. Frequency dependent and balancing selection favor preservation of variation.

4. **E**; Evolution is not goal driven and does not have a particular path or end point, nor does it always increase complexity, as seen in the loss of the digestive system in tapeworms. These are all common misconceptions of evolution.

5. **A**; Choices B, C, D, and E all are related to the theory of natural selection. The theory of acquired characteristics is synonymous with the idea that changes to an organism through its lifetime will be passed on to its offspring. Only changes to its gametes or cells producing gametes can be passed on.

6. **B**; Analogous structures serve the same function but are not constructed similarly and do not share a common ancestry. Homologous structures are anatomically similar because they are inherited from a common ancestor.

7. **E**; LaMarck recognized evolutionary changes in patterns of fossils and matched them to their environments. Unfortunately he is most remembered for his incorrect interpretation of the mechanism.

8. **C**; Cuvier was the father of paleontology and recognized that there were many mass extinctions in the fossil record.

9. **A**; Charles Darwin is recognized as the father of evolutionary theory and credited with first recognizing the correct mechanism of evolution.

10. **B**; Charles Lyell's ideas regarding the geology of the earth's surface, including the ideas that the earth is very old and small changes over long time periods add up to large-scale changes, were some of the ideas that prompted Darwin in his thinking.

Answer to FRQ

Part A (MAX 6)
Two points each, one point for idea, one point for contribution to Darwin

- Lyell
    - Geologist – principle of uniformitarianism, mechanisms are constant over time
    - *Gave Darwin idea that Earth is much older than previously thought and of gradual change over time
- Linneaus
    - Developed binomial nomenclature, father of taxonomy
    - Related similarities between creatures to patterns of creation, not evolution
    - *The hierarchy and relatedness to Darwin was evidence of relatedness through common descent.
- Malthus
    - Wrote on human suffering and human's potential to overpopulate the world leading to war, famine, disease
    - *Led to Darwin's idea of all organisms have capacity to overproduce
- Lamarck
    - Incorrect mechanism of evolution – theory of acquired characteristics
    - Idea of use and disuse and innate drive to become more complex
    - *Suggested life evolves as the environment changes

Part B (MAX 6)
- Descent with modification (internal max 4 pts)
    - Unity in life – all derived from a single common ancestor
    - Ancestral organisms lived in various habitats accumulating diverse modifications.
    - Over long periods of time, this led to the rich diversity of life today.
    - History of life as a tree with multiple branchings
- Natural Selection (internal max 5 pts)
    - Variation among a population which is inheritable
    - All species produce more offspring than the environment can support.

- o Limited resources mean not all survive.
- o Inference: Those with more favorable traits produce more offspring.
- o Inference: Unequal ability to survive and reproduce leads to an accumulation of favorable traits in the population over generations.

CHAPTER 16
HOW POPULATIONS EVOLVE

16.1   Population Genetics

A **population** consists of all of the members of a particular species that live together in the same area. **Population genetics** is the study of the allele differences that lead to variations in a population. These variations can even extend to **SNPs**, which are **single nucleotide polymorphisms**. SNPs are sequences of DNA where the two alleles of a gene differ by one nucleotide. Investigators are beginning to think that SNPs have significance because they may be an important source of variations in humans.

Microevolution is the evolutionary changes that occur in a population. Within a given population, the total number of alleles of all individuals is called the **gene pool**. The **Hardy-Weinberg equilibrium** is a constancy of allele frequencies in a gene pool that remains from generation to generation if certain conditions are met:

1. No mutations
2. No gene flow
3. Random mating
4. No genetic drift
5. No natural selection

Since these conditions are rarely met, a change in gene pool frequencies is likely. When gene pool frequencies change, microevolution has occurred. The changes in the allele frequencies of a gene pool can be determined by the **Hardy-Weinberg equilibrium**:

$$p^2 + 2pq + q^2 = 1$$

where: p^2 = frequency of the homozygous dominant genotype

p = frequency of the dominant allele

$2pq$ = frequency of the heterozygous genotype

q^2 = frequency of the homozygous recessive genotype

q = frequency of the recessive allele

Deviations from Hardy-Weinberg equilibrium allow us to determine when evolution has taken place. Mutations, gene flow, nonrandom mating, genetic drift, and natural selection all cause deviations from a Hardy-Weinberg equilibrium. Of these factors, the only one that results in adaptation is natural selection.

Mutations, changes in the DNA, are the raw material for evolutionary change. Recombinations in chromosomes help bring about adaptive genotypes. Mutations do not occur because an organism "needs" them. Rather, changes in the DNA occur by chance, and they just happen to result in the genetic variation that leads to adaptation.

Gene flow occurs when a breeding individual (in animals) migrates to another population or when gametes and seeds (in plants) are carried into another population. This results in a movement of alleles between two populations. Constant gene flow between two populations causes their gene pools to become similar. However, when gene flow brings a new or rare allele into the population, the allele frequencies change in the next generation.

Nonrandom mating occurs when relatives mate (inbreeding) or assortative mating takes place. **Assortative mating** occurs between individuals with the same phenotype with respect to a specific characteristic. Both of these cause an increase in the frequency of homozygotes and a decrease in the frequency of heterozygotes in a population.

Genetic drift occurs when allele frequencies are altered only because some individuals, by chance, contribute more alleles to the next generation. Genetic drift does not cause adaptation, but can cause the gene pools of two isolated populations to become dissimilar as some alleles are lost and others are fixed. Genetic drift is particularly evident after a bottleneck or when founders start a new population. The **bottleneck effect** occurs due to a natural disaster or some other event that greatly diminishes the number of individuals in a population, leading to severe inbreeding and extreme genetic similarity. The **founder effect** is seen when rare alleles occur at a higher frequency in a small population that is isolated from a larger population.

Take Note: *On the AP Biology exam, you may be asked to use the Hardy-Weinberg equilibrium to determine whether a population is evolving. Under what conditions would the allele frequencies in a population remain constant? What cellular, molecular, or environmental mechanisms introduce variety into the gene pool?*

16.2   Natural Selection

Most of the traits of evolutionary significance are polygenic; the many variations in population result in a bell-shaped curve. When a range of phenotypes is exposed to the environment, natural selection favors the one that is most adaptive for the present environment. Three types of selection occur: (1) **directional**—the curve shifts in one direction, as when bacteria become resistant to antibiotics; (2) **stabilizing**—the peak of the curve increases, as when there is an optimum clutch size for survival of Swiss starling young; and (3) **disruptive**—the curve has two peaks, as when *Cepaea* snails vary because a wide geographic range causes selection to vary.

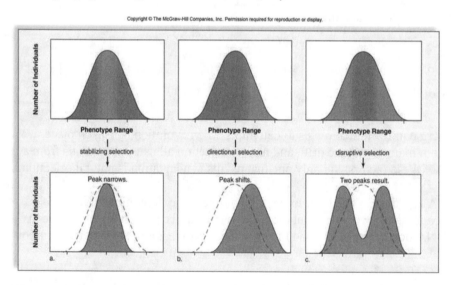

Three types of natural selection

The adaptive changes in each gender that lead to an increased ability to attract a mate lead to **sexual selection**. Traits that promote **fitness** (reproductive success) are expected to be advantageous overall, despite any possible disadvantage. Males produce many sperm and compete to inseminate females.

Females produce few eggs and are selective about their mates. In species where competition for mates is a behavioral adaptation, sexual dimorphism exists. It is possible that male competition and female choice also occur among humans. Biological differences between the sexes may promote certain mating behaviors because they increase fitness.

> **Take Note:** *For the AP Biology exam, you should be able to explain why sexual selection is a form of natural selection.*

16.3 Maintenance of Variations

Despite constant natural selection, genetic variation is maintained in a population due to several factors. Mutations and recombination still occur, creating new alleles that recombined in sexual reproduction. Gene flow among small populations can introduce new alleles, and natural selection itself sometimes results in variation.

In sexually reproducing diploid organisms, the heterozygote acts as a repository for recessive alleles whose frequency is low. The heterozygote is favored over either homozygote, resulting in a **heterozygote advantage** that maintains genetic diversity. An example of heterozygote advantage is seen in sickle-cell disease, where the heterozygote is more fit in areas where malaria occurs, and therefore both homozygotes are maintained in the population.

> **Take Note:** *For the AP Biology exam, you should be prepared to discuss the ways that cellular, molecular, or environmental mechanisms introduce variety into the gene pool.*

Multiple Choice Questions

1. Single nucleotide polymorphisms, or SNPs
    A. are similar RNA sequences in two organisms.
    B. are enzymes that are able to remove introns from mRNA.
    C. are DNA sequence variations.
    D. are components of splicesomes.

2. Which of the following is NOT a condition of the Hardy-Weinberg principle?
    A. Migration of alleles into or out of the population does not occur.
    B. Individuals pair by chance, not according to phenotypes.
    C. Changes of allele frequencies due to chance alone are insignificant.
    D. Selective forces do not favor one genotype over another.
    E. Allele changes are not balanced by changes in the opposite direction.

3. In a particular population of moths, the frequency of the DD genotype (dark) occurring in the population is 0.04. What is the frequency of the dd genotype (light) occurring in the population?
    A. 0.32
    B. 0.64
    C. 0.20
    D. 0.80
    E. 0.04

4. Mutations
    A. are the primary source of genetic difference in asexually reproducing prokaryotes.
    B. are temporary genetic changes.
    C. occur quickly in mitochondrial DNA.
    D. become significant only when a dominant genotype arises.
    E. present immediate phenotypic change in diploid organisms.

Questions 5–8 refer to the following five choices. For each question, select the choice that is most closely related. Each choice may be used once, more than once, or not at all.

    A.   bottleneck effect
    B.   gene flow
    C.   disruptive selection
    D.   heterozygote advantage
    E.   directional selection

5. The genotype $Hb^A Hb^S$ is favored over the extreme genotypes.

6. Prevents the majority off genotypes from participating in the production of the next generation

7. Extreme phenotype is favored.

8. A species has been pushed to near extinction due to natural disaster.

9. Subspecies help to maintain diversity because
    A.   there is interbreeding between species which introduces alleles that may keep each subspecies from fully adapting to their particular environment.
    B.   the alleles of one subspecies are the same as the other subspecies.
    C.   they are generally not located adjacent to one another and can therefore maximize their potential in a particular environment.
    D.   there is no gene flow between subspecies because each is particularly adapted to its own environment.
    E.   they increase the effects of the bottleneck effect.

10. The movement of alleles between populations is
    A.   genetic drift.
    B.   bottleneck effect.
    C.   gene flow.
    D.   stabilizing selection.
    E.   heterozygote advantage.

Free Response Question

Suppose investigators notice that on an island, when a species is subjected to a great deal of predation, the some organisms tend to have a different phenotypic appearance than those members of the population who are exposed to minimal or no predation.

    A. **Explain** the possible phenotypic characteristics of each population.

    B. **Identify** and **define** this type of selection.

    C. **Explain** when this type of selection may occur in an environment.

Annotated Answer Key (MC)

1. **C**; Single nucleotide polymorphisms (SNPs) generally have two alleles ("snips") and are DNA sequence variations occurring when a nucleotide (A, T, C, G) differs between members of a species or between paired chromosomes. SNPs are among the most useful of genetic markers. They occur in at least 1% of the population, are found in coding and noncoding DNA and occur on average about once in 100 to 300 bp in the human genome. Spliceosomes are composed of RNA and proteins (snRNPs or "snurps") that remove introns from a pre-mRNA strand.

2. **E**; The Hardy-Weinberg theory is a statement of theoretical genetic equilibrium that rarely occurs in nature. It provides a baseline to measure genetic change. In order for this to occur, a population must maintain all of the conditions listed except for the fact that allele changes are not balanced by changes in the opposite direction.

3. **B**; In the Hardy-Weinberg theory, $p^2 + q^2 + 2pq = 1$, where p^2 is the percentage of homozygous dominant individuals in a population, q^2 is the percentage of homozygous recessive individuals in a population, and $2pq$ is the percentage of heterozygous individuals in a population; $p + q = 1$, where p is the frequency of the dominant allele and q is the frequency of the recessive allele. Therefore, if the frequency of the dominant allele is 0.04, $p = 0.2$. If $p = 0.2$ then $q = 0.8$. Finally, $q^2 = (0.8)(0.8) = 0.64$.

4. **A**; Mutations are the only source of new variation in organisms. They can be spontaneous and occur in DNA that is then expressed as a phenotype. As such, mutations are not immediately expressed and can be very significant as part of a new genotype.

5. **D**; Often heterozygotes have a favorable advantage over extreme genotypes, and the superscript A and S indicate the individual is heterozygous.

6. **A**; In the bottleneck effect, a significant percentage of a population or species is prevented from reproducing. It often occurs due to an extreme environmental effect and increases genetic drift.

7. **E**; In directional selection, one extreme phenotype is favored.

8. **A**; Often the bottleneck effect is so significant that a species can be pushed towards extinction. Examples include bison, elephant seals, a species of antelope, and the giant panda.

9. **A**; Members of one subspecies differ morphologically or by different DNA sequences from members of other subspecies of the species.

10. **C**; Population bottlenecks occur when a population's size is reduced for at least one generation. Because genetic drift acts very quickly to reduce genetic variation in small populations, undergoing a bottleneck can reduce a population's genetic variation. However, normal movement of alleles between populations is called gene flow.

Answer to FRQ

PART A (MAX 4)
- Organisms with predators present will be duller (drab) in color.
- Organisms with decreased predation will be brighter in color, more spots…
- Example (2 pts – 1 point for drab coloration, 1 point for showy brighter coloration): guppies in ponds with or without a major predator (pike cichlid). Any viable example will work as long as there is a possible predator identified with one population.

PART B (MAX 3)
- Directional: one morph favored over another
- Stabilizing: extreme variants are removed from the population
- Disruptive: variants at either end of the continuum are favored
  Note: Labeled diagrams discussed in writing are acceptable.

PART C (MAX 3)
- Increased predation causes those more visible to be eaten, decreasing their fitness.
- Those with limited predation who are brighter will have more luck finding mates.
- Brighter coloration can lead to increased fitness with less predation.

CHAPTER 17
SPECIATION AND MACROEVOLUTION

17.1   Separation of the Species

The **evolutionary species concept** recognizes that every species has its own evolutionary history and can be recognized by certain diagnostic morphological traits. The **biological species concept** recognizes that a species is reproductively isolated from other species, therefore, the members of a species breed only among themselves. The use of DNA sequence data can also be used today to distinguish one species from another.

When two species are reproductively isolated, gene flow does not occur between them, preventing the production of fertile, viable offspring. Isolating mechanisms are reproductive barriers that prevent successful reproduction from occurring. **Prezygotic isolating mechanisms** (habitat, temporal, behavior, mechanical, and gamete isolation) prevent mating from being attempted or prevent fertilization from being successful if mating is attempted. **Postzygotic isolating mechanisms** (zygote mortality, hybrid sterility, and F_2 fitness) prevent hybrid offspring from surviving and/or reproducing. Take a moment to review these mechanisms before moving on to the next section.

TABLE 18.1

Reproductive Isolating Mechanisms

Isolating Mechanism	Example
Prezygotic	
Habitat isolation	Species at same locale occupy different habitats.
Temporal isolation	Species reproduce at different seasons or different times of day.
Behavioral isolation	In animal species, courtship behavior differs, or individuals respond to different songs, calls, pheromones, or other signals.
Mechanical isolation	Genitalia between species are unsuitable for one another.
Gamete isolation	Sperm cannot reach or fertilize egg.
Postzygotic	
Zygote mortality	Fertilization occurs, but zygote does not survive.
Hybrid sterility	Hybrid survives but is sterile and cannot reproduce.
F_2 fitness	Hybrid is fertile, but F_2 hybrid has reduced fitness.

Reproductive barriers

17.2   Modes of Speciation

During **allopatric speciation**, geographic separation precedes reproductive isolation. Geographic isolation allows genetic changes to occur so that the ancestral species and the new species can no longer breed with one another. This leads to one species splitting into two or more species over time, and requires a geographic barrier. For example, a series of salamander subspecies on either side of the Central Valley of California has resulted in two species that are unable to successfully reproduce when they come in contact.

Adaptive radiation, a type of allopatric speciation, occurs when multiple species evolve from an ancestral population because varied habitats permit varied adaptations to occur. Each species is adapted to a specific environment according to its ecological niche. Adaptive radiation is exemplified in the evolution of Hawaiian honeycreepers, a group of birds that comprises more than 20 species evolved from a goldfinch-like ancestor.

During **sympatric speciation**, no geographic separation precedes reproductive isolation, so a geographic barrier is not required. In this case, speciation is simply a change in genotype that prevents successful reproduction. In plants, the occurrence of polyploidy (chromosome number above 2n) reproductively isolates an offspring from the former generation. **Autoploidy** occurs within the same species. For example, if—due to nondisjunction—a diploid gamete fuses with a haploid gamete, a triploid plant results that cannot reproduce with 2n plants because some of the chromosomes would not pair during meiosis. **Alloploidy** occurs when two different, but related species hybridize. If an odd number of chromosomes results, the hybrid is sterile unless a doubling of the chromosomes occurs and the chromosomes can pair during meiosis. The sex chromosomes in animals make speciation by polyploidy highly unlikely.

> **Take Note:** *In preparation for the AP Biology exam, you should be able to compare and contrast allopatric speciation and sympatric speciation and give examples of both.*

17.3   Principles of Macroevolution

Macroevolution is evolution of new species and higher levels of classification. The fossil record, such as is found in the Burgess Shale, gives us a view of life many millions of years ago. The hypothesis that species evolve gradually after the isolation of a population is now being challenged by the hypothesis that speciation can occur rapidly. In that case, the fossil record could show periods of stasis interrupted by spurts of change, for example, a **punctuated equilibrium**. Transitional fossils would be expected with gradual change but not with punctuated equilibrium.

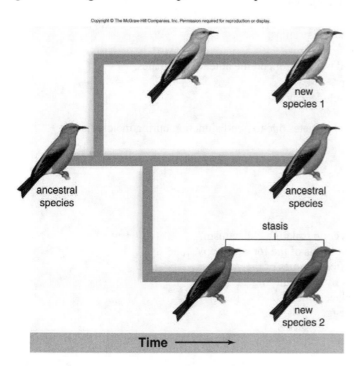

Gradualistic and punctuated equilibrium models

Take Note: *On the AP Biology exam, you may be asked to discuss gradualism and punctuated equilibrium and evaluate the role of each in the evolution of organisms.*

It could be that both models are seen in the fossil record, but rapid change can occur by differential expression of regulatory genes. The same regulatory gene (*Pax6*) controls the development of both the camera-type and the compound-type eye. The *Tbx5* gene controls development of limbs, whether the wing of a bird or the leg of a tetrapod. *Hox* genes control the number and appearance of a repeated structure along the main body axes of vertebrates. The same pelvic-fin genes control the development of a pelvic girdle. Changing the timing of gene expression, as well as which genes are expressed, can result in dramatic changes in shape.

Speciation, diversification, and extinction are seen during the evolution of *Equus*. These three processes are commonplace in the fossil record and illustrate that macroevolution is not goal-directed. The life we see about us represents adaptations to particular environments. Such adaptations have changed in the past and will change in the future.

Multiple Choice Questions

Questions 1–4 refer to the following five choices. For each question, select the choice that is most closely related. Each choice may be used once, more than once, or not at all.

    A.   behavioral isolation
    B.   mechanical isolation
    C.   gamete isolation
    D.   temporal isolation
    E.   habitat isolation

1. Florida maple and sugar maple trees do not exchange pollen.

2. Wood frogs breed in ponds and leopard frogs breed in swamps.

3. Female gypsy moths release pheromones detected by receptors on the antennae of males.

4. Courtship behaviors of the blue footed boobies include dances.

5. The process when a diploid plant produces diploid gametes due to nondisjunction during meiosis is
    A.   haploidy.
    B.   autoploidy.
    C.   polyploidy.
    D.   alloploidy.
    E.   diploidy.

6. Which of the following is an example of a postzygotic isolating mechanism?
    A.   Animal species are restricted to different levels of the forest canopy.
    B.   Several species of frogs reproduce at different times of the year.
    C.   Female crickets recognize male crickets by their chirping.
    D.   A male horse and a female donkey produce a mule.
    E.   The sperm cannot reach or fertilize the egg.

7. Sympatric speciation
    A. occurs when populations are separated by a geographic barrier.
    B. occurs when a single ancestral species gives rise to a variety of species.
    C. includes examples of polyploidy.
    D. allows the two species to mate with one another if there is no geographic barrier.
    E. limits neighboring populations to hybridize only if they are of the same species.

8. Within the biological species concept
    A. the members of a species have a single gene pool due to reproductive isolation.
    B. every species has its own evolutionary history and that is found in the fossil record.
    C. the members of a species share differing evolutionary pathways.
    D. organisms are looked at as the unit of evolution of new species.
    E. mutations create new species as they accumulate in the introns of a species.

9. An example of allopatric speciation would be
    A. two species of grasses in the same field hybridizing to make a new species.
    B. one species of grass duplicating its chromosomes to become triploid, forming a new species.
    C. two closely related neighboring species of birds hybridize to form a new species.
    D. two bacteria trading DNA through villi and hybridizing.
    E. two populations of squirrels on opposite sides of a canyon becoming two new species.

10. Strawberries are octoploid, having eight sets of their chromosomes. Strawberries most likely evolved by the process of
    A. coevolution.
    B. mutation.
    C. genetic drift.
    D. aneuploidy.
    E. polyploidy.

Free Response Question

Genes can bring about great changes in body shapes and organs.

    A. **Identify** and **define** two examples of organs whose genes provide evidence for a common ancestor.

    B. Choose **one** example from above and **explain** how this organ provides evidence of common ancestry.

    C. **Discuss** how new body plans in organisms can arise from fairly simple changes in the genes of an organism.

Annotated Answer Key (MC)

1. **E**; The Florida maple is found from Florida north coastally to Georgia, while the sugar maple is found from the Carolinas north.

2. **D**; The anatomy and size of the small wood frog prevents it from mating with the much larger leopard frog.

3. **A**; The release of pheromones, mating rituals, singing, and other rituals are all behavioral modifications that are used by organisms to recognize other members of the same species.

4. **A**; The release of pheromones, mating rituals, singing and other rituals are all behavioral modifications that are used by organisms to recognize other members of the same species.

5. **B**; The process of autopolyploidy produces more than two sets of chromosomes derived from a single species.

6. **D**; Mechanisms that occur after fertilization and a zygote forming are postzygotic. Mules are viable hybrids but cannot reproduce. Choices B and C are behavioral mechanisms and are prezygotic, as is sperm unable to reach the egg and animals in different canopy layers (habitat isolation).

7. **C**; Sympatric speciation occurs when new species arise in the same geographic area as the parent species, often by polyploidy in plants. Geographic barriers are an example of allopatric speciation and one species giving rise to many species is an example of radiation.

8. **A**; The biological species concept is the most widely accepted species concept. It defines species in terms of interbreeding. The biological species concept explains why the members of a species resemble one another (i.e., form phenetic clusters) and differ from other species. Members of a species share a common evolutionary heritage and the population is the unit of evolution.

9. **E**; Allopatric speciation involves the formation of new species surrounded by their parent species, as in choice A. Sympatric speciation involves a barrier preventing gene flow, while parapatric speciation is neighboring species hybridizing. Bacteria trading DNA via pilli is an example of conjugation.

10. **E**; Polyploidy is the result of an organism replicating an entire set of chromosomes generally through nondisjunction during meiosis. Aneuploidy is the occurrence of or deletion of an extra chromosome and is usually associated with cancer.

Answer to FRQ

Part A (MAX 4)

- One point – example of vestigial organ
- Definition vestigial organs: Remains of a structure once functional → no longer functional in the organism
- One point – example of a homologous organ
- Definition homologous organs: Similar in different types of organisms due to a common ancestor

Part B (MAX 4)

- Structures that were clearly designed for other complex purposes
- Perform relatively simple, minor, or inessential functions (also may have no function)
- No organism can have a vestigial structure that was not previously functional in one of its ancestors.
- If all living organisms descended from a common ancestor, then both functions and structures have been gained and lost in each lineage during macroevolution.
- Many organisms should retain vestigial structures as structural remnants of lost functions.
- Change is gradual.

Part C (MAX 5)

- Homeotic genes arose independently in the early evolution of animals and plants.
- Homeobox: DNA sequence found within genes that are involved in the regulation of patterns of development
- Mutations to homeobox genes can produce easily visible phenotypic changes.
- Hox genes (animals): Determine the identity of embryonic regions
- MADS-box genes (plants – not homologous to Hox)

CHAPTER 18
ORIGIN AND HISTORY OF LIFE

18.1   Origin of Life

The unique conditions of the early earth allowed a **chemical evolution** to occur. The atmosphere was very hot and contained little oxygen; it was a **reducing atmosphere**. The atmosphere consisted of many gases: H_2O, N_2, CO_2, H_2, CH_4, NH_3, H_2S, and CO. An **abiotic** synthesis of small organic molecules such as amino acids and nucleotides occurred, possibly either in the atmosphere or at hydrothermal vents. These monomers joined together to form polymers either on land (warm seaside rocks or clay) or at the vents. Stanley Miller's classic experiment demonstrated how this could have happened. His apparatus, shown below, contained an atmosphere that resembled the atmosphere of the early earth. He used a spark to simulate lightning, and discovered a variety of amino acids and organic molecules were produced.

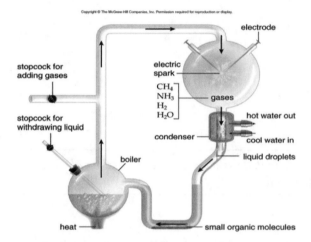

Stanley Miller's apparatus and experiment

The first polymers could have been proteins or RNA, or they could have evolved together. The aggregation of polymers inside a plasma membrane produced a **protocell** having some enzymatic properties such that it could grow. If the protocell developed in the ocean, it was a heterotroph; if it developed at hydrothermal vents, it was a chemoautotroph. A true cell had evolved once the protocell contained DNA genes. The first genes may have been RNA molecules, but later DNA became the information storage molecule of heredity. Biological evolution now began.

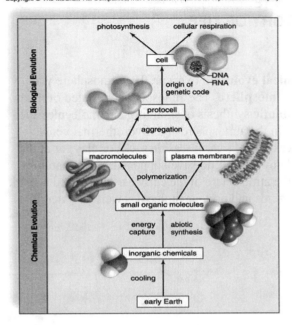

Origin of the first cells

Take Note: *For the AP Biology exam, you should be able to discuss the significance of the reducing atmosphere of the early earth to the evolution of the first cells.*

18.2 History of Life

The fossil record allows us to trace the history of life. The oldest prokaryotic fossils are cyanobacteria, dated about 3.5 bya, and they were the first organisms to add oxygen to the atmosphere because they conducted photosynthesis.

The eukaryotic cell evolved about 2.2 bya, gradually obtaining organelles over time. The **endosymbiotic theory** explains how eukaryotic cells evolved from prokaryotic cells. It says that mitochondria and chloroplasts were once free-living prokaryotes that were engulfed by a nucleated cell, and a symbiotic relationship formed between them. There are several pieces of evidence that support this theory (e.g., mitochondria and chloroplasts are about the same size as bacteria and divide by binary fission, as bacteria do). Both mitochondria and chloroplasts have their own DNA and make some of their own proteins. Though the outer membrane of both organelles resembles the membrane of a eukaryotic cell, their inner membranes are more similar to those of prokaryotic cells.

Multicellular animals (the Ediacaran animals) arose approximately 600 mya. Fossils reveal that these animals, which possibly reproduced sexually, were most likely soft-bodied invertebrates that lived in shallow marine waters. Most of them were flat and immobile with no mouths, and absorbed their nutrients into their bodies from the environment.

A rich animal fossil record starts at the Cambrian period of the Paleozoic era. The occurrence of external skeletons, which seems to explain the increased number of fossils at this time, may have been due to the presence of plentiful oxygen in the atmosphere, or perhaps it was due to predation. The fishes were the first vertebrates to diversify and become dominant. Amphibians are descended from lobe-finned fishes.

Plants also invaded land during the Ordovician period. These first plants were nonvascular plants. The swamp forests of the Carboniferous period contained seedless vascular plants, insects, and amphibians. This period is sometimes called the Age of Amphibians.

The Mesozoic era was the Age of Cycads and Reptiles. First mammals and then birds evolved from reptilian ancestors. During this era, dinosaurs of enormous size were present. By the end of the Cretaceous period, the dinosaurs were extinct.

The Cenozoic era is divided into the Tertiary period and the Quaternary period. The Tertiary is associated with the adaptive radiation of mammals and flowering plants that formed vast tropical forests. The Quaternary is associated with the evolution of primates; first monkeys appeared, then apes, and then humans. Grasslands were replacing forests, and this put pressure on primates, who were adapted to living in trees. The result may have been the evolution of humans—primates who left the trees.

> **Take Note:** *For the AP Biology exam, you should be prepared to explain how the origin of photosynthesis led to the development of the ozone layer and the evolution of eukaryotes. Can you explain the theory of endosymbiosis and describe the evidence that supports it?*

18.3 Factors That Influence Evolution

The continents are on massive plates that move, carrying the land with them. **Plate tectonics** is the study of the movement of the plates. **Continental drift** helps explain the distribution pattern of today's land organisms.

Mass extinctions have played a dramatic role in the history of life. It has been suggested that the extinction at the end of the Cretaceous period was caused by the impact of a large meteorite, and evidence indicates that other extinctions have a similar cause as well. It has also been suggested that tectonic, oceanic, and climatic fluctuations, particularly due to continental drift, can bring about mass extinctions.

Multiple Choice Questions

1. Earth's early atmosphere most likely consisted of
    A. ozone (O_3).
    B. ammonium (NH_4).
    C. carbon dioxide (CO_2).
    D. oxygen (O_2).
    E. methane (CH_4).

2. Which of the following is NOT a biological hypothesis for the evolution of early life on Earth?
    A. A protocell had enzymatic properties and developed into a heterotroph or chemoautotroph.
    B. Abiotic synthesis process created small organic molecules, including amino acids and nucleotides.
    C. Monomers joined together to form polymers of proteins or RNA.
    D. The first genes may have been RNA or DNA.
    E. The process of photosynthesis was one of the first metabolic processes in autotrophic prokaryotes.

3. Evidence for the endosymbiotic theory includes the fact that
    A. mitochondria reproduce by means of mitosis.
    B. chloroplasts do not make any of their own proteins.
    C. mitochondria and chloroplasts were larger than the nucleated cell.
    D. mitochondria and chloroplasts contain their own DNA.
    E. the outer membrane of mitochondria and chloroplasts are the same.

4. The first identifiable fossils are of
    A. sossilized stromatolites.
    B. angiosperms.
    C. unicellular prokaryotes.
    D. seedless plants.
    E. multicellular protists.

5. Which of the following appeared the earliest on the geologic time scale?
    A. flowering plants
    B. club mosses
    C. chordates
    D. protists
    E. seedless vascular plants

6. The first land plants
    A. contained water conducting tissues.
    B. were extremely tall.
    C. contained seeds.
    D. were similar to mosses.
    E. did not produce spores.

7. If Earth's early atmosphere was not a reducing environment, where would the most likely place have been for the early production of macromolecules?
    A. near submerged volcanoes
    B. in the upper atmosphere
    C. on the edges of volcanoes on continents
    D. inside electrical storms
    E. along continental geologic fault lines

Questions 8–10 refer to the following five choices. For each question, select the choice that is most closely related. Each choice may be used once, more than once, or not at all.
    A. Cenozoic
    B. Mesozoic
    C. Late Paleozoic
    D. Early Paleozoic
    E. Archaen

8. First prokaryotic cells

9. First insects and amphibians

10. Flowering plants evolve

Free Response Question

In cells, organic monomers join to form polymers in the presence of enzymes.

   A. **Explain** how the first organic polymers formed if there were no proteins.

   B. **Discuss** three characteristics of the protobiont.

Annotated Answer Key (MC)

1. **C**; Early models of Earth's atmosphere suggested it was a reducing atmosphere with ammonium and methane. Most scientists now do not believe the earth's early atmosphere was a reducing one and most likely contained the gases carbon dioxide and nitrogen. Oxygen and consequently ozone did not appear until much later with the advent of photosynthesis.

2. **E**; Photosynthesis evolved well after Glycolysis and aerobic respiration.

3. **D**; Mitochondria and chloroplasts contain their own DNA very similar to prokaryotes and manufacture most of their own proteins. Both of these organelles replicate on their own, inside the cell by a process similar to binary fission.

4. **A**; The earliest fossils are those of fossilized stromatolites (3.5 BYA), evidence of a photosynthetic microbial community. Unicellular prokaryotes must first have evolved before this time.

5. **D**; Plants, mosses, fungi, and chordates all evolved from early groups of protists.

6. **D**; The earliest land plants (liverworts, hornworts, and mosses) all still relied on water for their gametes to swim across the leaf surface. Seeds and spores evolved much later, as did water-conducting tissue allowing later plants to be tall.

7. **A**; Earth's early atmosphere was most likely not a reducing environment, thus the most likely candidate for a reducing environment is where minerals, sulfur, and other high energy compounds occur—near submerged volcanoes.

8. **E**; The first prokaryotic cells evolved prior to 3.5 BYA during the Archaen (Precambrian).

9. **C**; The first insects and amphibians appear in the fossil record during the late Paleozoic (Devonian, 400 mya).

10. **B**; The flowering plants first appear in the fossil record in the Mesozoic (Cretaceous, approximately 100 mya). They then become the dominant plant forms later in the Cenozoic.

Answer to FRQ

PART A (MAX 8)
Maximum three points each hypothesis

Protein first hypothesis
- Amino acids → collected → sun → proteinoids (small polypeptides with catalytic properties)
- Proteinoids → microspheres (protein with properties of a cell)
- Leads to selective advantage
- DNA genes came after protein enzymes arose.
- Protein enzymes are needed for DNA replication.

Simultaneous evolution
- Clay attracts inorganic catalysts for polypeptide formation.
- Collects energy from radioactive decay
- RNA nucleotides and amino acids →synthesis of RNA
- Polypeptides and RNA arose at the same time.

RNA first hypothesis
- RNA only was needed – can carry information

- Easily formed
- RNA can be both a substrate *and an enzyme.*
- First genes could have been RNA.
- Carried out the processes now associated with DNA and proteins

PART B (MAX 4)

- Chemically different internal environment
- Composed of organic molecules
- Membrane
  - Lipid-protein membrane possible
- No precise reproduction
- Heterotrophic OR chemoautotrophic depending upon environment
- Extraction of energy from carbohydrates (glycolysis)
- Carried on fermentation
- Conditioned responses to stimuli possible

CHAPTER 19
SYSTEMATICS AND PHYLOGENY

19.1   Systematics

Systematics is dedicated to understanding the evolutionary history of life on Earth. **Taxonomy,** a part of systematics**,** is the study of identifying and classifying organisms into groups based on the characteristics an organism shares with others of its same kind.

Taxonomists assign each organism an italicized binomial name in Latin that consists of the genus and the specific epithet. For instance, the scientific name of a lion is *Panthera leo*, where *Panthera* is the genus and *leo* is the specific epithet. Notice that the first name of an organism is capitalized, while the second name is lowercase.

Classification involves the assignment of species to categories. When an organism is named, a species has been assigned to a particular genus. From the least inclusive to the most inclusive category, or **taxa,** each **species** belongs to a **genus, family, order, class, phylum, kingdom,** and **domain.** Each higher taxa is more inclusive; species in the same kingdom share general characters, and species in the same genus share quite specific characters.

> **Take Note:** *On the AP Biology exam, you may be asked to describe the characteristics used to distinguish between members of one kingdom from the members of the other kingdoms. What similarities do the kingdoms share? How are they different?*

19.2   Phylogenetic Trees

Phylogenetic trees are diagrams that depict evolutionary relationships between different species. The tree shows how organisms evolved from common ancestors by showing the divergence (branching) between them due to their **derived traits** (traits not previously seen). Because of the hierarchical nature of Linnean classification, it is possible to draw up a phylogenetic tree based on classification categories.

Rather than using similarities to construct phylogenetic trees, **cladistics** offers a way to use shared derived traits to distinguish different groups of species from one another. When no original common ancestor can be found in the fossil record, the use of an outgroup allows us to determine with what trait to begin the tree. Based on the rest of the available data, it is possible to determine the sequence of **clades** in the tree. Clades are **monophyletic**; they contain the most recent common ancestor along with all its descendants. All the members of a clade share the same derived traits. Linnean classification permits the use of groupings other than those that are monophyletic, and therefore has come under severe criticism.

The fossil record, homology, and molecular data, in particular, are used to help decipher phylogenies. Because fossils can be dated, available fossils can establish the antiquity of a species. If the fossil record is complete enough, we can sometimes trace a lineage through time. Homology helps indicate when species belong to a monophyletic taxon (share a common ancestor); however, convergent evolution sometimes makes it difficult to distinguish homologous structures from analogous structures. DNA base sequence data are commonly used to help determine evolutionary relationships.

> **Take Note:** *In preparation for the AP Biology exam, you should be able to describe the evolutionary relationships between related groups of organisms. What structural and functional adaptations do the groups share? How did these adaptations lead to the success of the organisms?*

19.3 The Three-Domain System

The three domains of life are **Archaea, Bacteria,** and **Eukarya**. The first two domains contain prokaryotic organisms that are structurally simple but metabolically complex. They lack membrane-bound organelles, such as a nucleus, and they are both comprised of unicellular organisms that reproduce asexually. Archaea and bacteria are different in the organization of their DNA and composition of their cell walls. Also, archaea live in more extreme environments than bacteria. Neither archaea nor bacteria have been classified into kingdoms at this time.

Organisms in domain Eukarya have membrane-bound organelles and include protists, fungi, plants, and animals. Protists range from unicellular to multicellular organisms and include the protozoans and most algae. Among the fungi are the familiar molds and mushrooms. Plants are well known as the multicellular photosynthesizers of the world, while animals are multicellular and ingest their food. An evolutionary tree shows how the domains are related by way of common ancestors.

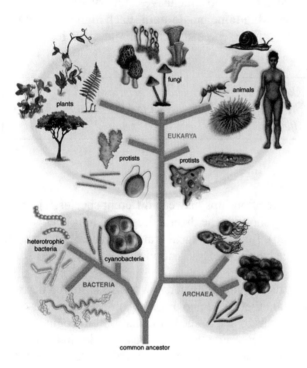

A tree of life showing the three domains

Multiple Choice Questions

1. Animals classified in the same phyla would also be in the same
   A. family.
   B. class.
   C. kingdom.
   D. order.
   E. genus.

2. An organism with hair, four limbs, vertebrae, and a notochord in an embryo would be
   A. finch.
   B. dog.
   C. crocodile.
   D. lizard.
   E. frog.

3. Echinoderms are closely related to vertebrates because
   A. they are acoelomate.
   B. they have bilateral symmetry.
   C. the embryonic blastopore becomes the anus.
   D. they are a protostome.
   E. they reproduce asexually.

4. Mitochondrial DNA is used to determine the phylogeny of closely related species because it
   A. contains over 100,000 genes.
   B. is found only in animals.
   C. is rearranged in the process of recombination.
   D. mutates very quickly.
   E. is inherited maternally and paternally.

5. Organisms classified in the domain Archaea posses
   A. a nuclear envelope.
   B. membrane bound organelles.
   C. ribosomes.
   D. a cellulose cell wall.
   E. only phospholipids in the cell membrane.

6. Fungi, being heterotrophic, multicellular eukaryotes, are most closely related to
   A. animals.
   B. plants.
   C. bacteria.
   D. archaea.
   E. protists.

7. Animals evolved from a(n)
   A. autotrophic fungus.
   B. heterotrophic plant.
   C. aquatic protist.
   D. photosynthetic protist.
   E. heterotrophic protist.

8. A nonmotile, multicellular, autotrophic organism with an alternation of generations is a(n)
   A. prokaryote.
   B. archaebacteria.
   C. fungus.
   D. plant.
   E. animal.

9. Fungi were placed in their own kingdom separate from plants because fungi are
   A. autotrophic and unicellular.
   B. multicellular and heterotrophic.
   C. both heterotrophic and autotrophic.
   D. unicellular and heterotrophic by ingestion.
   E. unicellular decomposers.

10. On a phylogenetic tree, a frog is most closely related to a(n)
    A. crocodile.
    B. finch.
    C. dog.
    D. lancelet.
    E. chimpanzee.

Free Response Question

Construction of a cladogram requires analysis of several traits to determine which are shared and which are not.

   A. **Discuss** how the following terms are used in constructing a cladogram.
      - outgroup
      - ancestral trait
      - shared derived traits

   B. **Explain** how three of the following traits are used to separate these species.

Traits	Species
mammary glands	chimpanzee
gizzard	tuna
hair	crocodile
epidermal scales	dog
vertebrae	lancelet
notochord in embryo	lizard

Annotated Answer Key (MC)

1. **C**; Kingdom is more inclusive than Phyla. Organisms in the same Phyla do not need to be part of the same Class, Order, Family, Genus or species.

2. **B**; Hair, vertebrae, and an embryonic notochord are features of mammals, of which the dog is an example.

3. **C**; In the Echinoderms and the Chordates, including ourselves, the blastopore is associated with the future anus.

4. **D**; The rate of evolution of the mitochondrial genome appears to exceed that of the single-copy fraction of the nuclear genome by a factor of about 10. This high rate may be due, in part, to an elevated rate of mutation in mitochondrial DNA. Because of the high rate of evolution, mitochondrial DNA is likely to be an extremely useful molecule to employ for high-resolution analysis of the evolutionary process.

5. **C**; Archaea organisms possess ribosomes, a cell membrane with branched lipids, a cell wall without peptidoglycan, and even contain some introns.

6. **A**; Molecular data suggests this even given the diversity of fungi and animals. Animals are also multicellular and entirely heterotrophic. Plants are autotrophic, bacteria are prokaryotic and unicellular, and protists are diverse in both their nutrition and structure.

7. **E**; Animals have true tissues and the organ system level of organization. They ingest their food. Plants evolved from an aquatic photosynthetic protist.

8. **D**; Plants are autotrophic, nonmotile multicellular organisms. Many algae, often classified as protists, are also

classified this way but protist is not a choice. Fungi and animals are heterotrophs, while prokaryotes and archaeans are unicellular.

9. **B**; Fungi are eukaryotes that form spores, lack flagella, and have cell walls containing chitin. They are multicellular with a few exceptions. Fungi are heterotrophic by absorption.

10. **A**; Phylogenetic trees show common ancestors and branches coming off the common ancestor. Each branch in a tree is a divergence that gives rise to two or more new groups. These shared derived traits indicate which species are closely related and which are distantly related. A frog is more distantly related to a mammal, with which it shares fewer traits, than to a crocodile, with which it shares more traits.

Answer to FRQ

PART A (MAX 6)

- Outgroup: species that define which trait is the oldest (vs. Ingroup: species that will be placed into clades in the cladogram)
- Ancestral trait: traits present in both the outgroup and the ingroup
- Shared derived traits: traits that distinguish a particular clade
- A cladogram contains several clades.
- Each clade contains a common ancestor
- The first clade consists of organisms with the largest number of ingroups.
- The last clade consists of organisms with the smallest number of ingroups.
- The outgroup is always separate from the ingroups.
- Each clade is separated by a different shared derived trait.

PART B (MAX 6)

	Chimpanzee	Dog	Crocodile	Lizard	Tuna	Lancelet
Mammary Glands	X	X				
Hair	X	X				
Gizzard			X			
Epidermal Cells			X	X		
Vertebrae	X	X	X	X	X	
Notochord in Embryo	X	X	X	X	X	X

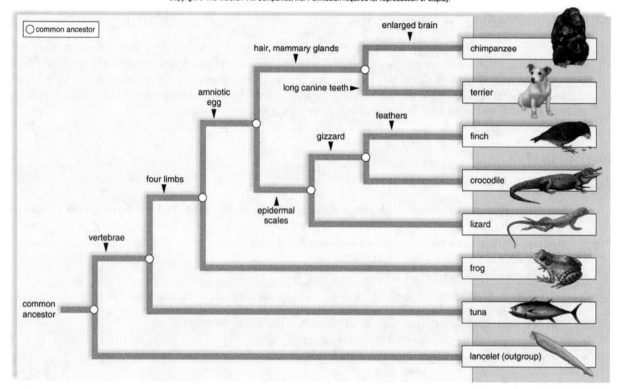

Constructing a cladogram: the phylogenetic tree

CHAPTER 20
VIRUSES, BACTERIA, AND ARCHAEA

20.1   Viruses, Viroids, and Prions

Viruses are noncellular, while prokaryotes are fully functioning organisms. All viruses have at least two parts: (1) an outer **capsid** composed of protein subunits and (2) an inner core of nucleic acid, either DNA or RNA, but not both. Some also have an outer membranous envelope made of plasma membrane from the host cell.

Viruses are **obligate intracellular parasites** that can reproduce only inside the living cells they infect. Unlike cellular organisms, they cannot replicate their own genetic material or synthesize their own proteins, but need the machinery of a host cell. Viruses can mutate, therefore, they do evolve. Viruses are able to infect cells by binding to specific receptors on the surface of a host cell, then injecting their nucleic acid into the cell. Once inside the cell, the genetic material of the virus is replicated and viral proteins are synthesized using the enzymes of the host cell. New viruses are reassembled then released when the host cell lyses.

Bacteriophages are viruses that infect bacteria and have two types of life cycles: the **lytic** cycle and the **lysogenic** cycle. The lytic cycle of a bacteriophage consists of attachment, penetration, biosynthesis, maturation, and release. In the lysogenic cycle of a bacteriophage, viral DNA is integrated into bacterial DNA for an indefinite period of time, becoming a **prophage**, but it can undergo the lytic cycle when stimulated.

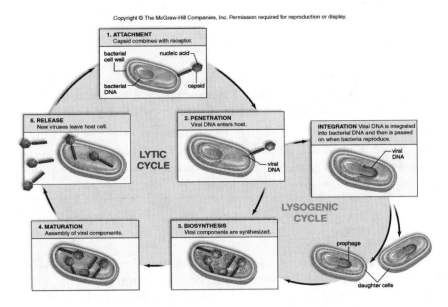

Lytic and lysogenic cycle in prokaryotes

The reproductive cycle differs for animal viruses because many of them have membranous envelopes surrounding the capsid. Uncoating is needed to free the genome from the capsid, and either budding or lysis releases the viral particles from the cell. RNA **retroviruses** have an enzyme called **reverse transcriptase** that produces one strand of DNA using viral RNA as a template, and then a complementary DNA strand. The resulting double-strand DNA becomes integrated into host DNA. The AIDS virus is a retrovirus.

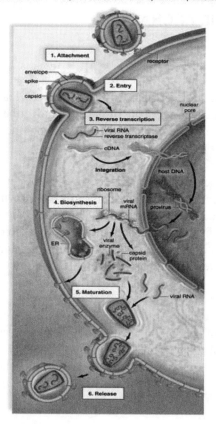

Reproduction of the retrovirus HIV

Viruses cause various diseases in plants and animals, including human beings. **Viroids** are naked strands of RNA (not covered by a capsid) that can cause disease in plants. **Prions** are protein molecules that have a misshapen tertiary structure. Prions cause diseases such as mad cow disease in cattle when they cause other proteins of their own kind also to become misshapen.

Take Note: *For the AP Biology exam, you should be able to describe how a DNA virus infects a cell and explain how this differs from the way RNA viruses infect cells.*

20.2   The Prokaryotes

The **bacteria** (domain Bacteria) and **archaea** (domain Archaea) are prokaryotes. Prokaryotic cells lack a nucleus and the membrane-bound organelles found in eukaryotic cells. The single chromosome of bacteria is found in the nucleoid region of the cell. Some prokaryotes have an additional small circular chromosome called a **plasmid** that is separate from the main chromosome. Prokaryotes have a cell wall outside of the cell membrane that protects them from any osmotic changes in their environment. Some prokaryotes have an additional coating of polysaccharides called a **glycocalyx** that forms a capsule or slime layer, while others have an S-layer made of glycoproteins. Some prokaryotes have flagella for movement. Those that adhere to surfaces have short, bristle-like fibers called **fimbriae** that extend from their surface.

Prokaryotes reproduce asexually by **binary fission**. In this process, the cell elongates as the chromosome replicates, eventually forming two identical daughter cells. Binary fission occurs more rapidly than mitosis, so any mutations in the chromosome are passed on to prokaryotic offspring faster than to eukaryotic offspring. Genetic variation in prokaryotes occurs in one of three ways: conjugation, transformation, and transduction. In **conjugation**, two prokaryotic cells are linked together via a **conjugation pilus**, through which the donor cell passes genetic information to the recipient cell. In **transformation**, a prokaryotic cell takes up naked DNA from the environment, which is incorporated into its genome. Bacteriophages are important for **transduction**, in which genetic material is transferred from one prokaryote to the next by viral infection.

> **Take Note:** *You should be able to describe the characteristics of prokaryotes and explain how they differ from eukaryotes in terms of structure, organization of genetic material, and ways of obtaining genetic variation.*

20.3   The Bacteria

Bacteria (domain Bacteria) are the more prevalent type of prokaryote. Though the classification of bacteria is still being developed, of primary importance at this time are the shape of the cell and the structure of the cell wall, which affects Gram staining. Bacteria occur in three basic shapes: spiral-shaped (**spirillum**), rod-shaped (**bacillus**), and round (**coccus**). The cell walls of bacteria consist of a complex of polypeptides and amino acids called **peptidoglycan**. The thickness of peptidoglycan makes bacteria **Gram-positive**; in other words, bacteria with a thick layer of peptidoglycan appear purple under a microscope when treated with Gram stain. **Gram-negative** bacteria appear ink under a microscope because they have a thin peptidoglycan layer.

Bacteria differ in their need (and tolerance) for oxygen. There are **obligate anaerobes** that must live in anaerobic environments, **facultative anaerobes** that can live in either an anaerobic or an aerobic environment, and **obligate aerobes** that live only in aerobic environments.

Some prokaryotes are autotrophic, being either **photoautotrophs** (photosynthetic) or **chemoautotrophs** (chemosynthetic). Some photosynthetic bacteria, such as cyanobacteria, give off oxygen, and some, like the purple and green sulfur bacteria, do not. Chemoautotrophs oxidize inorganic compounds such as hydrogen gas, hydrogen sulfide, and ammonia to acquire energy to make their own food. Surprisingly, chemoautotrophs support communities at deep-sea vents.

Many bacteria are **chemoheterotrophs** (aerobic heterotrophs) and are **saprotrophic decomposers** that are absolutely essential to the cycling of nutrients within ecosystems. Many heterotrophic bacteria are symbiotic. Mutualistic nitrogen-fixing bacteria live in nodules on the roots of legumes, where they convert atmospheric nitrogen to a form usable for plants in exchange for receiving nutrients. Other mutualistic bacteria live in the large intestines of humans, where they produce vitamins K and B12, again in exchange for nutrients. Parasitic bacteria are called pathogens because they cause disease. These bacteria can form **endospores** that protect them from destruction in harsh environments. Anthrax is an example of a bacteria that forms endospores, enabling them to lie dormant for hundreds of years.

Of special interest are the **cyanobacteria**, which were the first organisms to photosynthesize in the same manner as plants. When cyanobacteria are symbionts with fungi, they form lichens. Some bacterial symbionts, however, are parasitic and cause plant and animal, including human, diseases.

20.4   The Archaea

The archaea (domain Archaea) are a second type of prokaryote. The archaea appear to be more closely related to the eukarya than to the bacteria. They share more biochemical characteristics with the eukarya than do bacteria, such as ribosomal proteins, the initiation of transcription, and similar types of tRNA. Unlike bacteria, archaea do not have peptidoglycan in their cell walls. Instead, many of them have cell walls made mostly of polysaccharides or proteins. Additionally, the glycerol in their cell membranes is bound to branched-chain hydrocarbons that allow them to withstand the high temperatures at which many of them live.

Three types of archaea live under harsh conditions. The **methanogens** are obligate anaerobes that live in swamps, marshes, and animal intestines. These organisms are chemoautotrophs that produce ATP by reducing CO_2 to CH_4, a biogas that contributes to the greenhouse effect. **Halophiles** require high salt concentrations, such as those in the Dead Sea, to live and grow. They have special chloride pumps that prevent the osmotic loss of water. Halophiles are aerobic chemoautrophs, but can convert to photosynthesis under certain environmental conditions. **Thermoacidophiles** are found in hot, extremely acidic environments such as hot springs and near volcanoes. They are anaerobic chemoautotrophs that use H_2 and sulfur for ATP production.

Take Note: *Can you explain the similarities and differences between bacteria and Archaea? What adaptations do prokaryotes have that enable them to survive in their environments?*

Multiple Choice Questions

1. Viruses are NOT characterized by their
    A. type of nucleic acid.
    B. size.
    C. shape.
    D. number of cells.
    E. presence of an outer envelope.

2. The envelope of a virus usually contains
    A. glycoproteins.
    B. cellulose.
    C. cholesterol.
    D. phospholipids.
    E. peptidoglycan.

3. Viruses
    A. cannot mutate.
    B. are not host specific.
    C. do not contain a protein covering.
    D. can parasitize bacteria.
    E. can reproduce outside a living cell.

4. Which of the following diseases is NOT caused by bacteria?
    A. gonorrhea
    B. botulism
    C. HIV
    D. Lyme disease
    E. tetanus

5. A commensal relationship is seen between
   A. bacteria living in the human intestines.
   B. mycorrhizal associations between plant roots and fungi.
   C. ant protection of aphids.
   D. vertebrate hosts and tapeworms.
   E. epiphytic plants growing on trees.

Questions 6–9 refer to the following five choices. For each question, select the choice that is most closely related. Each choice may be used once, more than once, or not at all.
   A. conjugation
   B. transformation
   C. transcription
   D. transduction
   E. adaptation

6. Bacteria are temporarily linked together.

7. Bacteriophages carry portions of DNA from one bacterial cell to another.

8. Transfers chromosomal DNA from a donor cell to a recipient cell

9. Bacterial cells take up naked DNA molecules.

10. Archaea are prokaryotes with biochemical characteristics including
    A. diverse cell walls that do contain peptidoglycan
    B. plasma membranes with glycerol molecules linked to the lipids
    C. the lack of introns in genes
    D. one single type of RNA polymerase (domain bacteria)
    E. use of formyl-methionine as an initiator for protein synthesis (domain bacteria)

Free Response Question

Viruses can parasitize all forms of life and cause disease.

A. **Discuss** three characteristics of viruses that have contributed to their pathogenic success.

B. **Explain** why viruses are considered to be a nonliving organism.

C. **Discuss** how viruses are classified.

Annotated Answer Key (MC)

1. **D**; Viruses can contain RNA or DNA (and vary in the number of strands), vary tremendously in size and shape, and possess a capsid. The capsid may be surrounded by an outer membranous envelope (enveloped) or not (naked).

2. **A**; Glycoproteins interact with eukaryotic surface molecules.

3. **D**; Bacteriophages are viruses that parasitize bacteria. There are two types of bacteriophage life cycles: the lytic and lysogenic cycles.

4. **C**; HIV (Human immunodeficiency virus) is the cause of AIDS. It is a retrovirus that contains a special enzyme called reverse transcriptase, which carries out RNA → DNA transcription.

5. **E**; In a commensal relationship there are two organisms where one benefits and the other is not significantly harmed or gains a benefit. While epiphytes, which grow on plants, are often misclassified as parasites, they actually are commensal organisms, using the second plant for housing, but not harming it.

6. **A**

7. **D**

8. **A**

9. **B**

For Questions 6–9: During conjugation, two bacteria are temporarily linked together, often by means of a conjugation pillus. While they are linked, the donor cell passes DNA to a recipient cell. Transformation occurs when a cell picks up from the surroundings free pieces of DNA. During transduction, bacteriophages carry portions of DNA from one bacterial cell to another.

10. **B**; Archaea do possess some introns in their genome, have cell walls that do NOT contain peptidoglycan and have branched lipids in their membranes. Bacteria only possess one type of RNA polymerase and use formyl-methionine as an initiator for protein synthesis.

Answer to FRQ

PART A (MAX 4)
- Small size
- Capsid
- Diverse genetic material (either DNA or RNA)
- Varied shape
- Different modes of transmission
- Varied replication strategies

PART B (MAX 4)
Viruses do NOT
- metabolize or
- synthesize their own proteins (no ribosomes)
- replicate on their own
- respond to stimuli
- have a cell membrane/other organelles
- have varied genetic mutations
- go through lysogenic/lytic cycles

PART C (MAX 4)
Virus Classification
- Type of genetic material (DNA/RNA, single stranded/double stranded/mixture, sense)
- Replication mechanism
- Mechanism to express and regulate genes
- Morphology (size/shape/capsule)
- Size of genome
- Type of host/infection mechanism
- Mode of transmission/consequences of infection

CHAPTER 21
PROTIST EVOLUTION AND DIVERSITY

21.1   General Biology of Protists

Protists are in the domain Eukarya. They are generally unicellular, but they are still quite complex because they (1) have a variety of characteristics, (2) acquire nutrients in a number of ways, and (3) have complicated life cycles that include the ability to withstand hostile environments. They normally reproduce asexually by mitosis; sexual reproduction and spore formation occurs in unfavorable conditions.

Protists are of great ecological importance because in largely aquatic environments they are the producers (plankton) that support communities of organisms. Protists also enter into various types of symbiotic relationships. Their great diversity makes it difficult to classify protists, and as yet, there is no general agreement about their categorization. In this chapter, they have been arranged in six supergroups.

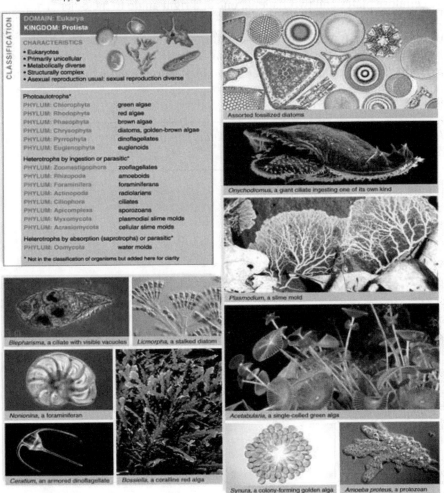

Protist diversity

21.2   Diversity of Protists

The Archaeplastids include the plants and all photosynthetic organisms that contain plastids, including green and red algae. **Green algae** possess chlorophylls *a* and *b,* store reserve food as starch, and have cell walls of cellulose as do land plants. Green algae are divided into chlorophytes and charophytes. **Chlorophytes** include forms that are unicellular *(Chlamydomonas)*, colonial *(Volvox)*, and multicellular *(Ulva)*. **Charophytes** include filamentous forms *(Spirogyra)*, as well as the stoneworts *(Chara)* and are thought to be the closest living relatives of land plants. The life cycle varies among the green algae. In most, the zygote undergoes meiosis, and the adult is haploid. *Ulva* has an alternation of generations like land plants, but the sporophyte and gametophyte generations are similar in appearance; the gametes look alike. **Red algae** are filamentous or multicellular seaweeds that are more delicate than brown algae and are usually found in warmer waters. Red algae have notable economic importance.

The supergroup Chromalveolates consists of stramenopiles such as brown algae, diatoms, golden brown algae, and water molds; and alveolates, including dinoflagellates, ciliates, and apicomplexans. **Brown algae** have chlorophylls *a* and *c* plus a brownish carotenoid pigment. The large, complex brown algae, commonly called **seaweeds**, are well known and economically important. **Diatoms**, which have an outer layer of silica, are extremely numerous in both marine and freshwater ecosystems, as are golden brown algae, which may have coverings of silica or organic material. **Water molds**, which are filamentous and heterotrophic by absorption, are unlike fungi in that they produce flagellated 2n zoospores. **Dinoflagellates** usually have cellulose plates and two flagella, one at a right angle to the other. They are extremely numerous in the ocean and, on occasion, produce a neurotoxin when they form red tides. The ciliates move by their many cilia. They are remarkably diverse in form, and as exemplified by *Paramecium*, they show how complex a protist can be despite being a single cell. The apicomplexans are nonmotile parasites that form spores; members of the genus *Plasmodium* cause malaria in humans, with Plasmodium vivax being the most common.

> **Take Note:** *For the AP Biology exam, you should have a clear understanding of the role of protists in the evolution of other eukaryotes.*

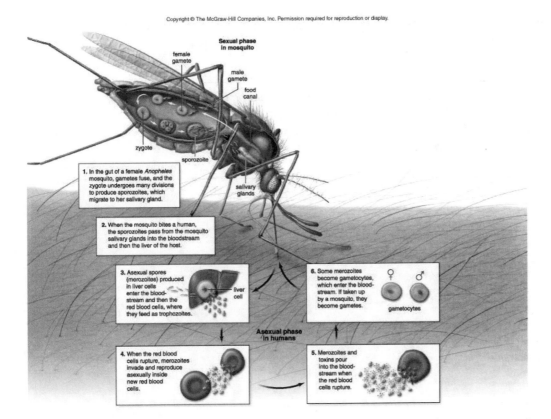

Sexual phase in mosquito

female gamete

male gamete

food canal

zygote

sporozoite

salivary glands

1. In the gut of a female *Anopheles* mosquito, gametes fuse, and the zygote undergoes many divisions to produce sporozoites, which migrate to her salivary gland.

2. When the mosquito bites a human, the sporozoites pass from the mosquito salivary glands into the bloodstream and then the liver of the host.

3. Asexual spores (merozoites) produced in liver cells enter the bloodstream and then the red blood cells, where they feed as trophozoites.

liver cell

6. Some merozoites become gametocytes, which enter the bloodstream. If taken up by a mosquito, they become gametes.

gametocytes

Asexual phase in humans

4. When the red blood cells rupture, merozoites invade and reproduce asexually inside new red blood cells.

5. Merozoites and toxins pour into the bloodstream when the red blood cells rupture.

Life cycle of Plasmodium vivax, a species that causes malaria

---

Take Note: *For the AP Biology exam, you should be able to explain the symbiotic relationships between different species as demonstrated by the life cycle of plasmodium. What adaptations have evolved in humans that are defenses against malaria?*

Supergroup Excavates is a diverse collection of single-celled, motile protists, including euglenids, parabasalids, diplomonads, and kinetoplastids. **Euglenids** are flagellated cells with a pellicle instead of a cell wall. Their chloroplasts are most likely derived from a green alga through endosymbiosis. Many of the kinetoplastids are parasites, including trypanosomes, such as those that cause Chagas disease and African sleeping sickness in humans. Parabasalids, such as *Trichomonas vaginalis*, and, diplomonads, such as *Giardia lamblia,* are common inhabitants of animal hosts. They thrive in low-oxygen conditions due to their lack of aerobic respiration.

The Amoebozoans supergroup contains amoeboids and slime molds, protists that use pseudopods for motility and feeding. **Amoeboids** move and feed by forming pseudopods. In *Amoeba proteus*, food vacuoles form following phagocytosis of prey. Contractile vacuoles discharge excess water. **Slime molds**, which produce nonmotile spores, are unlike fungi in that they have an amoeboid stage and are heterotrophic by ingestion.

The rhizarians are a supergroup that includes the foraminiferans and radiolarians, protists with thread-like pseudopods and skeletons called tests. The tests of foraminiferans and radiolarians form a deep layer of sediment on the ocean floor. The tests of foraminiferans are responsible for the limestone deposits of the White Cliffs of Dover.

The Opisthokonts supergroup includes kingdom Animalia, the animal-like protists known as choanoflagellates, kingdom Fungi, and the fungus-like protists called nucleariids.

Take Note: *In preparation for the AP Biology exam, you should be able to identify examples of plant-like protists, animal-like protists, and fungi-like protists. What characteristics make protists similar to these other organisms? What symbiotic relationships are seen between protists and other organisms? What adaptations do protists have that help them to survive in their environment?*

Multiple Choice Questions

1. Protists are NOT
   - A.   unicellular.
   - B.   colonial.
   - C.   filamentous.
   - D.   multicellular.
   - E.   prokaryotic.

Questions 2–5 refer to the following five choices. For each question, select the choice that is most closely related. Each choice may be used once, more than once, or not at all.
   - A.   brown algae
   - B.   water molds
   - C.   red algae
   - D.   amoeboids
   - E.   euglenoids

2. Unicellular with unique flagella and without pseudopodia

3. Colonial, multicellular, special carotenoid pigment

4. Multicellular seaweeds with phycoerythrin, phycocyanin, and chlorophyll

5. Cell walls with calcium carbonate

6. Dinoflagellates reproduce
   - A.   asexually by mitosis and sexually by meiosis.
   - B.   through binary fission.
   - C.   by sexual reproduction through conjugation.
   - D.   with the aid of a sex pillus.
   - E.   with an alternation of generation within different hosts.

7. Contractile vacuoles
   - A.   form pseudopods.
   - B.   discharge excess water.
   - C.   are essential in reproduction.
   - D.   contain carotenoid pigments.
   - E.   store reserve food as starch.

8. In alternation of generations, the sporophyte produces
   A. haploid spores by meiosis.
   B. haploid gametes.
   C. diploid gametophytes by mitosis.
   D. diploid zygotes.
   E. diploid spores by mitosis.

9. Which of the following accessory pigments is found in the golden brown algae?
   A. carotenoid
   B. chlorophyll a
   C. phycoerythrin
   D. chlorophyll b
   E. phycocyanin

10. Pseudopodia involve which of the following proteins?
    A. tubulin.
    B. pepsin.
    C. actin.
    D. keratin.
    E. myosin.

Free Response Question

Protists are an incredibly diverse collection of eukaryotic organisms that vary in many ways.

    A. **Describe** how three of the following characteristics vary in protists.
- form
- lifestyle
- nutrition
- locomotion
- reproduction

    B. **Discuss** the classification scheme of protists given these diverse characteristics.

Annotated Answer Key (MC)

1. **E**; Protists can be unicellular or multicellular, heterotrophic or autotrophic, and filamentous and even sometimes colonial. However, they are all eukaryotic.

2. **E**; Animal-like protists (amoeboids, euglenoids) are motile; euglenoids use flagella with pseudopods.

3. **A**; Brown algae possess a carotenoid pigment (brown tint).

4. **C**; Phycoerythrin gives red algae their characteristic color.

5. **C**; The red algae have cell walls with calcium carbonate and contribute as much to the growth of coral reefs as do coral animals.

6. **A**; Prokaryotes and some protists reproduce through binary fission (amoebas). However, dinoflagellates reproduce sexually and asexually based upon environmental conditions. When environmental conditions are favorable, meiosis with the zygospore produces haploid cells. During environmental stress, such as a shortage of nutrients, species may reproduce sexually.

7. **B**; contractile vacuoles discharge water in many protists for osmoregulatory purposes and a mode of propulsion.

8. **C**; The life cycle of slime molds is very similar to that of fungi. Alternation of generations occurs in almost all marine algae, most red algae, many green algae, and a few brown algae.

9. **A**; Accessory pigments work in conjunction with chlorophyll. Carotenoids are present in the golden brown algae.

10. **C**; Pseudopodia are flexible, cytoplasmic extensions that bulge from the cell surface, stretch outward, and anchor to a surface. They can surround and engulf food and are composed of the contractile protein actin.

Answer to FRQ

PART A (MAX 7)
One point each boldface square; max of 2 in each row

Form	**unicellular**	**multicellular**	**colonial**	
Lifestyle	**free living**	**parasitic**	**mutualistic**	
Nutrition	**autotrophic**	**heterotrophic**		
Locomotion	**cilia**	**flagella**	**pseudopodia**	**sessile**
Reproduction	**sexual**	**asexual**	**meiosis**	**mitosis**

PART B (MAX 6)
Two points for each classification

- Animal-like (single-celled, motile, phagocytosis – many exceptions)
- Plant-like (single/multicellular, autotrophic)
- Fungus-like (produce sporangia, chitin cell wall)
- New characterization includes genetic and biochemical similarities/differences.

CHAPTER 22
EVOLUTION AND DIVERSITY OF FUNGI

22.1 Evolution and Characteristics of Fungi

Fungi are multicellular, heterotrophic eukaryotes that are **saprotrophs**, that is, they obtain nutrients by absorption. After external digestion, they absorb the resulting nutrient molecules. Most fungi act as decomposers that aid the cycling of chemicals in ecosystems by decomposing dead remains. Some fungi are parasitic, especially on plants, and others are mutualistic with plant roots and algae.

Although some fungi such as yeasts are unicellular, most are multicellular. The body of a fungus is composed of thin filaments called **hyphae**, which make up a network called a **mycelium**. This network gives the fungus a large surface area for the absorption of nutrients. In some fungi, the hyphae are separated by **septae**, cross walls with pores that allow cytoplasm and organelles to pass from one hypha to another. **Nonseptate hyphae** have no cross walls; therefore, cytoplasm can freely flow between the multinucleated cells. The cell walls of fungi contain chitin, the same polysaccharide found in arthropods, and the energy reserve is glycogen. With the exception of the aquatic **chytrids**, which have flagellated zoospores and gametes, fungi do not have flagella at any stage in their life cycle and are thus nonmotile. They grow toward their food in order to obtain it.

Fungi produce spores during both asexual and sexual reproduction. During sexual reproduction, two haploid hyphae tips fuse so that **dikaryotic** (n + n) hyphae usually result, depending on the type of fungus. Following fusion of the two nuclei, the zygote undergoes meiosis, resulting in the production of spores. These spores will germinate and develop into a new organism. Fungi can also reproduce asexually by fragmentation or budding.

> **Take Note:** *In preparation for the AP Biology exam, you should have a full understanding of the ecological roles of fungi. What characteristics did they develop as they evolved? What adaptations do they have that helped them to survive on land? How is the alternation of fertilization and meiosis in fungi different from alternation of fertilization and meiosis in protists, plants, and animals?*

22.2 Diversity of Fungi

Five significant groups of fungi are the **chytrids** (chytridiomycota), **zygospore fungi** (zygomycota), **AM fungi** (glomeromycota), **sac fungi** (ascomycota), and **club fungi** (basidiomycota).

The **chytrids** are unique because they produce motile **zoospores** and gametes with flagella. Some chytrids have an alternation of generations life cycle similar to that of plants and certain algae. Most reproduce asexually through the production of zoospores that grow into new organisms. There are unicellular as well as hyphae-forming chytrids. When hyphae form, they are nonseptate. Most chytrids are involved in the decay and digestions of dead aquatic organisms, as are parasites on plants, animals, and protists. Chytriomyces is an example of a chytrid.

The **zygospore** fungi are saprotrophs that receive their nutrients from plant or animal remains in the soil. They are nonseptate, and during sexual reproduction they have a dormant stage consisting of a thick-walled zygospore. When the zygospore germinates, sporangia produce windblown spores. Asexual reproduction occurs when nutrients are plentiful and sporangia again produce spores. An example of a zygomycete is black bread mold (*Rhizopus stolonifer*).

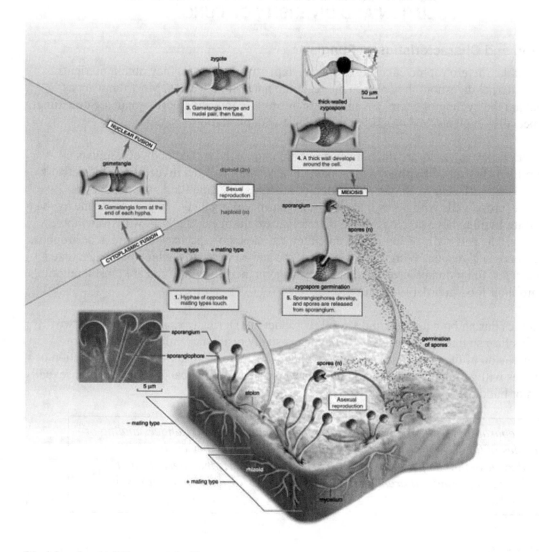

Black bread mold, Rhizopus stolonifer

The **AM** (arbuscular mycorrhizal) fungi were once classified with the zygospore fungi but are now viewed as a distinct group. AM fungi exist in mutualistic associations with the roots of land plants. **Arbuscles** are the branching invaginations that the fungus uses to invade the roots of plants.

The **sac fungi** are septate, and the body is typically a mycelium. Asexual reproduction, which is dependent on the production of **conidiospores**, is more common. These spores vary in shape, size, and number of cells. During sexual reproduction, saclike cells called **asci** produce spores. Asci are sometimes located in **fruiting bodies** called **ascocarps**. Sexual reproduction is unknown in some sac fungi. Sac fungi include *Talaromyces*, *Aspergillus*, *Candida*, morels and truffles, and various yeasts and molds, some of which cause disease in plants and animals, including humans.

The **club fungi** consist of mycelium bodies that are septate. Most are saprotrophs; some are parasites. During sexual reproduction, club-shaped structures called **basidia** produce spores called **basidispores**. Basidia are located in fruiting bodies called **basidiocarps**. Club fungi have a prolonged dikaryotic stage, and asexual reproduction by conidiospores is rare. A dikaryotic mycelium periodically produces fruiting bodies. Mushrooms and puffballs are examples of club fungi.

> **Take Note:** *For the AP Biology exam, you should be able to give an example from each group of fungi. How does each group differ with respect to their life cycle and the structures used to produce spores?*

22.3   Symbiotic Relationships of Fungi

Lichens are an association between a fungus, usually a sac fungus, and a cyanobacterium or a green alga. A lichen has three layers: a tough upper layer, a photosynthetic layer, and a loosely packed lower layer. They reproduce asexually by fragmentation, releasing a complex consisting of hyphae and an algal cell. Traditionally, this association was considered mutualistic, where the fungus received nutrients from the algae and algae received protection from water loss. Experimentation, however, suggests a controlled parasitism by the fungus on the alga. Lichens may live in extreme environments and on bare rocks; they allow other organisms that will eventually form soil to establish.

The term **mycorrhizae** refers to an association between a soil fungus and the roots of a plant. The fungus helps the plant absorb minerals, and the plant supplies the fungus with carbohydrates. Plants with mycorrhizae grow better in conditions where the soil is of poor quality than plants without them. **Endomycorrhizae** penetrate the cell walls of the plant root and **ectomycorrhizae** grow between cell walls, forming a mantle on the exterior of the root. In either situation, mycorrhizae increase the surface area for absorption so that the plant receives the minerals it needs.

Multiple Choice Questions

1. Which of the following is a primary difference between plants and fungi?
    A.   Fungal cells lack chloroplasts.
    B.   Plant cell walls are composed of peptdoglycan.
    C.   Fungi lack motility at any stage in their life cycle.
    D.   Many plant cells are multinucleate.
    E.   Plants convert $C_6H_{12}O_6$ to CO_2 and H_2O.

2. Penicillin is produced by a
    A.   bacteria.
    B.   fungus.
    C.   yeast.
    D.   lichen.
    E.   mycorrhizae.

3. The infection caused by ringworm is due to
    A.   the secretion of toxins that suppress the immune system.
    B.   the production of lesions that produce penicillin.
    C.   the release of enzymes that degrade keratin and collagen in the skin.
    D.   the digestion of cellulose in the cell walls.
    E.   the production of ethanol and CO_2.

4. What adaptation ensures that terrestrial fungal offspring will proliferate?
    A.   Fungi reproduce only asexually.
    B.   Fungi reproduce asexually through the use of runners.
    C.   Fungi produce nonmotile gametes during meiosis.
    D.   Fungi produce spores during both sexual and asexual reproduction.
    E.   Fungi produce a network of filaments called a mycelium.

5. The relationship between fungal and algal cells (lichen) is
   A. mutualistic, in which the fungus receives nutrients from the algal cells and the algal cells are protected from dessication by the fungus.
   B. controlled parasitism, with the fungus receiving nutrients and the algae not benefitting at all from the interaction.
   C. commensal, with the fungi getting no benefits from the algae.
   D. a predator-prey relationship, as the fungi feeds on the algae.
   E. mutualistic, with the fungus feeding on the algae and the lichen receiving nutrients.

6. Mycorrhizae
   A. give the plant a larger absorptive surface for the intake of minerals.
   B. grow on moist seeds and secrete pathogenic toxins.
   C. release windborn spores that cause allergic reactions.
   D. release spores that parasitize cereal crops.
   E. protect algae from dessication by fungi.

Questions 7–10 refer to the following five choices. For each question, select the choice that is most closely related. Each choice may be used once, more than once, or not at all.
   A. archae
   B. bacteria
   C. green algae
   D. lichen
   E. mycorrhizae

7. Form a mutualistic relationship with the roots of plants

8. Reproduce asexually by releasing hyphae

9. Reproduce asexually by mitosis

10. Often parasitize algae

Free Response Question

At one time, fungi were considered a part of the plant kingdom, and then later, they were considered a protist.

   A. **Explain** why fungi could have been classified as a plant or a protist.

   B. **Discuss** the reproductive strategies of fungi and explain how these diverse reproductive strategies have contributed to the success of fungi.

Annotated Answer Key

1. **A**; Fungi are not autotrophic, they are saprophytic and feed on decaying organic matter.

2. **B**; Penicillin is a group of antibiotics derived from *Penicillium* fungi.

3. **C**; Ringworm is not a worm at all. It is caused by a parasitic fungus. Parasitic fungi often feed on keratin, the material found in the outer layer of skin, hair, and nails and are often found in moist, warm, dark locations.

4. **D**; Terrestrial fungi are dependent upon spores to reproduce. Spores are well adapted for dispersal and can survive for long periods in unfavorable conditions, even without stored food reserves, as compared to seeds.

5. **B**; In the past, lichens were assumed to be mutualistic relationships in which the fungus received nutrients from the algal cells and the algal cells were protected from dessication by the fungus. However, this controlled parasitic relationship is supported by experiments in which the fungal and algal components are removed and grown separately. The algae grow faster when they are alone than when they are part of a lichen. It is also difficult to cultivate the fungus, which does not grow naturally on its own.

6. **A**; Mycorrhizae are mutualistic relationships between soil fungi and the roots of most plants. The presence of the fungus gives the plant a greater absorptive surface for the intake of minerals. The fungus receives carbohydrates.

7. **E**; Plants whose roots are invaded by mycorrhizae grow more successfully in poor soils than do plants without mycorrhizae.

8. **D**; Hyphae penetrate the photosynthetic cells, transferring nutrients directly to the rest of the fungus.

9. **C**; Unicellular green alga reproduce both sexually and asexually. During environmental stress, the haploid cell divides first by mitosis to produce haploid gametes, and then a pair of gametes from a different individual fuse to form a pair.

10. **D**; Lichens are an association between a fungus and a cyanobacterium or a green alga.

Answer to FRQ

PART A (MAX 6)
Three points for each category

- Not monophyletic; diverse in form and function

Similarities to protists
- Fungi can reproduce both sexually and asexually.
- Fungi are unicellular, multicellular, or multinucleate.
- Cell walls of chitin

Similarities to plants
- Possess cell walls (previously thought cellulose; now chitin)
- Produce spores
- Morphological resemblance
- Lack of locomotion

PART B (MAX 6)
- Asexually or sexually
- Asexual means: fragmentation, fission, budding, spore formation
- Sexual means: fusion of gamete with or without sex organs

Production of spores
- Small size
- Protective coating
- Wide dispersal
- Numerous
- Dormancy of spores

Sexual reproduction
- Increased variation
- Adapt to new environments/can live in a rapidly changing environment

CHAPTER 23
PLANT EVOLUTION AND DIVERSITY

Take Note: *As plants evolved, they developed adaptations that helped them to survive in their environments. You should be able to identify these adaptations and explain how they enabled plants to overcome the challenges of living on land.*

23.1 The Green Algal Ancestor of Plants

Plants are multicellular, photosynthetic eukaryotes. Land plants evolved from a common ancestor with multicellular, freshwater green algae called **charophytes**. These algae share many common characteristics with the land plants of today: 1) they have cell walls made of the same type of cellulose; 2) they have apical tissue that allows them to grow in length; 3) their cells communicate via plasmodesmata; and 4) both provide nutrients for the zygote. During the evolution of plants, protecting the embryo, apical growth, vascular tissue for transporting water and organic nutrients, possession of megaphylls, using seeds to disperse offspring, and having flowers were all adaptations to a land existence.

All plants have a life cycle that includes an **alternation of generations**. In this life cycle, a haploid gametophyte alternates with a diploid sporophyte. The **sporophyte generation** produces haploid spores by meiosis, and the **gametophyte generation** produces the haploid gametes that will fuse to form the sporophyte generation. During the evolution of plants, the sporophyte gained in dominance, while the gametophyte became microscopic and dependent on the sporophyte.

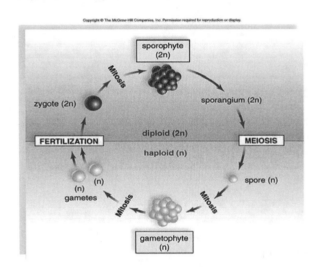

Alternation of generations in land plants

23.2 Evolution of Bryophytes: Colonization of Land

Ancient **bryophytes** were the first plants to colonize land. Today, the bryophytes consist of liverworts, hornworts, and mosses that lack well-developed vascular tissue. Because they are **nonvascular** plants, they are small, and are found in moist areas. The sporophytes of bryophytes are nutritionally dependent on the gametophyte, which is more conspicuous and photosynthetic. They are the only land plants where the gametophyte generation is dominant. Consult your textbook to locate a picture of the life cycle of the moss. As you study this picture, you will see that it demonstrates reproductive strategies such as flagellated sperm and dispersal by means of windblown spores. Sperm produced in the **antheridium** travel through water to reach the **archegonium**, where fertilization of the egg occurs. The zygote develops into the sporophyte, which grows out of the gametophyte. Spores will be released from the capsules at the top of the sporophyte.

23.3   Evolution of Lycophytes: Vascular Tissue

In **vascular** plants, such as the rhyniophytes, the sporophyte has two kinds of well-defined conducting tissues. **Xylem** is specialized to conduct water and dissolve minerals, and **phloem** is specialized to conduct organic nutrients. The cell walls of the vascular tissue are strengthened by a substance called **lignin**. The lycophytes are descended from these first plants, and they have vascular tissue. Ancient lycophytes also had the first leaves, which were **microphylls**; their life cycle is similar to that of the fern.

23.4   Evolution of Pteridophytes: Megaphylls

Pteridophytes (ferns and their allies, horsetails and whisk ferns) and also lycophytes, are seedless **vascular** plants where the sporophyte is dominant and is separate from the tiny gametophyte, which produces flagellated sperm. Windblown spores are dispersal agents. Consult your textbook to locate a picture of the life cycle of the fern. As you study this picture, you will see that the gametophyte arises from the germination of spores that were produced by meiosis in the sporophyte generation. Notice that the sporophyte grows from the gametophyte, and as it does, the gametophyte decreases in size. The leaves of a fern are called **fronds**, and they are divided into leaflets. Notice that the fronds start out as **fiddleheads** that grow from the **rhizome**, a horizontal underground stem.

23.5   Evolution of Seed Plants: Full Adaptation to Land

Seed plants are also vascular plants that use seeds to protect and nourish the embryo. Seed plants also have an alternation of generations, but they are **heterosporous**, meaning they produce both microspores and megaspores. **Microspores** become the windblown or animal-transported male gametophytes, or the pollen grains. **Pollen** grains carry sperm to the egg-bearing female gametophyte. **Megaspores** develop into the female gametophyte within the ovule. Following fertilization, the ovule becomes the seed. A seed contains a sporophyte embryo, and therefore seeds disperse the sporophyte generation. Fertilization no longer uses external water, and sexual reproduction is fully adapted to the terrestrial environment.

The **gymnosperms** (cone-bearing plants) and also possibly angiosperms (flowering plants) evolved from woody seed ferns. There are four groups of gymnosperms: conifers, cycads, ginkgoes, and gnetophytes. The conifers, represented by the pine tree, are the most abundant of these plants. Gymnosperms have "naked seeds" because the seeds are not enclosed by fruit, as are those of flowering plants. Consult your textbook to locate a picture of the life cycle of gymnosperms. As you study this picture, notice that the microspores and megaspores are located in separate cones, pollen cones and seed cones, respectively. The pollen grain has wings to make it more aerodynamic. The zygote is formed when the pollen lands on the seed cone and a **pollen tube** forms, allowing sperm to travel to the egg for fertilization. The **ovule** matures and becomes the **seed**, which contains the embryo and a food source, protected by a seed coat.

In **angiosperms**, the reproductive organs are found in flowers; the ovules, which become seeds, are located in the ovary, which becomes the fruit. Therefore, angiosperms have "covered seeds." In many angiosperms, pollen is transported from flower to flower by insects and other animals. Both flowers and fruits are found only in angiosperms and may account for the extensive colonization of terrestrial environments by the flowering plants. Consult your textbook to locate a picture of the life cycle of angiosperms. Note that sperm are formed in the **anthers** and the eggs are formed in the ovule. A pollen tube is formed when pollen lands on the stigma and **double fertilization** occurs to form a diploid zygote and a triploid nucleus that becomes the **endosperm**. The sporophyte generation develops when the seed germinates.

> **Take Note:** *For the AP Biology exam, you should be prepared to describe the life cycles of the four major groups of plants. How are they similar? How are they different? Why does the life cycle of bryophytes cause them to be less widely distributed than other groups of plants?*

Multiple Choice Questions

1. Flowering plants and gymnosperms are divided by
   A. the type of fruit.
   B. mode of embryo protection.
   C. apical growth.
   D. the type of vascular tissue.
   E. possession of megaphylls or microphylls.

Questions 2–5 refer to the following five choices. For each question, select the choice that is most closely related. Each choice may be used once, more than once, or not at all.
   A. Bryophytes
   B. Charophytes
   C. Gymnosperms
   D. Angiosperms
   E. Club Mosses

2. Low-lying nonvascular plants

3. Sporophyte bears cones which produce male gametophytes and seeds.

4. Possess double fertilization

5. Fruit develops from an ovary.

6. Bryophytes do NOT share which of the following characteristics with vascular plants?
   A. alternation of generations
   B. covered by a cuticle
   C. possess stomata
   D. possess a dominant gametophyte stage
   E. multicellular

7. Microphylls differ from megaphylls in that they
   A. are more broad.
   B. have a single strand of vascular tissue.
   C. are found in ferns.
   D. can gather more sunlight.
   E. produce more organic food.

8. Ginkgoes are dioecious, which means that they
   A. produce microspores and megaspores.
   B. have male and female reproductive structures.
   C. have separate trees producing seeds and pollen.
   D. are pollinated by insects and via wind.
   E. produce two gametes.

9. Flowers that bloom at night are usually
   A. large and showy.
   B. colorful.
   C. aromatic.
   D. light colored.
   E. dark or red colored.

10. In double fertilization
    A. the megasporangium divides by meiosis.
    B. the pollen sacs produce microspores.
    C. meiosis produces four megaspores.
    D. one sperm fertilizes the egg, the other forms an endosperm nucleus.
    E. the cytoplasm differentiates into and egg and two polar nuclei.

Free Response Question

Plants evolved from freshwater green algae and over a long period of time became fully adapted to living on land.

   A. **Explain** the challenges of a terrestrial existence and the adaptations that are required for the success of plants.

   B. **Discuss** how the alternation of generations is different for bryophytes, gymnosperms, and angiosperms.

Annotated Answer Key (MC)

1. **B**; Angiosperms differ from gymnosperms in that they produce seeds which are enclosed in ovaries as they develop.

2. **A**; Bryophytes also prefer moist locations and are limited in size due to a lack of conducting vessels.

3. **C**; The male gametophyte is the pollen and the seed is the cone.

4. **D**; Following double fertilization, ovules become seeds that enclose a sporophyte embryo and endosperm.

5. **D**; The fruit may provide a fleshy covering (apples, tomatoes) or a dry covering for seeds (pea pods, acorns).

6. **D**; The gametangia are called antheridia and archegonia. Antheridia produce flagellated sperm.

7. **B**; Megaphylls are large leaves with several to many veins; microphylls are small leaves with one vein.

8. **C**; Male trees are usually preferred for planting because the fleshy seeds give off such a foul odor.

9. **D**; Light colored flowers are more visible at night.

10. **D**; Double fertilization in angiosperms results in an egg and an endosperm.

Answer to FRQ

PART A (MAX 6)
Three points for challenges, three points for adaptations

Challenges
- Obtaining minerals from the soil
- Obtaining water from the soil/possibility of desiccation
- Photosynthesis products require transport.
- Require internal supporting structure
- Less water for sperm OR spores to pass through

Adaptations
- Stems and branches → support (must support its own weight)
- Vascular tissue (xylem, tracheids, vessels, heartwood, phloem)
- Vines or tendrils
- Extensive root system → anchor
- Cell wall (lignin)
- Waxy cuticle → prevent water loss
- Stomata → closed when water is depleted → wilting
- Sporophyte as dominant generation/reduced gametophyte → variation
- Protected gametophyte (megaspore/megasporangium)
- Woody tissue
- Pollen (light weight for transport)
- Seeds and seed dormancy
- Flowers (attract pollinators – shape, color, smell, chemical)
- Fruit and seed dispersal
- Flagellated → nonflagellated cells

PART B
- Gymnosperms and angiosperms have a dominant sporophyte (2n) generation.
- Gametophyte (1n) is reduced further than the ferns and bryophytes.
- Produce two kinds of spores that vary in size
- Flagellated sperm in bryophytes

CHAPTER 24
FLOWERING PLANTS: STRUCTURE AND ORGANIZATION

> **Take Note:** *In preparation for the AP Biology exam, you should be able to describe the structure and function of the major organs of a plant, and explain how each organ aids in the survival of plants.*

24.1   Organs of Flowering Plants

Most flowering plants have a **root system**, the underground parts of the plant, as well as a **shoot system**, which consists of the above ground parts of the plant. A flowering plant has three vegetative organs which allow it to live and grow. **Roots** anchor a plant, absorb water and minerals, and store the products of photosynthesis. **Stems** produce new tissue, support leaves, conduct materials to and from roots and leaves, and help store plant products. **Leaves** are specialized for gas exchange, and they carry on most of the photosynthesis in the plant.

Flowering plants are divided into the **monocots** and **eudicots** according to the number of cotyledons (embryonic leaves) in the seed; the arrangement of vascular tissue in roots, stems, and leaves; and the number of flower parts. Examples of monocots are grasses, lilies, and palm trees. Eudicots include dandelions and oak trees. Review the differences between monocots and eudicots below before moving on to the next section.

	Seed	Root	Stem	Leaf	Flower
Monocots	One cotyledon in seed	Root xylem and phloem in a ring	Vascular bundles scattered in stem	Leaf veins form a parallel pattern	Flower parts in threes and multiples of three
Eudicots	wo cotyledons in seed	Root phloem between arms of xylem	Vascular bundles in a distinct ring	Leaf veins form a net pattern	Flower parts in fours or fives and their multiples

Flowering plants are either monocots or eudicots

> **Take Note:** *You may be asked on the AP Biology exam to compare and contrast monocots and eudicots.*

24.2   Tissues of Flowering Plants

Flowering plants are able to grow because they have meristematic tissue. Plants have **apical meristem** plus three types of primary meristem. The apical meristem is located at the tips of stems and roots and allows plants to grow in length. All of the meristematic tissue develops into three types of specialized tissues as plants grow: 1) epidermal tissue, 2) ground tissue, and 3) vascular tissue.

Protoderm is the meristem that produces epidermal tissue. **Epidermal tissue** contains tightly packed cells, and it has a waxy covering called the **cuticle** that protects the plant. In the roots, epidermal cells bear root hairs; in the leaves, the epidermis contains guard cells. In a woody stem, epidermis is replaced by **periderm**.

Ground meristem produces ground tissue, which composes most of the plant body. **Ground tissue** is composed of **parenchyma** cells, which are thin-walled and capable of photosynthesis when they contain chloroplasts. **Collenchyma** cells have thicker walls for flexible support. **Sclerenchyma** cells are hollow, nonliving support cells with secondary walls fortified by **lignin**.

Procambium produces vascular tissue. **Vascular tissue** consists of xylem and phloem. **Xylem** contains two types of conducting cells: vessel elements and tracheids. **Vessel elements**, which are larger and have perforation plates, form a continuous pipeline from the roots to the leaves. In elongated **tracheids** with tapered ends, water must move through pits in end walls and side walls. Xylem transports water and minerals. In **phloem**, sieve tubes are composed of **sieve-tube members**, each of which has a **companion cell**. Phloem transports sucrose and other organic compounds including hormones.

24.3 Organization and Diversity of Roots

A root tip has a zone of cell division, a zone of elongation, and a zone of maturation. The apical meristem in the **zone of cell division** is protected by a **root cap**. The newly formed cells lengthen and become specialized in the **zone of elongation**. Cells are fully formed and differentiated in the **zone of maturation**, where root hairs are found on the outer surface of the roots.

A cross section of an herbaceous eudicot root reveals the epidermis, which protects; the cortex, which stores food; the endodermis, which regulates the movement of minerals; and the vascular cylinder, which is composed of vascular tissue. In the vascular cylinder of a eudicot, the xylem appears star-shaped, and the phloem is found in separate regions, between the points of the star. In contrast, a monocot root has a ring of vascular tissue with alternating bundles of xylem and phloem surrounding the pith.

Roots are diversified. Taproots are specialized to store the products of photosynthesis; a fibrous root system covers a wider area. Prop roots are adventitious roots specialized to provide increased anchorage. Roots have mycorrhizae and root nodules, symbiotic relationships with fungi and bacteria respectively, that enable them to obtain nutrients and minerals from the soil.

> **Take Note:** *You should be able to describe how the structure of monocot roots differs from that of eudicot roots. What adaptations do roots have that help them to better perform their functions?*

24.4 Organization and Diversity of Stems

The activity of the shoot apical meristem within a terminal bud accounts for the primary growth of a stem. A **terminal bud** contains internodes and **leaf primordia** at the nodes. When stems grow, the **internodes**, spaces between the nodes, lengthen.

In a cross section of a nonwoody eudicot stem, epidermis surrounds the **cortex**, vascular bundles arranged in a ring, and an inner **pith**. The epidermis is covered by a cuticle that prevents the stem from drying out. Monocot stems have scattered vascular bundles, and the cortex and pith are not well defined.

Secondary growth of a woody stem is due to **vascular cambium**, which produces new xylem and phloem every year, and **cork cambium**, which produces new cork cells when needed. **Cork**, a part of the bark, replaces epidermis in woody plants. In a cross section of a woody stem, the bark is all the tissues outside the vascular cambium. It consists of secondary phloem, cork cambium, and cork. **Wood** consists of secondary xylem, which builds up year after year and forms the annual rings.

Stems are diverse. There are horizontal aboveground and underground stems. Stolons, found in strawberries, are aboveground horizontal stems. Ferns have rhizomes, which are underground horizontal stems. Corms and some tendrils are also modified stems.

> **Take Note:** *You should be able to describe how the structure of monocot stems differs from that of eudicot stems. What adaptations do stems have that help them to better perform their functions? Can you describe how secondary growth occurs in woody stems?*

24.5   Organization and Diversity of Leaves

The **leaves** are very important for plants, as they are the location of most of photosynthesis. The bulk of a leaf is **mesophyll** tissue bordered by an upper and lower layer of epidermis. There are two types of mesophyll: **spongy mesophyll** and **palisade mesophyll**. Spongy mesophyll has many air spaces for the gas exchange required for photosynthesis. The epidermis is covered by a cuticle and may bear **trichomes**, tiny epidermal hairs that decrease the flow of air across the surface of the leaf. **Stomata**, openings for gas exchange, tend to be in the lower epidermis. They are surrounded by **guard cells**, which control the opening and closing of the stomata. Vascular tissue (xylem and phloem) is present within the leaf veins. The vascular tissue is surrounded by **bundle sheath cells**.

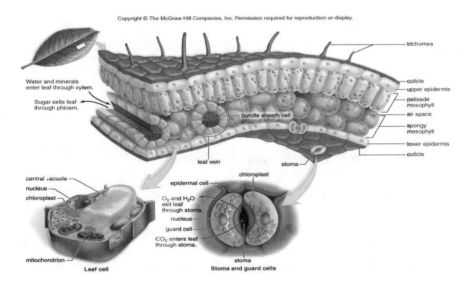

Leaf structure

Leaves are diverse and are adapted to their environment. The spines of a cactus are leaves. Other succulents have fleshy leaves. An onion is a bulb with fleshy leaves, and the tendrils of peas are leaves. The Venus's flytrap has leaves that trap and digest insects.

> **Take Note:** *Describe the adaptations of leaves, and how these adaptations aid in the survival of plants. How is the structure of leaves suitable to their function?*

Multiple Choice Questions

1. The function of the Casparian strip is to
   A. divide and develop lateral roots.
   B. prevent the passage of water between adjacent cell walls.
   C. conduct water and minerals across the width of a plant.
   D. support the xylem of the vascular cylinder.
   E. contain chloroplasts and photosynthesize.

2. A climbing ivy that is not rooted in the ground would have which of the following types of root(s)?
   A. taproot
   B. fibrous roots
   C. pneumatophores
   D. prop root
   E. aerial roots

3. Which of the following is NOT true about the bark of a tree?
   A. Secondary phloem is not produced every year.
   B. Phloem builds up from season to season.
   C. Girdling is not harmful to a tree.
   D. Cork cells have a waxy layer that impedes gas exchange.
   E. Cork cell division does not disrupt the epidermis.

4. A plant that grows in the shade will tend to have
   A. spiny leaves.
   B. broad, wide leaves.
   C. fleshy leaves.
   D. sunken stomata.
   E. hinged leaves.

5. Most photosynthesis occurs in the
   A. mesophyll.
   B. endodermis.
   C. cork cambium.
   D. tracheids.
   E. epidermis.

6. Which of the following plants have stems that are adapted for support and climbing?
   A. strawberries
   B. potatoes
   C. sugar cane
   D. grapes
   E. cacti

Questions 7–10 refer to the following five choices. For each question, select the choice that is most closely related. Each choice may be used once, more than once, or not at all.
   A. parenchyma
   B. meristem
   C. epidermis
   D. xylem
   E. schlerynchema

7. Transports water and minerals from the roots to the leaves

8. Many have root hairs.

9. Nonliving cells without pits that function in support of the plant

10. Consists of only an exterior single layer of cells

Free Response Question

Roots have various adaptations and associations to better perform their functions.

      A. **Describe** the specialized tissues of the dicot root.

      B. **Explain** how the structure of roots is specialized to their function.

      C. **Describe** how plants absorb nitrogen in a usable fashion from the environment.

Annotated Answer Key (MC)

1. **B**; Due to the Casparian Strip, the only access to the vascular cylinder is through the endodermal cells. This regulates the entrance of minerals into the vascular cylinder.

2. **E**; As the vines climb, the rootlets attach the plant to any available vertical structure.

3. **D**; The bark of a tree contains periderm (cork and cork cambium) and phloem. Secondary phloem is produced every year, but phloem does not build up from season to season. Girdling of a tree is harmful. The cork cells are impregnated with suberin, a waxy layer that makes them waterproof but also causes them to die. This makes the stem less edible.

4. **B**; The large, wide leaves provide the greatest surface for photosynthesis.

5. **A**; The mesophyll contains elongated and irregular cells bounded by air spaces. There are many chloroplasts in this region, therefore the greatest amount of photosynthesis occurs here.

6. **D**; The tendrils of grape plants, which are stem branches, allow them to climb, as do the stems of morning glories. These help plants expose their leaves to the sun.

7. **D**; Xylem tissue is composed of tracheids and vessel elements.

8. **C**; Root hairs increase the surface area of the root for absorption of water and minerals and help to anchor the plant firmly in place.

9. **E**; Schlerenchyma cells are impregnated with lignin, which makes the walls tough and hard. Their primary function is to support the mature regions of the plant.

10. **C**; In most plants, the epidermis is a single layer of cells set close together to protect the plant from water loss, invasion by fungi, and physical damage. The epidermis that is exposed to air is covered with a protective substance called cuticle.

Answer to FRQ

PART A (MAX 6)

- Epidermis: outer layer, single layer of cells, thin-walled, may have root hairs
- Cortex: thin walled, irregular shaped, loosely packed, contain starch (food storage)
- Endodermis: single layer, rectangular, boundary/snug fitting, Casparian strip (lignin and suberin)
- Vascular tissue/pericycle: ability to divide, can differentiate, xylem and phloem, star shaped in dicots

PART B (MAX 4)

Must tie structure to function; can get structural points above without functional significance

- Epidermis: protection
- Cortex: water and mineral transport, food storage
- Endodermis: boundary, regulatory for mineral transport
- Vascular tissue: differentiate, efficient transport

PART C (MAX 6)

- Plants can use nitrogen as NH_4^+ and NO_3^-.
- Most plants obtain nitrogen as inorganic nitrates from the soil.
- Ammonium toxicity is an issue.
- Denitrifying bacteria $N_2 \rightarrow NO_3^-$.
- Nitrogen fixing bacteria in root nodules of legumes or in soil
- Bacterial conversion of NH_4^+ to NO_2^-.
- NO_2^- to NO_3^-
- Decomposers (bacteria/fungi) convert NH_4^+ to NO_2^-.

CHAPTER 25
FLOWERING PLANTS: NUTRITION AND TRANSPORT

25.1   Plant Nutrition and Soil

Plants need both essential and beneficial inorganic nutrients. Carbon, hydrogen, and oxygen make up 95% of a plant's dry weight. The other necessary nutrients, including phosphorus, potassium, sulfur, calcium, and magnesium, are taken up by the roots as mineral ions. Nitrogen (N), which is present in the atmosphere, is most often taken up as NO3.

You can determine mineral requirements by **hydroponics**, in which plants are grown in a nutrient solution. The solution is varied by the omission of one mineral. If the plant dies when treated with the modified solution, then the missing mineral must be essential for growth. If the plant grows poorly with the missing mineral, then the mineral is beneficial.

Terrestrial life is dependent on soil, which forms by the weathering of rock. Organisms contribute to the formation of humus and soil. Soil is a mixture of mineral particles, humus, living organisms, air, and water. Soil particles are of three types from the largest to the smallest: sand, silt, and clay. **Loam**, which contains about equal proportions of all three types, retains water but still has air spaces. **Humus,** decaying organic matter, contributes to the texture of soil and its ability to provide inorganic nutrients to plants. Topsoil contains humus, and this is the layer that is lost by erosion, a worldwide problem.

25.2   Water and Mineral Uptake

Water, along with minerals, can enter a root by passing between the porous cell walls until it reaches the **Casparian strip**. The Casparian strip is a waterproof band made of **suberin** and **lignin** that surrounds the **endodermal** cells and facilitates the one-way flow of water into the root. Once water has passed through the endodermal cells, it enters the xylem. Water can also enter root hairs and then pass through the cells of the **cortex** and endodermis to reach xylem. Both methods of water entry occur due to a lower osmotic pressure in the root than in the soil.

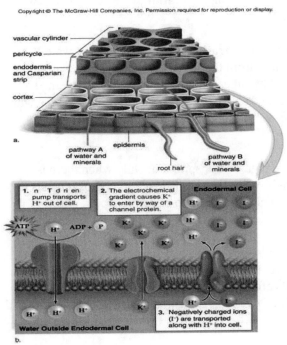

Water and mineral uptake

Mineral ions cross the plasma membranes of endodermal cells by a chemiosmotic mechanism. A proton pump transports H^+ out of the cell using energy from ATP. This establishes an electrochemical gradient that causes positive ions such as K^+ to flow through channels into the cells. Negative ions are carried across the membranes by a cotransporter in conjunction with H^+, which is moving along its concentration gradient.

Plants have various adaptations that assist them in acquiring nutrients. Legumes have **root nodules** infected with the bacterium *Rhizobium,* which makes nitrogen compounds available to these plants. These bacteria fix atmospheric nitrogen, converting it from N_2 to NH_4^+, so that it can be transported into the plant. Most plants have **mycorrhizae**, or fungus roots. The fungus gathers and breaks down nutrients from the soil, and the root provides the fungus with sugars and amino acids. Some plants have poorly developed roots. Most **epiphytes** live on, but do not parasitize trees, whereas dodder and some other plants parasitize their hosts. The Venus flytrap obtains nitrogen and minerals from the insects it digests.

> **Take Note:** *In preparation for the AP Biology exam, you should be able to identify the nutritional needs of plants and explain how they are different from the nutritional needs of animals. What adaptations (structural and relational) do plants have to help them obtain the nutrients they need?*

25.3 Transport Mechanisms in Plants

As an adaptation to life on land, plants have a vascular system that transports water and minerals from the roots to the leaves and must also transport the products of photosynthesis to various organs of the plants. Vascular tissue includes xylem and phloem. **Xylem** transports water and minerals, while **phloem** transports sugars and other organic molecules.

In xylem, vessels composed of **vessel elements** aligned end to end form an open pipeline from the roots to the leaves. Xylem also contains smaller conducting vessels called **tracheids**. As described in the previous section, water enters a plant via the root. The water creates a positive pressure called **root pressure**, which pushes the xylem sap upward. Particularly at night, root pressure can build in the root. This does not contribute significantly to xylem transport, but can cause **guttation**, when droplets of water are forced out of the vein endings on the edges of leaves.

The **cohesion-tension model** of xylem transport states that **transpiration** (the evaporation of water from the leaves) creates a tension that pulls water upward in xylem from the roots to the leaves. As water molecules evaporate through the stomata, they are replaced by other water molecules in the xylem. This means of transport works only because water molecules are cohesive with one another and adhesive with xylem walls. This creates a pulling force that allows water to travel upward through the plant against gravity.

Most of the water taken in by a plant is lost through the stomata by transpiration. Only when there is plenty of water do stomata remain open, allowing carbon dioxide to enter the leaf and photosynthesis to occur. The consequence, however, is that when the stomata close in order to conserve water, gas exchange—and therefore photosynthesis—cannot occur.

Stomata open when guard cells take up water. The guard cells are anchored at their ends. They can only stretch lengthwise because **microfibrils** in their walls prevent lateral expansion. Therefore, guard cells buckle out when water enters. Water enters the guard cells after potassium ions (K^+) have entered. Light signals stomata to open, and a high carbon dioxide (CO2) level may signal stomata to close. Abscisic acid produced by wilting leaves also signals for closure.

In phloem, sieve tubes composed of **sieve-tube members** aligned end to end form a continuous pipeline from the leaves to the roots. Sieve-tube members have sieve plates through which plasmodesmata allow the contents of these cells to extend from one to the other. The **pressure-flow model** of phloem transport proposes that a positive pressure drives the phloem contents in sieve tubes from one cell to the next. Sucrose is actively transported into sieve tubes—by a chemiosmotic mechanism—at a source, and water follows by osmosis. The resulting increase in pressure creates a flow that moves water and sucrose to a sink. A sink can be at the roots or any other part of the plant that requires organic nutrients.

> **Take Note:** *For the AP Biology exam, you should be able to describe the flow of water through the xylem and the flow of sugars through the phloem. What mechanisms are involved? What adaptations do plants have that aid in the transport of water and nutrients?*

Multiple Choice Questions

1. Which of the following stimuli will cause the stomata to remain open?
    A. the secretion of abscisic acid
    B. decrease in turgor pressure due to the exit of K^+ ions
    C. low humidity levels
    D. high solute concentration outside of the guard cells
    E. high light intensity

2. The presence of a vascular transport system does NOT
    A. allow plants to increase in height.
    B. facilitate the dispersal of pollen and seeds.
    C. make it easier for animals to recognize flowers.
    D. promote easier distribution of nutrients to fast growing tissues.
    E. require plants to live near an underground water source.

3. The cells of a wilted plant have a
    A. higher water potential inside the cell.
    B. higher concentration of solutes outside of the cell.
    C. lower osmotic potential outside of the cell.
    D. lower pressure potential in the extracellular fluid.
    E. higher concentration of water inside the cell.

4. The primary function of companion cells is to
    A. provide proteins to sieve tube members.
    B. transport organic nutrients to all parts of the plant.
    C. allow water to pass from cell to cell.
    D. provide internal support.
    E. digest organic nutrients before they reach the roots.

5. Which of the following is NOT a factor in the movement of water throughout a plant?
    A. cohesion of the water molecules throughout the stem
    B. adhesion of water molecules to the xylem walls
    C. higher water potential at the surface of the leaf than in the air
    D. higher concentration of solutes in the xylem sap than in the root cells
    E. higher water potential in the root hairs than the surrounding soil

6. Mycorrhizae
    A. are interactions between bacteria and plant roots.
    B. decrease mineral uptake in the roots.
    C. are involved in a parasitic relationship with a fungus on plant leaves.
    D. cause a plant to be more susceptible to disease.
    E. increase the surface area available for water uptake.

7. Calcium is an essential inorganic nutrient in plants because it
   A. regulates responses to stimuli and transport across the cell membrane.
   B. is essential in cellular respiration as part of cytochrome.
   C. is required for photosynthesis in the porphyrin ring.
   D. plays a role in chlorophyll formation.
   E. splits water in photosynthesis.

8. Sucrose is moved into the phloem by
   A. passive transport.
   B. active transport.
   C. phagocytosis.
   D. diffusion.
   E. facilitated diffusion.

9. Plants acquire nutrients in all of the following ways EXCEPT
   A. engaging in a mutualistic relationship with fungi.
   B. sending out root-like projections that tap into vessels of a host plant.
   C. possessing root nodules containing nitrogen-fixing bacteria.
   D. obtaining nitrogen from digesting insects.
   E. breaking down dead organisms extracellularly for sugars.

10. Phytoremediation is effective because
    A. it is much quicker than conventional clean-up of sites.
    B. it can be effective at extremely deep sites by plants with taproots.
    C. plants have unique abilities to absorb and tolerate toxic substances.
    D. toxins absorbed by the plant increase reproductive rates of the plants.
    E. plants convert the toxins to usable substances and release them in the soil.

Free Response Question

Plants are well adapted to living in terrestrial environments and transporting water from the roots through the stem to the leaves.

A. **Discuss** how water is transported against gravity to reach the leaves.

B. **Explain** the mechanism of stomatal opening and how the following factors influence the opening and closing of the stomata.
   - increased humidity
   - increased temperature
   - wind

Annotated Answer Key (MC)

1. **E**; Stomata open with the concentration of K^+ inside of the cell increases, thus causing an influx of H_2O. This occurs under conditions of high humidity and high light intensity. Abscisic acid causes the stomata to close.

2. **E**; Vascular plants have the ability to be terrestrial and live in areas of very low water availability.

3. **B**; Water flows from an area of high water potential to low water potential. Solutes contribute to this potential. If there are more solutes outside the cell, water will follow, causing the cells of the plant to lose turgidity and the plant will wilt.

4. **A**; Companion cells move sugars and amino acids into and out of the sieve elements. Transmembrane proteins actively transport sugars and amino acids, and water follows by osmosis.

5. **E**; Water travels from a higher water potential to lower potential.

6. **E**; Mycorrhizae are mutualistic relationships between soil fungi and the roots of most plants. The presence of the fungus gives the plant a greater absorptive surface.

7. **A**; Cytochromes contain iron, porphyrin rings contain magnesium (which is in chlorophyll), and light is required for the photolysis of water in photosynthesis.

8. **B**; Sucrose is drawn in to the phloem actively at the source.

9. **E**; Plants use diverse means to acquire nutrients; however, fungi are decomposers.

10. **C**; Phytoremediation uses plants to treat environmental problems. The cost is low, plants can be monitored, there is the possibility of recovering valuable metals, and it is least harmful to the environement.

Answer to FRQ

PART A (MAX 6)

- Water → polar molecule
- Hydrogen bonds (− charged oxygen atom with a + charged hydrogen atom)
- Water enters roots → osmosis
- Capillarity
- High water potential → low water potential
- Cohesion: water attracted to water
- Adhesion: water attracted to xylem walls
- Transpirational pull (water evaporates form the surface of a leaf; replacement molecule) OR cohesion tension theory
- Environmental driven water movement (sun, wind, humidity)

PART B (MAX 6)
Three points for mechanism; three points for effects

- The increase in osmotic pressure in the guard cells is caused by an uptake of **potassium** ions (K^+).
- The concentration of K^+ in open guard cells far exceeds that in the surrounding cells. This is how it accumulates:
  o Blue light is absorbed by phototropin.
  o Activation of a proton pump (an H^+-ATPase) in the plasma membrane of the guard cell.
  o ATP drives the pump.
  o As protons (H^+) are pumped out of the cell, its interior becomes increasingly negative.
  o This attracts additional K^+ ions into the cell, raising its osmotic pressure.
  o Water fills the vacuoles of the guard cells

Environmental Condition	Plant Response
Increased humidity	Stomata remain open
	Higher water potential outside the guard cells
Increased temperature	Stomata will close
	Tend to open during the night; plants have photosynthetic adaptations to accommodate this temporal change
Wind	Stomata close

CHAPTER 26
FLOWERING PLANTS: CONTROL OF GROWTH RESPONSES

26.1   Plant Hormones

All organisms, including plants, respond to stimuli in the environment. Like animals, flowering plants use a reception–transduction–response pathway when they respond to a stimulus. The process involves receptor activation, transduction of the signal by relay proteins, and a cellular response, which can consist of the turning on of a gene or an enzymatic pathway.

The most common hormones found in plants are **auxins, gibberellins, cytokinins, abscisic acid,** and **ethylene.** Their effects on plants are summarized in the table below:

Hormone	Effects
Auxins (IAA)	• Cause apical dominance • Produce growth of roots and fruit • Prevents loss of leaves and fruit • Triggers cell elongation in phototropism and gravitropism
Gibberellins	• Promote stem elongation between nodes • Interrupt dormancy of seeds and buds
Cytokinins	• Facilitate cell division • Prevent leaf senescence
Abscisic acid	• Initiates and maintains dormancy of seeds and buds • Causes stomata to close in response to water stress
Ethylene	• Causes fruit to ripen • Initiates abscission • Suppresses stem and root elongation

> **Take Note:** *For the AP Biology exam, you should be prepared to describe the effects of hormones on plant growth and development. How do these responses compare to the effects that hormones have on animal growth and development? For example, what is the role of hormones in the regulation of water balance in plants and animals?*

26.2   Plant Responses

Like all other organisms, plants respond to their environment. Plants can respond to the abiotic and biotic factors of their environment. When flowering plants respond to stimuli, growth and/or movement occurs. Responses to the abiotic environment include tropisms, nastic movements, and photoperiodism.

Tropisms are growth responses toward or away from unidirectional stimuli. Growth toward a stimulus is called a **positive tropism**, while growth away from a stimulus is called a **negative tropism**. The three most common tropisms are **phototropism** (movement in response to light), **gravitropism** (movement in response to gravity), and **thigmotropism** (movement in response to touch). The positive phototropism of stems results in a bending toward light, and the negative gravitropism of stems results in a bending away from the direction of gravity. Roots that bend toward the direction of gravity show positive gravitropism. Thigmotropism occurs when a plant part makes contact with an object, as when tendrils coil about a pole.

Nastic movements are not directional, and include **turgor movements** and **sleep movements**. Some plants, like the *Mimosa pudica*, respond to touch when motor cells lose K+, and water, resulting in the loss of turgor pressure. This causes the leaves to collapse when an electrical signal is passed from one leaflet to another. Plants exhibit **circadian rhythms**, which are believed to be controlled by a biological clock. The sleep movements of prayer plants (closing of the leaves at night), the closing of stomata, and the daily opening of certain flowers have a 24-hour cycle.

Phytochrome is a pigment that is involved in **photoperiodism**, the ability of plants to sense the length of the day and night during a 24-hour period. This sense leads to seed germination, shoot elongation, and flowering during favorable times of the year. Daylight causes phytochrome to exist as P_{fr}, but during the night, it is reconverted to P_r by metabolic processes. Phytochrome in the P_{fr} form leads to a biological response such as flowering. Short-day plants flower only when the days are shorter than a critical length, and long-day plants flower only when the days are longer than a critical length. In fact, research has shown that it is the length of darkness that is critical. Interrupting the dark period with a flash of white light prevents flowering in a short-day plant and induces flowering in a long-day plant.

Flowering plants have defenses against predators and parasites. The first line of defense is their outer covering, the bark and cuticle. They also routinely produce **secondary metabolites** such as tannins and alkaloids that protect them from herbivores, particularly insects. Wounding causes plants to produce **systemin**, which travels though the plant and causes cells to produce proteinase inhibitors that destroy the digestive enzymes of an insect. During **a hypersensitive response**, an infected area is sealed off, allowing a plant to protect itself from bacteria, viruses, or fungi. As an indirect response, plants temporarily attract animals that will destroy predators, and going one step further, plants have permanent symbiotic relationships with animals, such as ants, that will attack predators.

> **Take Note:** *On the AP Biology exam, you may be asked to explain how plants respond to their environment in order to survive. How does the way plants respond to their environment compare to the way that animals respond to their environment?*

Multiple Choice Questions

Questions 1–4 refer to the following five choices. For each question, select the choice that is most closely related. Each choice may be used once, more than once, or not at all.

   A. gibberellin
   B. auxin
   C. cytokinin
   D. abscisic acid
   E. ethylene

1. Closes stomata

2. Responsible for apical dominance

3. Seed development and germination

4. Ripens fruit

5. A researcher discovers that cocklebur will not flower if a required long dark period is interrupted by a brief flash of white light. Which of the following conclusions can be made?
   A. A cocklebur is a long-day plant.
   B. When the night is longer than a critical length, cocklebur will flower.
   C. The length of the light period controls flowering.
   D. Interrupting the light period with darkness will prevent flowering.
   E. A flash of far-red light during the daylight hours will promote flowering.

6. To qualify as a circadian rhythm the activity must
    A. occur in the morning and the evening.
    B. occur only in the presence of external stimuli.
    C. be able to reset if external cues are provided.
    D. affect seasonal changes.
    E. impact the photoperiod.

7. Plants do NOT possess which of the following defense mechanisms?
    A. production of tannins on the epidermis of leaves
    B. secretion of secondary metabolites that break down to cyanide
    C. production of alkaloids with astringent properties
    D. release of volatiles to attract insects during times of distress
    E. teeth-like barbs that sting predators

8. Unripe bananas could be ripened by an external treatment of
    A. ethylene.
    B. abscisic acid.
    C. auxin.
    D. cytokinins.
    E. gibberillin.

9. Which of the following stimuli would activate a phototropin in the cell membrane?
    A. auxin
    B. defense hormones
    C. gibberrillin
    D. blue light
    E. attack by a herbivore

10. Which of the following is true regarding plant hormone concentrations?
    A. Root production by the callus requires a higher concentration of auxin to cytokinin.
    B. Auxin alone causes floral shoots in a plant.
    C. A specific ratio of auxin to cytokinin is required for the production of leaves.
    D. Gibberellin prevents senescence and promotes cell division
    E. Genetically modified tomaotes produce excess ethylene to stay ripe longer.

Free Response Question

You hypothesize that abscisic acid (ABA) brings about the closing of the stomata when a plant is under water stress.

   A. **Design** an experiment to test this hypothesis, and explain the expected results.

   B. **Explain** how protein molecules may play a role in stomatal closing.

Annotated Answer Key (MC)

1. **D**; During times of environmental stress, stomata will close to preserve water. Abscisic acid is also applied to nursery plants prior to shipping to induce dormancy and make them more resistant to damage.

2. **B**; Auxins respond to gravity and light and induce lengthening of stems.

3. **A**; In the spring, gibberellins assist in budding and seeds breaking dormancy.

4. **E**; Ethylene causes aging responses including fruit ripening.

5. **C**; Plant growth and development are dependent upon the number of hours of light available each day. Long-day plants require the daylength to exceed a critical value. Short-day plants require a daylength shorter than a critical value. Day-neutral plants flower regardless of daylength.

6. **C**; Circadian rhythms persist in constant conditions, are maintained over a range of temperatures, and can be reset by external stimuli.

7. **E**; Plants have many defense mechanisms.

8. **A**; Ethylene contributes to the aging response and therefore causing ripening.

9. **D**; Phototropins are proteins in the membranes of higher plants. When blue light hits the prototropin protein, an enzyme cascade will occur to cause stems to bend towards light and stomata to open.

10. **D**; Gibberellins have diverse functions in stem elongation, dormancy of seeds and buds, flowering, bud, root and leaf growth and development, and in embryos.

Answer to FRQ

PART A (MAX 3)

Design Structures
- Tests appropriate variable (ABA)
- Experimental constants maintained
- Control group
- Quantitative experiment
- Multiple trials/repeatability

PART B (MAX 5)

Results
- Analysis of results: Interpret hypothetical results

Explanation of results
- ABA $\rightarrow$ binds to receptors at the plasma membrane of the guard cells
- Receptors initiate cascade $\rightarrow$
    - Increased pH
    - $\rightarrow$ transfer of Ca^{2+} from the vacuole to the cytosol
    - $\rightarrow$ Ca^{2+} blocks the uptake of K^+
    - loss of Cl^-
    - $\rightarrow$ reduces osmotic pressure $\rightarrow$ decreased turgor pressure $\rightarrow$ stomata close
- G Protein Coupled Receptors (GPCR)
- Transmembrane
- Binding of a ligand $\rightarrow$ activates a G protein
- $\rightarrow$ second messenger (cAMP) $\rightarrow$ enzyme cascade

CHAPTER 27
FLOWERING PLANTS: REPRODUCTION

27.1   Sexual Reproductive Strategies

Sexual reproduction in plants leads to genetic variation in offspring due to meiosis and fertilization, just as it does in animals. Angiosperms exhibit an alternation of generations life cycle (see Chapter 23). Flowers (the reproductive structures of angiosperms) produced by the sporophyte produce microspores and megaspores by meiosis. Microspores develop into a male gametophyte, and megaspores develop into the female gametophyte. The gametophytes produce gametes by mitotic cell division. Following fertilization, the sporophyte is enclosed within a seed covered by fruit.

A typical flower has several parts: **Sepals**, which are usually green in color, form an outer whorl; **petals**, often colored, are the next whorl; and **stamens**, each having a **filament** and **anther**, form a whorl around the base of at least one carpel. The **carpel**, in the center of a flower, consists of a **stigma**, **style**, and **ovary**, which contains **ovules**. The stamens are the male parts of the flower, and the carpel is the female part of the flower.

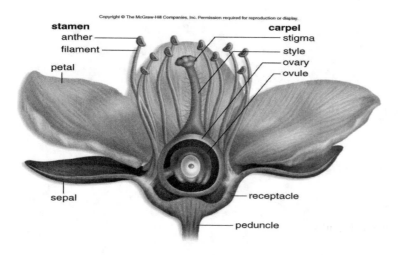

Anatomy of a flower

Flowers can vary in structure. Those that have sepals, petals, stamens, or carpels are called **complete** flowers, while those that do not are **incomplete**. Flowers that have both stamens and carpels are **perfect**, and while flowers that have one or the other (**staminate** or **carpellate**) are called **imperfect** flowers. **Monoecious** plants have staminate and carpellate flowers on the same plant. **Dioecious** plants have separate plants for staminate and carpellate flowers.

The flowering plant life cycle is adapted to a land existence. The microscopic gametophytes are protected from desiccation by the sporophyte; the pollen grain has a protective wall and fertilization does not require external water. The seed has a protective seed coat, and seed germination does not occur until conditions are favorable.

Each ovule contains a megaspore mother cell, which divides meiotically to produce four haploid megaspores, only one of which survives. This megaspore divides mitotically to produce the female gametophyte (embryo sac), which usually has seven cells. One is an egg cell and another is a central cell with two **polar nuclei**.

The anthers contain microspore mother cells, each of which divides meiotically to produce four haploid microspores. Each of these divides mitotically to produce a two-celled **pollen grain**. One cell is the **tube cell**, and the other is the **generative cell**. The generative cell later divides mitotically to produce two sperm cells. The pollen grain is the male gametophyte. After pollination (the transfer of pollen from anther to stigma) the pollen grain germinates, and as the pollen tube grows, the sperm cells travel to the embryo sac. Flowering plants experience **double fertilization**. One sperm nucleus unites with the egg nucleus, forming a 2n zygote, and the other unites with the polar nuclei of the central cell, forming a 3n endosperm cell.

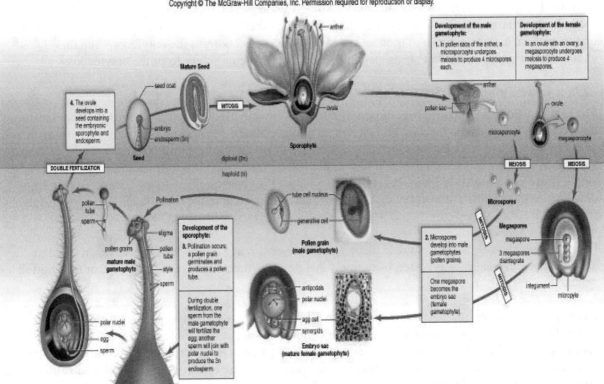

Life cycle of flowering plants

After fertilization, the endosperm cell divides to form multicellular **endosperm**. The zygote becomes the sporophyte embryo. The ovule matures into the seed (its integuments become the seed coat). The ovary becomes the fruit.

> **Take Note:** *On the AP Biology exam, you should be able to explain how flowers are an adaptation for survival in a terrestrial environment. What are the adaptive advantages of the reproductive structures of flowering plants? Can you describe the life cycle of a flowering plant? Why are flowering plants better adapted to life on land than mosses?*

Genetic variation is ensured by cross pollination between different plants of the same species. Some angiosperms produce large amounts of pollen that are carried by the wind. Most angiosperms rely on animals (insects, birds, or mammals) for pollination. In fact, plants and their pollinators have **coevolved**— that is, as one species changed, the other did as well—and ultimately the two species are suited to each other. In other words, the reproductive parts of some flowers are positioned such that the pollinator has to pick up pollen from one plant and deliver to another; likewise, the mouthparts of the pollinator are suited for the collection of nectar from these particular flowers.

27.2 Seed Development

As the ovule is becoming a seed, the zygote is becoming an embryo. After the first several divisions, it is possible to discern the embryo and the suspensor. The **suspensor** attaches the embryo to the ovule and supplies it with nutrients. The eudicot embryo becomes first heart-shaped and then torpedo-shaped as rapid cell division leads to the development of the **cotyledons**, or seed leaves. Once you can see the two cotyledons, it is possible to distinguish the shoot tip and the root tip, which contain the apical meristems. In eudicot seeds, the cotyledons frequently take up the endosperm. As the embryo develops inside the seed, the external protective seed coat develops from the outer layers of the ovule. Monocot seeds have one cotyledon, and the embryo of monocot seeds obtains nutrients from the cotyledons and the endosperm, whereas in eudicot seeds the embryo obtains nutrients solely from the cotyledons.

27.3 Fruit Types and Seed Dispersal

The seeds of flowering plants are enclosed by fruits. There are different types of fruits. **Simple fruits** are derived from a single ovary (which can be simple or compound). Some simple fruits are fleshy, such as a peach or an apple. Others are dry, such as peas, nuts, and grains. **Compound fruits** are produced from several groups of ovaries and consist of **aggregate fruits**, which develop from a number of ovaries of a single flower, and **multiple fruits** develop from a number of ovaries of separate flowers. During fruit development, the walls of the ovary thicken to become the **pericarp**, the three-layered covering of the fruit.

Flowering plants have several ways to disperse seeds. Seeds may be blown by the wind, attached to animals that carry them away, eaten by animals that defecate them some distance away, or transported by water.

Once seeds are dispersed, they will germinate to form a seedling under favorable conditions—sufficient water, warmth, and oxygen to sustain growth. If conditions are not favorable for growth, seeds will remain dormant. Prior to germination, you can distinguish a bean (eudicot) seed's two cotyledons and **plumule**, which is the shoot that bears leaves. Also present are the **epicotyl**, **hypocotyl**, and **radicle**. In a corn kernel (monocot), the endosperm, the cotyledon, the plumule, and the radicle are visible.

Take Note: *In preparation for the AP Biology exam, you should be able to describe the structure of a either a bean or a corn seed and explain how it germinates to form a seedling.*

27.4 Asexual Reproductive Strategies

Many flowering plants reproduce asexually, eliminating the need for pollen or seeds. Asexual reproduction occurs when the nodes of stems (either aboveground or underground) give rise to entire plants, such as when strawberries arise from **stolons** (aboveground stems) or when irises grow from **rhizomes** (underground stems). In some plants, like fruit trees, roots can also produce new shoots.

Somatic embryogenesis is the development of adult plants from protoplasts in tissue culture. **Micropropagation**, the production of clonal plants as a result of meristem culture in particular, is now a commercial venture. Flower meristem culture results in somatic embryos that can be packaged in gel for worldwide distribution. Anther culture results in homozygous plants that express recessive genes. Leaf, stem, and root culture can result in cell suspensions that allow plant chemicals to be produced in large tanks.

Multiple Choice Questions

1. In flowering plants
    A. the diploid sporophyte is dominant and it is the generation that bears flowers.
    B. the embryo contains a seed which then develops into a flower.
    C. a diploid male gametophye is produced through meiosis and produces pollen.
    D. the microspore undergoes mitosis to form an egg.
    E. there is only one multicellular stage.

2. In all land plants
    A. there are flagellated sperm.
    B. pollen grains are released when a thick wall is present.
    C. gametes are produces by meiosis.
    D. megaspores give rise to sperm.
    E. the sporophyte produces haploid spores by meiosis.

Questions 3–6 refer to the following five choices. For each question, select the choice that is most closely related. Each choice may be used once, more than once, or not at all.
    A. simple fruit
    B. accessory fruit
    C. legume
    D. aggregate fruit
    E. multiple fruit

3. Contains fleshy fruits from simple ovaries of the same flower

4. A tomato; contains seed produced from a compound ovary

5. A grape (a true berry), which develops from a singly ovary

6. A dry fruit from a nitrogen-fixing plant

7. Double fertilization in angiosperms
    A. requires one sperm to unite with an egg and the other to form an endosperm.
    B. produces two seeds from the fusion of two sperm and two ova.
    C. produces a triploid seed when two sperm cells unite with an egg cell.
    D. occurs when the pollen grain lands on the flower and begins development.
    E. occurs only between dioecious plants.

8. A typical flower does NOT consist of
    A. sepals.
    B. petals.
    C. calyx.
    D. stamen.
    E. pericarp.

9. Which of the following examples of pollinators and plants is NOT correctly matched?
    A. bee – brightly colored flowers
    B. moth – lightly shaded flowers
    C. bird – flowers with strong odors
    D. bat – large, sturdy, musky flowers
    E. butterfly - flat landing platforms

10. Which of the following allows flowering plants to not require water for fertilization?
    A. pollen tube
    B. flagellated sperm
    C. filament
    D. style
    E. carpel

Free Response Question

Angiosperms are the most diverse and widespread of all plants, including in their means of reproduction.

A. **Discuss** the advantages and disadvantages of sexual and asexual reproduction in plants.

B. **Discuss** how plants can asexually reproduce without fertilization.

Annotated Answer Key (MC)

1. **A**; Sporophytes are vegetative bodies that grow by mitosis from a fertilized egg and bear flowers.

2. **E**; When that sporophyte bears flowers, haploid spores form and develop into gametophytes. Sperm form inside male gametophytes and eggs from inside female gametophytes.

3. **D**

4. **A**

5. **A**

6. **C**

    For Questions 3–6: Fruits are divided into several categories: Simple dry or fleshy fruits (from one simple or compound ovary), accessory fruits (from simple ovary plus other tissues), aggregate fruits (from several separate ovaries of one flower), and multiple fruits (combined from numerous separate flowers). A legume is a simple dry fruit.

7. **A**; Double fertilization takes place in flowering plants.. An embryo and nutritive tissue (endosperm) form in the ovule, which becomes a seed.

8. **E**; The typical flower consists of whorls of modified leaves called petals, stamens, and carpels. The sepals enclose the other parts of the flower inside the bud. The pericarp is the outer fleshy layer of the fruit.

9. **C**; Bees are attracted to brightly colored flowers, and moths are attracted to light shaded flowers as they can see them in the dark. Bats' sense of smell contributes to their attraction to musky flowers that also are large to support the bats, and butterflies require a flat landing platform. Birds have a capacity to smell but it usually plays little or no part in their ability to locate food.

10. **A**; Flagellated sperm are required in bryophytes. The pollen tube grows down through the tissues and carries the sperm nuclei with it and therefore eliminates the need for a water environment.

Answer to FRQ

PART A (MAX 8)

Three points + one example asexual; three points + one example sexual

Asexual advantages

- Less energy investment for asexual (only requires one parent)
- Faster
- Best in stable environments
- Examples: stolons, runners, rhizomes, tubers, bulbs, corms, plantlets, suckers

Sexual advantages

- Genetic recombination of maternal and paternal chromosomes
- Increased variation in offspring
- variation →increased resistance
- Examples: various

PART B (MAX 4)

- Formation of sporophytes by parthenogenesis of gametophyte cells.
- The embryo arises from an unfertilized egg that was produced without meiosis.
- Seeds are genetically identical to the parent plant.
- Apomixis is an abnormal sexual reproduction in which the embryo develops from the egg cells associated with it without fertilization.
- Examples:
  - Alternates between a diploid sporophyte and a haploid gametophyte generation
  - Meiosis produces two types of spores →microspores and megaspores
  - Multicellular haploid AND multicellular diploid
  - Gametes formed by differentiation
  - Growth by mitosis

CHAPTER 28
INVERTEBRATE EVOLUTION

28.1   Evolution of Animals

Animals are multicellular, eukaryotic organisms that are heterotrophic and ingest their food. They have the diploid life cycle; the diploid adult produces haploid gametes by meiosis. The gametes fuse during fertilization to form a zygote that develops into a diploid adult. Most animals reproduce sexually, though some can reproduce asexually.

Animals have two tissues that other organisms do not: **muscles** and **nerves**. Typically, they have the power to move by means of contracting muscle fibers. Vertebrates have a spinal cord or backbone that runs down the center of the back, while invertebrates do not.

It is hypothesized that animals evolved from a protist that resembles the **choanoflagellates** of today. The colonial flagellate theory states that the ancestor of animals resembles a hollow spherical colony of flagellated cells. As the ancestor evolved, the cells become specialized and infolded to form multiple layers of tissues.

Molecular data used to construct an evolutionary tree of the animals can be substantiated by morphological data. All animals exhibit one of the three types of symmetry: asymmetry, radial symmetry or bilateral symmetry. **Asymmetric** animals have no particular body shape. **Radially symmetric** animals are organized circularly and are sometimes **sessile**, meaning they are attached to a substrate. Animals that exhibit **bilateral symmetry** move forward with an anterior end and show **cephalization**, the localization of a brain and sensory organs at the anterior end. Most animals have true tissues that arise from germ layers during embryonic development. **Diploblastic** animals have two germ layers (ectoderm and endoderm), while **triploblastic** animals have three (**ectoderm, mesoderm, and endoderm**). Triploblastic animals are either protostomes or deuterostomes.

There are three major events that determine whether an animal with three germ layers will be a protostome or deuterostome: 1) cleavage, 2) blastopore fate, and 3) coelom formation. During **protostome** development, cleavage is spiral and determinate, the blastopore becomes the mouth, and the coelom forms by splitting of the mesoderm. During **deuterostome** development, cleavage is radial and determinate, the blastopore becomes the anus, and the coelom forms by outpocketing of the gut.

> **Take Note:** *In preparation for the AP Biology exam, you should be prepared to discuss the similarities and differences between animals and other organisms. How do they compare to plants and fungi? What adaptive advantage do animals gain by having muscles?*

Echinoderms and chordates make up the deuterostomes. The protostomes consist of two groups: **ecdysozoa** (molting animals) and **lophotrochozoa**. Lophotrochozoa also consists of two groups: **lophophores** and **trochophores**.

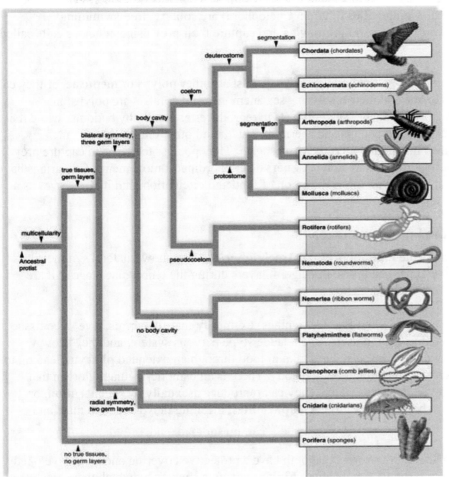

Phylogenetic tree of animals

Take Note: *For the AP Biology exam, you should be able to describe the differences between radial and bilateral symmetry, coelomates and acoelomates, and protostomes and deuterostomes. Why are these differences important in grouping animals in the phylogenetic tree shown in the figure above? Can you identify the fate of the three germ layers during animal development?*

28.2 Introducting the Invertebrates

Sponges (phylum Porifera) are aquatic animals that resemble colonial protozoans. They have the cellular level of organization, lack tissues, and have various symmetries. They do not have nerves or muscles. Sponges are sessile **filter feeders (suspension feeders)** and depend on water that enters through the **osculum** and flows through the body to acquire food, which is digested in vacuoles within **collar cells (choanocytes)** that line a central cavity. Sponges have few predators because of their spicules, needle-like rays that make up their endoskeleton. Sponges reproduce asexually by fragmentation or budding. They reproduce sexually when eggs and sperm are released into their central cavity, forming a zygote that develops into a flagellated larva that swims to a new location.

Comb jellies and cnidarians have two tissue layers derived from the germ layers ectoderm and endoderm. They are radially symmetrical. Comb jellies (phylum Cnetophora) are solitary, free-swimming invertebrates. They are made of a transparent **mesoglea** and capture their prey using adhesive cells called **colloblasts**.

Cnidarians (phylum Cnidaria) have a sac body plan. They exist as either **polyps** or **medusae**, or they can alternate between the two. Hydras and their relatives—sea anemones and corals—are polyps; in jellyfishes, the medusan stage is dominant. The polyp stage reproduces asexually by budding, while the medusa stage reproduces sexually using eggs and sperm. In *Hydra* and other cnidarians, an outer epidermis is separated from an inner gastrodermis by mesoglea. They possess tentacles to capture prey and **cnidocytes** armed with **nematocysts** to stun it. A **nerve net** coordinates movements. Cnidarians have a **gastrovascular cavity** that is involved in the digestion of food and circulation, and it also serves as a **hydrostatic skeleton**.

28.3   Variety Among the Lophotrochozoans

The **lophotrochozoans** include the **lophophores** and **trochophores**. They show bilateral symmetry at some point during their development and have three germ layers during the embryonic stage. All are protostomes, and some are true coelomates.

Free-living flatworms (planarians, phylum Platyhelminthes) exemplify that flatworms have three tissue layers and no coelom. **Planarians** have muscles and a ladder-type nervous system, and they show cephalization, with a brain and two eyespots. They take in food through an extended **pharynx** leading to a gastrovascular cavity, which extends throughout the body. There is an osmotic-regulating organ that contains **flame cells** which remove excess water. They can reproduce asexually by **regeneration**, or sexually by cross-fertilization because they are **hermaphrodites**, meaning they have both male and female sex organs.

Flukes and tapeworms are parasitic flatworms that both have a protective covering and well-developed reproductive systems but do not show cephalization. **Flukes** have two suckers by which they attach to and feed from their hosts. They can invade a variety of organs, such as the liver and the lungs, as well as blood. Find the figure of the life cycle of flukes in your textbook, and review it before moving on. **Tapeworms** have a **scolex** with hooks and suckers for attaching to the host intestinal wall. The body of a tapeworm is made up of **proglottids**, which, when mature, contain thousands of eggs. If these eggs are taken up by pigs or cattle, larvae become encysted in their muscles. If humans eat this meat, they too may become infected with a tapeworm. Most tapeworms have a life cycle that involves several hosts. Review the life cycle of tapeworms found in your textbook before you continue to the next section.

Rotifers (phylum Rotifera) are microscopic and have a **corona**, a crown of cilia that resembles a spinning wheel when in motion, that serves as a mouth and for locomotion. Most are transparent, but some are quite colorful. Most are aquatic organisms, living in both freshwater and marine environments, and some are terrestrial.

The body of a **mollusc** (phylum Mollusca) typically contains a **visceral mass** (the internal organs), a **mantle** (the covering of the visceral mass), and a foot. Many also have a head and a **radula**, a tongue-like organ with many rows of teeth that is used to obtain food. The nervous system consists of several ganglia connected by nerve cords. There is a reduced coelom and an open circulatory system. Clams (bivalves) are adapted to a sedentary coastal life, squids (cephalopods) to an active life in the sea, and snails (gastropods) to life on land.

Annelids (phylum Annelida) are segmented worms. They have a well-developed coelom, body segments divided by **septa**, a closed circulatory system, a ventral solid nerve cord, and paired **nephridia** for waste removal. They have a hydrostatic skeleton, and movement is assisted by **setae**—bristles on the body wall that anchor the worm. Earthworms are oligochaetes ("few bristles") that use the body wall for gas exchange. Polychaetes ("many bristles") are marine worms that have **parapodia** that are used for movement and gas exchange. They may be predators, with a definite head region, or they may be filter feeders, with ciliated tentacles to filter food from the water. Leeches are also part of this phylum.

28.4 Quantity Among the Ecdysozoans

Ecdysozoans are protostomes that molt, meaning they shed their outer covering in order to grow. **Roundworms** (phylum Nematoda) have a **pseudocoelom**, a body cavity not completely surrounded by mesoderm. Roundworms are usually small and very diverse; they are present almost everywhere in great numbers. Many are significant parasites of humans. The parasite *Ascaris* is representative of the group. Infections can also be caused by *Trichinella,* whose larval stage encysts in the muscles of humans. Elephantiasis is caused by a filarial worm that blocks lymphatic vessels.

Arthropods (phylum Arthropoda) are the most varied and numerous of animals. Their success is largely attributable to five major characteristics: 1) a rigid, but jointed exoskeleton, 2) segmentation, 3) a high degree of cephalization, 4) a variety of respiratory organs, and 5) reduced competition through metamorphosis. Crustaceans (crayfish, lobsters, shrimps, copepods, krill, and barnacles) live in aquatic and marine environments. They have a head that bears compound eyes, antennae, antennules, and mouthparts such as mandibles and maxillae. Crayfish illustrate other features, such as an open circulatory system, respiration by gills, and a ventral solid nerve cord. Centipedes and millipedes have many segments and differ by the number of pairs of legs on the segments (one pair for centipedes, two pairs for millipedes).

Insects live a primarily terrestrial life and include butterflies, grasshoppers, bees, and beetles. The anatomy of the grasshopper illustrates insect anatomy and the ways they are adapted to life on land. Like other insects, grasshoppers have three segments (head, thorax, abdomen), wings, and three pairs of legs attached to the **thorax**. Grasshoppers have a **tympanum** for sound reception, a digestive system specialized for a grass diet, **Malpighian tubules** for excretion of solid nitrogenous waste, **tracheae** for respiration, internal fertilization, and incomplete metamorphosis.

Chelicerates (horseshoe crabs, spiders, scorpions, ticks, and mites) live in a variety of environments. They have chelicerae, pedipalps, and four pairs of walking legs attached to a cephalothorax.

28.5 Invertebrate Deuterostomes

Echinoderms (phylum Echinodermata) (e.g., sea stars, sea urchins, sea cucumbers, and sea lilies) have radial symmetry as adults (bilateral as larvae) and internal calcium-rich plates with spines. They have an internal **water vascular system** made of several canals that is involved in locomotion, gas exchange, feeding, and sensory reception. Typical of echinoderms, sea stars have tiny skin gills, a central nerve ring with branches, and a water vascular system for locomotion. Each arm of a sea star contains branches from the nervous, digestive, and reproductive systems.

> **Take Note:** *For each group of invertebrates described in this chapter, you should be able to compare and contrast them with respect to their nervous, digestive, and circulatory systems, as well as their modes of reproduction.*

Multiple Choice Questions

1. Which of the following organisms exhibits bilateral symmetry, cephalization, and is a deuterostome?
    A. sponge
    B. cnidarians
    C. chordate
    D. roundworm
    E. rotifer

2. Homeotic genes
    A. are only inherited maternally.
    B. may be responsible for development differences in animals.
    C. are DNA sequences capable of randomly moving from one site to another.
    D. signal where transcription is to begin at the promoter sequence.
    E. have remained unchanged through animal evolution.

3. Which of the following statements provides evidence that echinoderms evolved from bilaterally symmetrical animals?
    A. Echinoderms have an excretory and circulatory system for exchange of waste products.
    B. Sea stars reproduce only asexually through fragmentation.
    C. Bilateral larva undergo metamorphosis to become radially symmetrical adults.
    D. Echinoderms are acoelomate as are other bilaterally symmetrical animals.
    E. The spines on echinoderms are only on the dorsal side of the sea star.

4. Which of the following pairs is incorrectly matched?
    A. beetle – first pair of wings hardened
    B. Llafhopper – thickened forewings
    C. dragonly – elongated forewings
    D. wasp – ovipositor/stinger
    E. louse – chewing mouthparts

5. Grasshoppers
    A. undergo complete metamorphosis involving drastic changes in form.
    B. possess Malpighian tubules to secrete solid nitrogenous waste.
    C. use different food sources as larvae and adults.
    D. do not have wings.
    E. are composed of a head and an abdomen.

6. Planarians can reproduce asexually because they
    A. are hermaphroditic.
    B. practice cross fertilization.
    C. can regenerate.
    D. contain flame cells.
    E. are parasitic.

Questions 7–10 refer to the following five choices. For each question, select the choice that is most closely related. Each choice may be used once, more than once, or not at all.
    A. sponges
    B. flatworm
    C. mollusc
    D. cnidarian
    E. annelida

7. Utilize nematocysts for capturing prey

8. Filter feeder

9. Elongated foot; may be marine or terrestrial

10. Contain spicules for support

Free Response Question

There are several evolutionary ties that bind major groups of animals together while allowing them to adapt to particular habitats.

    A. **Discuss** why animals are classified separately from fungi and plants.

    B. **Explain** and give an example of each of the morphological data below used to construct the phylogenetic tree of animals.
- types of symmetry
- embryonic development

Annotated Answer Key (MC)

1. **C;** Cephalization is dependent upon bilateral symmetry. Deuterostome development occurs only in Chordates and Echinoderms; all other bilaterally symmetrical animals contain three layers and possess a body cavity, but are protostomes.

2. **B;** Homeotic genes control the overall body plan by controlling the fate of groups of cells during development. Slight shifts in Hox (homeotic) genes are responsible for the major differences between animals that arise during development. Mitochondrial DNA is inherited maternally; transposons are DNA sequences capable of randomly moving from one site to another.

3. **C;** Echinoderms do not have excretory and circulatory systems. They can also reproduce sexually and asexually. Echinoderms are coelomate animals.

4. **E;** A louse has piercing-sucking mouthparts.

5. **B;** Grasshoppers undergo an incomplete metamorphosis—a gradual change in form as the animal matures—and therefore use similar food sources as larvae and adults. They have wings and are composed of a head, thorax, and abdomen.

6. **C;** Planarians are hermaphroditic and practice cross fertilization but these are characteristics of sexually reproducing organisms. They contain flame calls for excretory purposes.

7. **D;** Nematocysts are stinging structures unique to cnidaria for capturing prey.

8. **A;** Sponges are also known as suspension feeders because they filter suspended particles from the water by means of a straining device.

9. **C;** The foot is a muscular organ that may be adapted for locomotion, attachment, food capture, or a combination of those functions.

10. **A;** Spicules are small, needle-shaped structures. The type of spicule has been used to classify sponges.

Answer to FRQ

PART A (MAX 4)

Similarities
- Multicellular
- Eukaryotic

Differences
- Heterotrophs
- Injest whole food and digest it internally

PART B (MAX 8 – Four each)

Symmetry
- Assymetrical: no particular body shape
- Examples: sponges
- Radial symmetry: organized circularly, two identical halves
- May be sessile or floating → allows them to reach out in all directions with one center
- Examples: cnidarians and comb jellies
- Bilateral symmetry: left and a right half
- Examples: Flatworms, mollusks, annelids, roundworms, arthropods, echinoderms, chordates
- Accompanied by cephalization

Embryonic development
- No tissues: i.e., sponges → cellular level of organization
- True tissues: two (diploplastic) or three (triploblastic)
- Two layers have the tissue level of organization
- Three layers have the organ level of organization
- Protostomes:
  - cleavage (spiral and determinate),
  - fate of blastopore (mouth)
  - coelom development (splitting of the mesoderm)

CHAPTER 29
VERTEBRATE EVOLUTION

29.1   The Chordates

Chordates (sea squirts, lancelets, and vertebrates) have an internal skeleton with muscles attached to the outer surface of the bones. They have four defining characteristics: 1) a notochord, 2) a dorsal tubular nerve cord, 3) pharyngeal pouches, and 4) a postanal tail at some time in their life history. The **notochord** is located below the nerve cord and is replaced by the vertebral column in most vertebrates. The anterior part of the **dorsal nerve cord** becomes the brain and will develop into the spinal cord in most vertebrates. The **pharyngeal pouches** will become the gills in aquatic vertebrates and various other organs in terrestrial vertebrates, such as the auditory tubes in humans. The **postanal tail** extends past the anus in the embryo and in some adult chordates.

Lancelets and sea squirts are the nonvertebrate chordates. Lancelets are the **cephalochordates** and are the only chordate to have all four characteristics in the adult stage. These marine organisms also show segmentation. Sea squirts, the **urochordates**, lack chordate characteristics (except gill slits) as adults, but they have a larva that could be ancestral to the vertebrates. They are bilaterally symmetric animals that live a sessile adult lifestyle on the ocean floor.

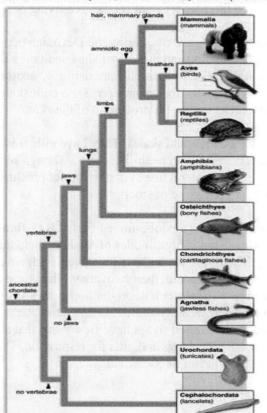

Phylogenetic tree of the chordates

29.2   The Vertebrates

Vertebrates have the four chordate characteristics as embryos. As adults, the notochord is replaced by the **vertebral column**, which is composed of individual vertebrae. Vertebrates undergo cephalization and have a skull that protects the brain, which has regions that are highly specialized for specific functions. Vertebrates also have an endoskeleton made of cartilage or bone that protects internal organs and serves as a place for muscle attachment. Paired appendages are characteristic of vertebrates, as are well-developed internal organs that are specialized and organized into organ systems.

Vertebrate evolution is marked by the evolution of vertebrae, jaws, a bony skeleton, lungs, limbs, and the amniotic egg. The earliest vertebrates were the fishes that lived an aquatic life. Amphibians were the first vertebrates to have limbs, but are not fully adapted to life on land like reptiles and some birds, which have a terrestrial lifestyle due to the use of an **amniotic egg** for reproduction. Most **placental mammals** also live on land because of the ability of the fertilized egg to develop within the body of the female.

> **Take Note:** *For the AP Biology exam, you should be able to explain how animals adapted to a terrestrial environment as they evolved. How do the ways that animals adapted compare to the ways that plants adapted to a terrestrial environment?*

29.3   The Fishes

The first vertebrates lacked jaws and paired appendages. They are represented today by the hagfishes and lampreys. They have a skeleton made of cartilage and have a cylindrical body with smooth, nonscaly skin. Hagfishes are scavengers, while many lampreys are filter feeders that can be free-living or parasitic.

Ancestral bony fishes, which had jaws and paired appendages, gave rise during the Devonian period to two groups: today's cartilaginous fishes (skates, rays, and sharks) and the bony fishes, including the ray-finned fishes and the lobe-finned fishes. These fishes are typified by four characteristics: 1) **ecothermy**, dependent on the environment for temperature regulation; 2) **gills**, for respiration; 3) an **endoskeleton** made of cartilage or bone, for support and protection; and 4) **scales**, for protection of the skin.

The **cartilaginous fishes** (Chrondrichthyes) include sharks, rays, and skates. They have gills slits on both sides of the pharynx, but lack a gill cover. Their skin is covered with teeth-like scales. Though some of these fishes are predators, others are filter feeders. They use a lateral line system to detect pressure changes in the environment around them and they have a keen sense of smell.

The **bony fishes** (Osteichthyes) include the ray-finned fishes and the lobe-finned fishes. **Ray-finned fishes** have fan-shaped fins supported by bony rays. They have various modes of feeding, including filter-feeding, opportunistic feeding, and predatory feeding. Their gills are protected by an external covering called the **operculum**, and they have **swim bladders** to help maintain their buoyancy. They have a single-looped closed circulatory system and a well-developed nervous system. Reproduction occurs sexually when males and females deposit their gametes into the water. The ray-finned fishes (actinopterygii) have become the most diverse group among the vertebrates. **Lobe-finned fishes** have fleshy fins that are supported by bones. Lungfishes are lobe-finned fishes that have lungs and gills for respiration. The lobe-finned fishes (sarcopterygii), are represented today by two species of coelacanths.

29.4   The Amphibians

Amphibians are **tetrapods**, meaning they have four limbs, and are represented primarily today by frogs and salamanders. They have a well-developed skeleton and smooth, nonscaly skin kept moist by mucus. Respiration occurs by lungs and across the surface of the skin (**cutaneous respiration**). Amphibians have a three-chambered heart with a closed double-looped circulatory system. They have specialized sense organs and are ectothermic. Most frogs and some salamanders return to the water to reproduce and then

metamorphose into terrestrial adults. Amphibians are thought to have evolved from the lobe-finned fishes.

29.5   The Reptiles

Reptiles (today's alligators and crocodiles, birds, turtles, tuataras, lizards, and snakes) lay a shelled **amniotic egg**, which allows them to reproduce on land. Inside the amniotic egg, the embryo is surrounded by three membranes: 1) the **chorion**, which is responsible for gas exchange, 2) the **allantois**, which removes wastes, and 3) the **amnion**, which acts as a shock absorber and prevents dessication. The egg also contains a **yolk sac**, which provides nourishment for the embryo.

Reptiles have paired limbs with five toes and a thick, dry skin with tails that prevents dessication. Respiration occurs with lungs, and the kidney excretes urine. Reptiles have a three- or four-chambered heart with a septum and are ectothermic. One main group of ancient reptiles, the stem reptiles that presumably evolved from amphibian ancestors, produced a line of descent that evolved into both dinosaurs and birds during the Mesozoic era. A different line of descent from stem reptiles evolved into mammals.

Birds are feathered, which helps them maintain a constant body temperature. Feathers (modified scales) also help birds fly; the tail feathers are used for steering. They are adapted for flight: their bones are hollow, their shape is compact, their breastbone is keeled, and they have well-developed sense organs. Their lungs are connected to air sacs that allow the one-way flow of air through the lungs and make them lighter for flight. Birds have a four-chambered heart and a double-looped circulatory system. Unlike their reptilian predecessors, birds are **endothermic**, meaning they use metabolic heat to maintain a constant body temperature. Birds reproduce sexually, laying eggs that are incubated externally.

29.6   The Mammals

Mammals remained small and insignificant while the dinosaurs existed, but when dinosaurs became extinct at the end of the Cretaceous period, mammals became the dominant land organisms.

Mammals are vertebrates with hair and **mammary glands**. Hair helps them maintain a constant body temperature, preventing heat loss, and aids in camouflage and sensory reception. The mammary glands allow them to feed and establish an immune system in their young. Mammals have an internal skeleton with a larger skull to accommodate the brain. Mammals breathe using lungs, and like birds have a four-chambered heart with a double-looped circulatory system. Mammals reproduce sexually. **Monotremes** lay eggs, while **marsupials** have a pouch in which the newborn crawls and continues to develop. The **placental mammals**, which are the most varied and numerous, retain offspring inside the uterus until birth.

Take Note: *For the AP Biology exam, you should be able to compare and contrast each group of vertebrates with respect to their modes of nutrition, nervous systems, respiratory systems, circulatory systems, excretory systems, and reproduction. What adaptations does each group have that allow them to survive in their environment?*

Multiple Choice Questions

1. Vertebrates do NOT possess
   A. gills or lungs to obtain oxygen from the environment.
   B. a high degree of cephalization and complex sense organs.
   C. a pseudocoelom and incomplete digestive tract.
   D. remnants of the notochord in the intervertebral disks.
   E. two pairs of appendages.

2. The three-chambered heart of amphibians and most reptiles
   A. contains two ventricles, to pump blood to the lung and the skin capillaries, and one atria to receive blood from the body.
   B. sends blood to the lungs and body via the ventricle and returns blood from the lungs and body via the atria.
   C. returns oxygen-poor blood from the body and the lungs to the ventricle.
   D. separates deoxygenated blood from oxygenated blood throughout the body.
   E. is a single-loop pathway, with the blood flowing from the ventricles to the lungs, directly to the body and returning to the atrium.

3. Lancelets exhibit which of the following features in addition to the four features of all chordates?
   A. segmentation
   B. limbs
   C. mammary glands
   D. bony skeleton
   E. ventral nerve cord

4. The evolution of amphibians from lobe-finned fishes occurred because
   A. the supply of food on land and the absence of predators promoted further adaptations to a terrestrial existence.
   B. lobe finned fishes had a disadvantage over others and could not move from pond to pond.
   C. amphibians needed to move on to land to find new food sources.
   D. as fishes experienced new phenomena during their lifetime they modified their behaviors and adapted to new environments.
   E. individual fishes evolved new adaptations for life on land, which led to the evolution of amphibians.

5. Ectothermic animals
   A. can maintain a warm body temperature by warming themselves in the sun.
   B. work to maintain temperatures matching that of the ambient temperature of the environment.
   C. often have less enzymes than homeothermic animals, to operate at lower temperatures.
   D. generate their own body heat.
   E. do not need to control their heat loss.

6. The amniotic egg in reptiles
   A. provides a fluid-filled sac in which the egg develops.
   B. keeps the sexes separate and promotes external fertilization.
   C. allows the passage of nutrients through a placenta.
   D. provides nutrients necessary for asexual reproduction.
   E. provides a protected internal environment for the embryo.

Questions 7–10 refer to the following five choices. For each question, select the choice that is most closely related. Each choice may be used once, more than once, or not at all.

    A. birds
    B. amphibians
    C. reptiles
    D. vertebrates
    E. fish

7. Modified respiration; lobed lungs connect to air sacs for continuous air flow

8. Cutaneous respiration across a moist membrane, aided by lungs

9. Scaly skin is impermeable to water; excretion of uric acid

10. Single-loop circulatory pathway

Free Response Question

Mammals include the largest animal and the largest animal ever to live. Given this diversity, there are several distinguishing characteristics for all mammals.

    A. **Describe** the adaptive advantage of three of these distinguishing characteristics.

    B. **Explain** the evidence to show that mammals share a remote ancestor with reptiles and how this adaptive radiation occurred.

Annotated Answer Key (MC)

1. **C**; Vertebrates are coelomate animals and possess a complete digestive system.

2. **B**; The three-chambered heart of the amphibian and that of most reptiles is a double-loop pathway with a limited amount of mixing of oxygenated and deoxygenated blood. The single ventricle is where this mixing occurs. Blood enters the right atrium from the body, passes to the ventricle and is pumped to the lung/skin capillaries and to the other capillaries. Blood from the lungs and skin capillaries returns to the heart via the left atrium (which is totally oxygenated) but then mixes in the ventricle with the blood returning from the body.

3. **A**; The four basic characteristics occurring in all chordates are: notochord, dorsal tubular nerve cord, pharyngeal pouches, and a postanal tail. In lancelets, the notochord extends from the tail to the head, and this accounts for their group name, the cephalochordates. In addition to the four features of all chordates, segmentation is present, as witnessed by the fact that the muscles are segmentally arranged and the dorsal tubular nerve cord has periodic branches.

4. **A**; There are two suggested hypotheses for this phenomenon: 1.) Lobe-finned fishes had an advantage over others because they could use their lobed fins to move from pond to pond. 2.) The supply of food on land and the absence of predators promoted further adaptations. Recent fossil evidence supports these theories.

5. **C**; Ectotherms work to maintain homeostasis by external means. They often have more enzymes (to work at multiple temperatures) and control their heat loss by physical changes in their location and behavior.

6. **D**; Amniotic eggs contain extra-embryonic membranes which protect the embryo, remove nitrogenous wastes, and provide the embryo with oxygen, food, and water.

7. **A**; The air sacs of birds permit a unidirectional flow of air through the lungs. It contains a higher oxygen content.

8. **B**; Amphibians breathe primarily through their skin, with assistance from the lungs.

9. **C**; Reptiles conserve water due to their impermeable skin and excreting dry uric acid.

10. **E**; Fish have a single-loop pathway with blood traveling from the ventricle to the gills to the body and then to the atrium, returning to the ventricle.

Answer to FRQ

PART A (MAX 6)
Two points each characteristic: one for identification, one for explanation

- Hair: insulation against heat loss, camouflage, ornamental, sensory functions
- Mammary glands: nurse young, ensures parental care, provides antibodies
- Skeleton: larger brain due to increased skull size, teeth differentiation, arched vertebral column
- Internal organs: efficient respiratory and circulatory systems for oxygen demands, ready supply of oxygen to muscles that produce heat as endotherms, double-loop circulatory pathway, four-chambered heart, conservation of water by kidneys, highly developed nervous system, special senses
- Internal development: development in the uterus for shelter and allows the mother to move during gestation

PART B (MAX 6)
Two for each group

Monotremes
- Egg-laying mammals
- Hard-shelled amniotic eggs
- Cloaca

Marsupials
- Internal development with pouch/mammary glands
- Born very immature

Placental
- Organ of exchange between maternal blood and fetal blood
- Long period of dependency on parents: increased frontal lobe of the brain

CHAPTER 30
HUMAN EVOLUTION

30.1   Evolution of Primates

Primates, in contrast to other types of mammals, are adapted for an **arboreal** (living in trees) life. The evolution of primates is characterized by trends toward mobile limbs; grasping hands; a flattened face; stereoscopic vision; a large, complex brain; and a reduced reproductive rate. These traits are particularly useful for living in trees. The grasping hands of mammals have five digits, including a thumb, which is **opposable** in Old World monkeys, great apes, and humans. Along with mobile forelimbs, grasping hands allow primates to grasp tree limbs for ease of locomotion and food for ease of feeding. **Stereoscopic (three-dimensional) vision** allows primates to accurately judge the distance and position of tree limbs. A flattened face indicates a decreased reliance on smell as the primary sense and an increased reliance on vision. Primates show cephalization, allowing the brain to specialize for processing sensory information (important for eye-hand coordination) and communication. A reduced reproductive rate is associated with increased sexual maturity and gestational periods and longer life spans. Movement through trees is made easier with fewer offspring to carry and care for.

The term **hominin** is now used for humans and their closely related, but extinct, relatives. A hominin is a member of the group **hominines** that also includes the chimpanzee and gorilla, which are the apes most closely related to hominins on the basis of molecular data. These groups were preceded in evolutionary time by the anthropoids and prosimians, which include the lemurs and the tarsiers.

Proconsul is a transitional link between monkeys and the **hominoids**, which include the gibbons, organutans, and the hominines.

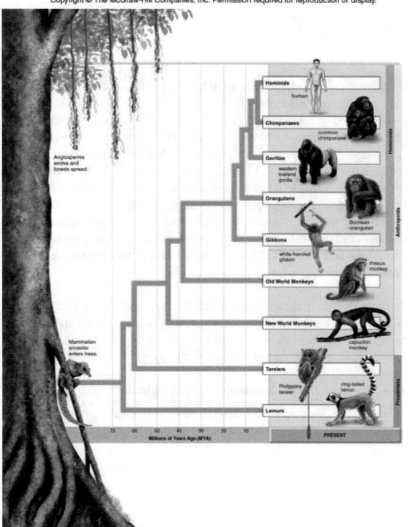

Evolution of primates

Take Note: *For the AP Biology exam, you should be able to describe the characteristics that make primates different from other vertebrates. What evidence points to the fact that primates evolved from a common ancestor?*

30.2   Evolution of Early Hominins

Fossil and molecular data tell us we shared a common ancestor with certain of the apes (chimpanzees and gorillas) until about 7 mya. The split between their lineage and human lineage occurred around this time.

Hominins walk erect, and this causes our anatomy to differ from the other hominines. In humans, the S-shaped spinal cord curves and exits from the center of the skull, rather than from its rear. The human pelvis is broader and more bowl-shaped to place the weight of the body over the legs. Humans are **bipedal**, using only the longer, heavier lower limbs for walking; in apes, all four limbs are used for walking, and the upper limbs are longer than the lower limbs.

To be a hominin, a fossil must have an anatomy suitable to standing erect. Perhaps bipedalism developed when hominins stood on branches to reach fruit overhead, and then they continued to use this stance when foraging among bushes. An upright posture reduces exposure of the body to the sun's rays, and leaves the hands free to carry food, perhaps as a gift to receptive females.

Several early hominin fossils, such as *Sahelanthropus tchadensis,* have been dated around the time of a shared ancestor for apes and humans (7 mya). The ardipithecines appeared about 4.5 mya. All the early hominins have a chimp-sized braincase but are believed to have walked erect.

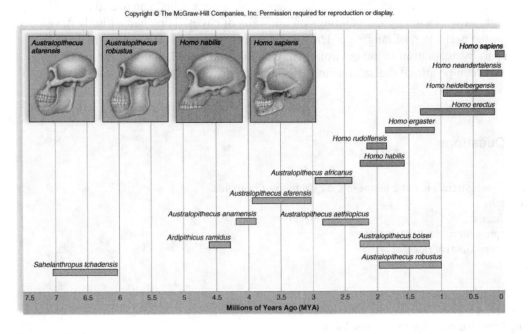

Human evolution

30.3 Evolution of Later Hominins

It is possible that an **australopith** (4 mya–1 mya) is a direct ancestor for humans. These hominins walked upright and had a brain size of 370–515 cc. In southern Africa, hominins classified as australopiths include *Australopithecus africanus,* a gracile form, and *Paranthropus robustus,* a robust form. In eastern Africa, hominins classified as australopiths include, *A. afarensis* (Lucy), a gracile form, and also robust forms. Many of the australopiths coexisted, and the species *A. garhi* is the probable ancestor to the genus *Homo.*

30.4 Evolution of Early *Homo*

Early *Homo,* such as *Homo habilis* (handyman) and *Homo rudolfensis,* dated around 2 mya, is characterized by a brain size of at least 600 cc, a jaw with teeth that resembled those of modern humans, and the use of tools. They had an organized social structure, but were scavengers rather than hunters.

Homo ergaster and *Homo erectus* (1.9–0.3 mya) had a striding gait, made well-fashioned tools, and could control fire. *Homo ergaster* migrated into Asia and Europe from Africa between 2 and 1 mya, and had a larger brain and flatter face than *Homo habilis. Homo ergaster* was taller than the any of the previously mentioned hominins. *Homo erectus* evolved in Asia and gave rise to *H. floresiensis,* which used tools and fire, but had a brain that was one-third the size of that of modern humans.

30.5 Evolution of Later *Homo*

The **replacement model** or **out-of-Africa hypothesis** of human evolution says that modern humans originated only in Africa and, after migrating into Europe and Asia, replaced the archaic *Homo* species found there. This hypothesis is supported by the fossil record and DNA data.

The **Neandertals**, a group of archaic humans, lived in Europe and Asia. Their chinless faces, squat frames, and heavy muscles are apparently adaptations to the cold. They were culturally advanced, living in caves or constructed dwellings, and they manufactured and used tools for hunting and food preparation. **Cro-Magnon** is a name often given to modern humans. Their tools were sophisticated, and they definitely had a culture, as witnessed by the paintings on the walls of caves. They had a highly developed brain that allowed for the perfection of language. The human ethnic groups of today differ in ways that can be explained in part by adaptation to the environment. Genetic studies tell us that there are more genetic differences between people of the same ethnic group than between ethnic groups. We are one species.

Multiple Choice Questions

1. Which of the following trends is not a factor in the evolution of primates?
    A. mobile limbs
    B. grasping hands
    C. stereoscopic vision
    D. increased reproductive rate
    E. flattened face

2. The human skeletal system is modified by
    A. supporting quadripedal locomotion with angled knee.
    B. walking long distances due to an arched foot.
    C. the spine exiting from the center of the skull to allow the body to remain upright.
    D. allowing swaying while walking due to narrow hip and pelvic joints.
    E. the femur being larger at the bottom to support large weights.

3. Bipedalism was NOT associated with the need to
    A. climb trees to escape predators.
    B. carry food.
    C. collect overhead fruit.
    D. travel from woodland to woodland.
    E. carry helpless infants from place to place.

4. In an effort to regulate body temperature, animals in colder regions of their range may have
    A. shorter fingers and toes.
    B. larger ears.
    C. taller than average height.
    D. longer arms and legs.
    E. longer noses.

5. Which of the following is NOT true regarding genetic evidence for a common ancestry?
    A. Mitochondrial DNA was used to determine differences.
    B. Genotypes of different modern populations are similar.
    C. More genetic variation occurs within ethnic groups than among them.
    D. Results are consistent with their having a common ancestor 10 million years ago.
    E. Genetic variation is large between Caucasians, but not between Asians or Africans.

6. Anatomical differences among ethnic groups such as hair texture, a fold on the upper eyelid, or the shape of lips are due to
    A. environmental adaptations.
    B. genetic drift.
    C. natural selection.
    D. large population size.
    E. geographical differences.

Questions 7–10 refer to the following five choices. For each question, select the choice that is most closely related. Each choice may be used once, more than once, or not at all.
    A. *Homo habilis*
    B. Australopithecines
    C. *Homo erectus*
    D. *Homo floresiensis*
    E. *Proconsul*

7. Quadripedal lifestyle; flat vertebral column; short forelimbs

8. Great differences in anatomy (gracile/robust); small brain; walked erect

9. First people to use tools; omnivorous; small height

10. May have been the first to cook meat; largest brain

Free Response Question

Most researchers believe that modern humans evolved from *Homo ergaster* but they differ as to the details.

    A. Compare and contrast the replacement model and the out-of-Africa hypothesis.

    B. Explain the following characteristics of the Cro-Magnons and discuss how these advancements were possible:
        *   tool use
        *   language

Annotated Answer Key (MC)

1. **D**; The trend in primate evolution is a general reduction in the rate of reproduction associated with increased age of sexual maturity and extended life plans. This allows for long gestation periods and time for forebrain development.

2. **B**; Humans have an arched foot to walk long distances and run with less chance of injury.

3. **A**; A quadripedal existence would be better for climbing trees.

4. **A**; Animals in colder regions of their range have shorter limbs, digits, and ears. Both of these effects help regulate body temperature by increasing the surface area to volume ratio in hot climates and decreasing the ratio in cold climates.

5. **D**; Mitochondrial DNA results are consistent with a common ancestor about 1 million years ago.

6. **B**; These features became fixed in different populations due simply to genetic drift. These changes cannot be explained by adaptations to the environment. The effects of genetic drift are strongest in small populations. There is no evidence to support the fact that these are environmental adaptations.

7. **E**; *Proconsul* is a transitional link between the monkeys and the hominoids. Fossil evidence suggests that the species didn't have a monkey tail, but its limb proportions suggest it was quadripedal and a tree dweller.

8. **B**; The australopithecines are a group of hominins that evolved and diversified in Africa. They give mosaic evidence for evolution, meaning that different body parts changed at different rates.

9. **A**; *Homo habilis* were very socially organized and were scavengers. Their brain was larger than that of australopithecines but not as large as *Homo erectus*.

10. **C**; This extensive population movement is a first in the history of humankind and a tribute to the intellectual and physical skills of these peoples.

Answer to FRQ

PART A (MAX 6)
Replacement Model
- Supported by the fossil record
- The earliest remains are at 130,000 years bp; until older fossils are found, this hypothesis is supported.
- Mitochondrial DNA of Africans is more diverse than those of Europeans.
- Mitochondrial DNA has a constant rate of mutations.
- Africans should show the greatest diversity since human existence has been the longest in Africa.

Out of Africa hypothesis
- Modern humans arose from archaic humans in the same manner in Africa, Asia, and Europe.
- Genetic continuity will be found between modern and archaic populations.

PART B (MAX 6)
Three per part

TOOL USE
Mechanism
- Larger brain
- Free hands
- Opposable thumbs

Characteristics
- Compound tools
- Wooden handles
- Knifelike blades
- Weapons thrown in the air
- Increased sophistication

LANGUAGE
Mechanism
- Larger brain

Characteristics
- Cooperation
- Social organization
- Sculptures
- Religious significance
- Think symbolically
- Spoken language

CHAPTER 31
ANIMAL ORGANIZATION AND HOMEOSTASIS

> **Take Note:** *Chapter 1 addressed the hierarchy of organization. This chapter covers the organization of animals as well as the maintenance of homeostasis. In the coming chapters, each organ system will be discussed, so be sure to have a solid understanding of the information in this chapter before moving on.*

31.1   Types of Tissues

All organisms start out as a single cell—a fertilized egg, or zygote. During development, the zygote divides to produce cells that become tissues composed of similar cells specialized for a particular function. Tissues make up organs, organs make up organ systems, and organ systems make up the organism. This sequence describes the levels of organization within an organism. Cells were studied in depth in Chapter 4. It may help you review

There are four types of tissues in the body:

1. Epithelial tissue
2. Connective tissue
3. Muscular tissue
4. Nervous tissue

Epithelial tissue is made of tightly packed cells and functions to cover the body and line cavities. The cells of epithelial tissue are connected to each other by one of the three types of cell junctions discussed in Chapter 5: tight junctions, adhesion junctions, or gap junctions. Epithelial cells are typically attached to a basement membrane that anchors them to the extracellular matrix. Epithelial tissue can be classified as being simple, stratified, or pseudostratified. **Simple** epithelium consists of a single layer of cells, whereas **stratified** epithelium consists of several layers of cells. **Pseudostratified** epithelium appears to be layered, but in fact, all of the cells are in contact with the **basement membrane**. Additionally, the cells of epithelial tissue can have a variety of shapes: squamous, cuboidal, and columnar. **Squamous** cells are flat, with no defined shape; **cuboidal** and **columnar** cells are shaped like cubes and columns, respectively. Epithelial tissue can also be glandular or have modifications, such as cilia. Epithelial tissue protects, absorbs, secretes, and excretes.

The most abundant tissue in the body is **connective tissue**. Connective tissue has a generous extracellular matrix between its cells. Loose fibrous connective tissue and dense fibrous connective tissue contain fibroblasts and fibers. Loose fibrous connective tissue has both collagen and elastic fibers. Dense fibrous connective tissue, like that of tendons and ligaments, contains closely packed collagen fibers. In adipose tissue, the cells enlarge and store fat. Both cartilage and bone have cells within lacunae, but the matrix for cartilage is more flexible than that for bone, which contains calcium salts. In bone, the lacunae lie in concentric circles within an osteon (or Haversian system) about a central canal. Blood is a connective tissue in which the matrix is a liquid called plasma.

There are three types of **muscular tissue**: skeletal muscle, smooth muscle, and cardiac muscle. In humans, **skeletal** muscle is attached to bone. It is striated in appearance and voluntary. **Smooth** muscle is found in the wall of internal organs, and is nonstriated and involuntary. **Cardiac** muscle makes up the heart. It is striated and involuntary.

Nervous tissue has two types of cells, **neurons** and **neuroglia**. Nervous tissue is responsible for communication between different parts of the body. The neurons are the cells that send the signals, while the neuroglia support and nourish the neurons. The majority of neurons have dendrites, a cell body, and an axon. The brain and spinal cord contain complete neurons, while nerves contain only axons. Axons are specialized to conduct nerve impulses.

31.2   Organs and Organ Systems

Organs contain two or more tissues that work together to perform a common function. The integumentary system—the skin and its appendages—is the largest organ of the body. All animals have skin, though its characteristics vary among the different phyla. For example, fish and reptiles have scales, while amphibians have smooth skin. The skin protects the internal organs from trauma, invading pathogens, and water loss. Human skin has two major layers, the epidermis (stratified squamous epithelium), and the dermis (fibrous connective tissue containing sensory receptors, hair follicles, blood vessels, and nerves). The subcutaneous layer, also known as the hypodermis, is composed of loose connective tissue and is found between the skin and either muscles or bone. Human skin has several accessory organs, which include the nails, hair, and sweat glands.

Organ systems contain several organs. The organ systems of humans have specific functions and carry out the life processes that are common to all organisms. The organ systems of the body and their functions are as follows:

Life Processes	Human Systems
Coordinate body activities	Nervous system Endocrine system
Acquire materials and energy (food)	Skeletal system Muscular system Digestive system
Maintain body shape	Skeletal system Muscular system
Exchange gases	Respiratory system
Transport materials	Cardiovascular system
Excrete wastes	Urinary system
Protect the body from disease	Lymphatic system Immune system
Produce offspring	Reproductive system

The human body has two main cavities. The dorsal cavity contains the brain and spinal cord. The ventral cavity is divided into the thoracic cavity (heart and lungs) and the abdominal cavity (most other internal organs).

31.3   Homeostasis

Homeostasis is the dynamic equilibrium of the internal environment (blood and tissue fluid). It is important that an organism maintain homeostasis, because the external environment is constantly changing. If the internal environment also changed, it would be difficult for organisms to survive. So, organisms must maintain a constant internal environment in order to survive in a changing environment. This means that all of the body's characteristics, such as temperature, pH, and blood glucose levels, need to be maintained within a narrow range.

The two organ systems that are primarily responsible for controlling homeostasis are the endocrine system and the nervous system, though all organ systems contribute to homeostasis. Special contributions are made by the liver, which keeps the blood glucose constant, and the kidneys, which regulate the pH levels of the blood. The nervous and endocrine systems regulate the other body systems. Both systems are controlled by negative feedback mechanisms, which result in slight fluctuations above and below desired levels.

Negative feedback mechanisms reverse changes that occur in the body. For instance if the pH of the blood increases above a certain level, negative feedback will bring it back down. Conversely, if the pH of the blood decreases, negative feedback will restore it to acceptable levels. Let's look at an example of a thermostat to understand how negative feedback works:

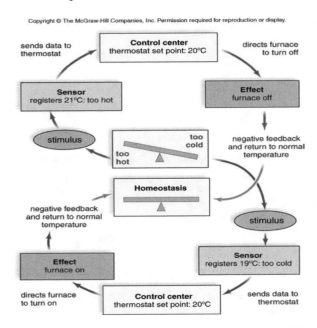

Regulation of room temperature

Notice that when room temperature exceeds the set point, a control center recognizes the change and sends a signal through the system to turn the furnace off, allowing the temperature to return to the original set point. On the other hand, when the temperature is too low, the control center sends a signal through the system to turn the furnace on, causing the room to heat up and return to the original set point. The body works in the same way. The control center is the brain, and when changes occur in the body, it sends signals through the nervous and/or the endocrine system to reverse the change and restore the set point. Body temperature, for example, is regulated by a center in the hypothalamus that maintains human temperatures at 98.6 °C.

Positive feedback mechanisms are not as numerous in the body as negative feedback, yet they are still important. Positive feedback mechanisms amplify a change that has occurred in the body. In positive feedback, a control center (i.e., the brain) recognizes a change and causes that change to increase in the same direction. An example of this is seen in childbirth. Positive feedback helps the mother give birth to the baby by causing her contractions to occur with greater frequency and intensity.

Multiple Choice Questions

Questions 1–4 refer to the following five choices. For each question, select the choice that is most closely related. Each choice may be used once, more than once, or not at all.

    A. epithelial tissue
    B. blood
    C. nervous tissue
    D. muscular tissue
    E. bones

1. Responsible for communication between different parts of the body

2. Found inside the walls of internal organs

3. Can be glandular and contain cilia

4. Can be striated and voluntary

5. While a person consumes a hearty Thanksgiving dinner,
    A. blood glucose levels decrease.
    B. glucagon is secreted.
    C. insulin is secreted by the pancreas.
    D. glycogen is converted to glucose.
    E. bile converts glucose into glycogen.

6. Which of the following effectors is incorrectly matched with a response in a negative feedback loop?
    A. salivary glands – secretion of saliva
    B. smooth muscle in blood vessels – dilation to carry blood to skin
    C. adrenal glands – secretion of insulin
    D. skeletal muscles – shivering or uncontrolled contractions
    E. uterine contractions – labor

7. Negative feedback loops
    A. are associated with instability in a system.
    B. Sst in motion a chain of events that intensify changes.
    C. keep the internal environment at homeostasis.
    D. decrease body temperatures to avoid hypothermia.
    E. trigger responses in the uterus to initiate child birth.

8. Which of the following physiological responses is an example of positive feedback?
    A. temperature regulation of a dog on a hot day
    B. blood clotting after injury
    C. a turtle basking on a rock
    D. sodium ions pumped into a nerve cell
    E. oxygen diffusing into a fish's gills

9. Tight junctions are present in
    A. nervous tissue at the synapse – release of neurotransmitters.
    B. cardiac muscle of the heart – conduction of impulses.
    C. epithelial tissue of the skin – prevention of blisters.
    D. skeletal muscle – prevention of muscle tears and damage.
    E. epithelial tissue of the nephron – conservation of water.

10. Transitional epithelia that can contract and expand are found in the
    A. heart.
    B. liver.
    C. urinary bladder.
    D. lungs.
    E. skeletal muscle.

Free Response Question

Animals utilize a diverse array of mechanisms to maintain body heat.

   A. For **each** the following, **describe** how the organism maintains or generates body heat:
      * reptiles
      * mammals

   B. **Describe** how either birds **OR** fish utilize a counter-current exchange mechanism to preserve body heat.

Annotated Answer Key (MC)

1. **C**

2. **D**

3. **A**

4. **D**

   For Questions 1–4: Nervous tissue is responsible for communication and is generally found in the brain, spinal cord, and peripheral nerves. Muscular tissue is diverse (skeletal, smooth, and cardiac) and is found inside the walls of internal organs (smooth) and can be striated and voluntary (skeletal). Epithelial tissue can take on many structures adapted for many functions and can be glandular and contain cilia.

5. **C**; During Thanksgiving dinner, blood glucose levels will increase. This will cause a secretion of insulin (not glucagon) by the pancreas. Shortly after dinner, the conversion of glucose to glycogen will occur. When blood glucose levels drop long after dinner, glucagon will stimulate the conversion of glycogen to glucose. This is an example of a negative feedback loop to maintain homeostasis.

6. **C**: Insulin is secreted by the pancreas, not the adrenal glands.

7. **C**; Negative feedback loops are associated with stability and homeostasis. They actually counteract, not intensify changes, as in positive feedback loops.

8. **B**; Positive feedback mechanisms intensify an action, as with blood clotting. The mechanism will continue until the platelet plug and fibrin threads are complete.

9. **E**; Tight junctions function to prevent leaking of solutes from one cell to another (stomach cells, parts of the kidney). Nervous tissue at the synapse and the cardiac muscle cells in the heart contain gap junctions and the epithelial tissue of the skin contains desmosomes.

10. **C**; Transitional epithelia are unique in that they can contract and expand as needed in the urinary bladder.

Answer to FRQ

Part A (MAX 8)
Four points each

- Reptiles
  - o Ectothermic
  - o Behavorial modifications – move to warmer areas (basking)
  - o Darken skin
  - o Flatten bodies
  - o Open mouth/pant
  - o Shuttle between warm and cold locations
  - o Underground
  - o Torpor
  - o Gain/lose heat through (2 pts)
    - ▪ Convection
    - ▪ Conduction
    - ▪ Panting
    - ▪ Radiation
- Mammals
  - o Endothermic
  - o Physiological mechanisms (3 pts internal max)
    - ▪ Vasoconstriction/vasodilation
    - ▪ Shivering
    - ▪ Metabolic heat
    - ▪ Hair standing up/laying down
    - ▪ Elaboration of nervous response, temperature receptors (hypothalamus)
  - o Behavorial mechanisms
    - ▪ Move to warmer/cooler areas
    - ▪ Putting on clothes/taking of clothes
    - ▪ Huddle together
    - ▪ Burrow

Part B (MAX 4)
- Location of mechanism
  - o Birds: mechanism is in legs
  - o Fish: mechanism is between body core and periphery
- Examples – other correct examples are permissible
  - o Birds: penguins standing on ice
  - o Fish: tunas diving into deep colder waters
- Purpose same in each: to preserve body heat, allows the organism to diffuse more than 50% of body heat back to blood returning to the body
- Anatomy: arteries carrying blood away from the body lie adjacent to veins carrying blood back to the body

CHAPTER 32
CIRCULATION AND CARDIOVASCULAR SYSTEMS

> **Take Note:** *To successfully prepare for the AP Biology exam, you should understand how the circulatory system functions in the maintenance of the overall homeostasis of an animal.*

32.1   Transport in Invertebrates

Some invertebrates do not have a transport system to circulate oxygen and nutrients throughout their bodies. The presence of a thin body wall and a **gastrovascular cavity** allows diffusion alone to supply the needs of cells in sponges, cnidarians, and flatworms. Roundworms make use of their pseudocoelom in the same way that echinoderms use their coelom to circulate materials. They each use the coelomic fluid in their body cavity for circulation.

Other invertebrates do have a transport system—either an open circulatory system or a closed circulatory system. An **open circulatory system** consists of a heart, blood vessels, and open spaces. The circulatory fluid is called **hemolymph**, which consists of blood and tissue fluid. The heart pumps the hemolymph through the blood vessels into the open spaces of the body (the hemocoel), where the organs obtain their cellular needs as the hemolymph washes over them. A **closed circulatory system** consists of a heart and blood vessels. The circulatory fluid is **blood**, which travels in one direction through the body due to valves in the blood vessels. As blood flows through smaller vessels called capillaries, exchange between the blood and the tissue occurs, allowing the cells to obtain oxygen and nutrients, and to remove wastes. Insects have an open circulatory system, and earthworms have a closed one.

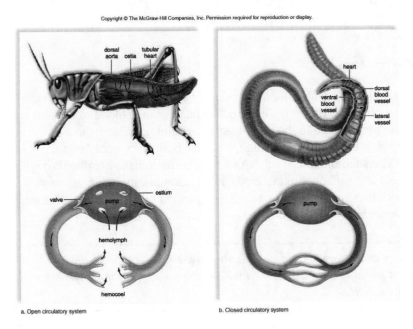

Copyright © The McGraw-Hill Companies, Inc. Permission required for reproduction or display.

a. Open circulatory system

b. Closed circulatory system

Open versus closed circulatory systems

32.2   Transport in Vertebrates

Vertebrates have a closed circulatory system in which **arteries** carry blood away from the heart to **capillaries**, where exchange takes place, and **veins** carry blood back to the heart. Arteries and veins are thick blood vessels that consist of layers of connective tissue, smooth muscle, and epithelial tissue. Arteries branch into smaller vessels called **arterioles**, which branch into capillaries. Capillary beds consist of interconnected capillaries that are very numerous, so that every cell in the body is within close proximity of a capillary. **Sphincters** control whether capillary beds are open; when they are, blood flows through and exchange can occur. When blood leaves a capillary bed, it flows through **venules**, small vessels that converge into veins. Veins have valves to ensure the one-way flow of blood back to the heart.

Vertebrates have either a one-circuit (single-loop) or two-circuit (double-loop) circulatory system. Fishes have a **one-circuit** circulatory pathway because the heart, with a single atrium and ventricle, pumps blood to the gills, where gas exchange occurs. Blood then flows through the dorsal aorta, which distributes blood through the rest of the body. Blood returns to the atrium via an enlarged chamber called the **sinus venosus**.

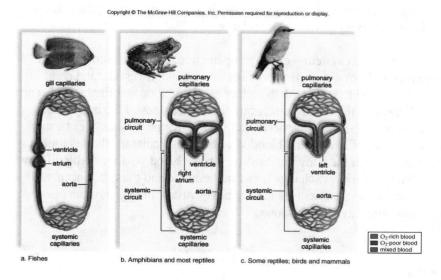

Comparison of circulatory circuits in vertebrates

The other vertebrates have a **two-circuit** (double-loop) circulatory system consisting of both a pulmonary and a systemic circuit. In the **pulmonary circuit**, the heart pumps blood to the lungs, while in the **systemic circuit**, it pumps blood to the body tissues. Amphibians have three-chambered hearts that have two atria but a single ventricle. Because both atria empty into the ventricle, oxygenated and deoxygenated blood can mix. As a result, amphibians breathe through their skin as well as through their lungs. Crocodilians, birds, and mammals, including humans, have a four-chambered heart with two atria and two ventricles separated by a **septum**; therefore, O_2-rich blood is always separate from O_2-poor blood.

> **Take Note:** *For the AP Biology exam, you should be able to compare and contrast the circulatory systems of different phyla of animals. How does an animal's circulatory system enable it to survive in its environment?*

32.3 Transport in Humans

The human heart is a cone-shaped organ consisting mostly of cardiac muscle called **myocardium**. It has four chambers, the left and right **atria**, and the left and right **ventricles**. The left and right sides of the heart are separated by a muscular wall called the **septum**. Between the atria and ventricles are **atrioventricular valves** that ensure the one-way flow of blood through the heart. **Semilunar valves** are located between the ventricles and their corresponding arteries.

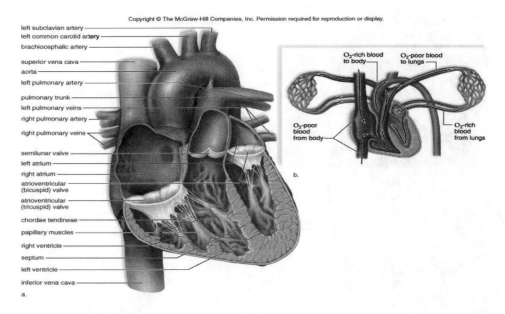

Internal view of the heart

Blood flows through the heart in the following manner: Vena cavae → right atrium → right ventricle → pulmonary arteries → lungs → pulmonary veins → left atrium → left ventricle → aorta → body. Blood that enters the right atrium is deoxygenated, while blood that enters the left atrium is oxygenated. Oxygenated and deoxygenated blood never come in contact due to the presence of the septum.

The heartbeat in humans begins when the **SA (sinoatrial) node** (the cardiac pacemaker) causes the two atria to contract, and blood moves through the atrioventricular valves to the two ventricles. The SA node also stimulates the **AV (atrioventricular) node**, which in turn causes the two ventricles to contract. Ventricular contraction sends blood through the semilunar valves to the pulmonary trunk and the aorta. Now all chambers rest. The heart sounds, **lub-dub**, are caused by the closing of the valves. Contraction of the heart is called **systole**, while **diastole** is the relaxation of the heart.

In the pulmonary circuit, blood can be traced to and from the lungs. Deoxygenated blood leaves the heart and goes to the lungs, where gas exchange occurs. Oxygenated blood returns to the heart. In the systemic circuit, the aorta divides into blood vessels that serve the body's cells. The venae cavae return O_2-poor blood to the heart.

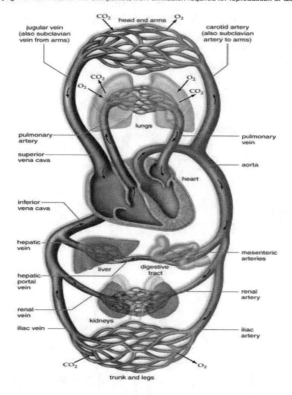

Path of blood

Blood pressure created by the beat of the heart accounts for the flow of blood in the arteries, but skeletal muscle contraction is largely responsible for the flow of blood in the veins, which have valves that prevent a backward flow.

Hypertension and atherosclerosis are two circulatory disorders that lead to heart attack and stroke. Following a heart-healthy diet, getting regular exercise, maintaining a proper weight, and not smoking cigarettes are protective against the development of these conditions.

> **Take Note:** *For the AP Biology exam, you should be able to describe how oxygen and nutrients are transported through the body. What adaptations of the cardiovascular system facilitate the transport and absorption of oxygen and nutrients?*

32.4   Blood, a Transport Medium

Blood plays an important role in homeostasis. It has two main parts: **plasma** and **formed elements**. Plasma contains mostly water (90–92%) and proteins (7–8%) but it also contains nutrients and wastes. The proteins and salts in plasma act as buffers to maintain the pH of the blood at approximately 7.4. They also assist in maintaining the osmolarity of blood so the body can retain water. The proteins in plasma have a variety of functions such as blood clotting, immune response, and the transport of substances throughout the body.

There are three types of formed elements: red blood cells (**erythrocytes**), white blood cells (**leukocytes**), and platelets. The red blood cells do not have nuclei, but contain **hemoglobin** and function in oxygen transport. This is described in detail in Chapter 35. An insufficiency of red blood cells or hemoglobin leads to anemia. Erythrocytes are manufactured in the red bone marrow, which is stimulated by the hormone erythropoietin produced by the kidneys.

Defense against disease depends on the various types of leukocytes, or white blood cells. There are several types of white blood cells, including neutrophils, monocytes, lymphocytes, eosinophils, and basophils. **Neutrophils** and **monocytes** are phagocytic and are very active during the inflammatory reaction. **Lymphocytes** are involved in the development of immunity to disease. **Eosinophils** are important in allergic reactions and parasitic infections, and **basophils** contain the anticoagulant heparin, which prevents blood from clotting too quickly.

The **platelets** and two plasma proteins (prothrombin and fibrinogen) function in blood clotting, an enzymatic process that results in the formation of fibrin threads. Blood clotting is a complex process that includes three major events: (1) platelets and injured tissue release prothrombin activator, which (2) enzymatically changes prothrombin to thrombin, an enzyme; (3) thrombin causes fibrinogen to be converted to fibrin threads. The fibrin threads then block the area of damage, resulting in a blood clot.

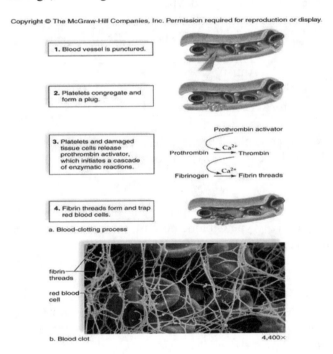

Blood clotting

When blood reaches a capillary, water moves out and into the **tissue fluid** at the arterial end due to blood pressure. At the venous end, water moves in due to osmotic pressure. In between, where the blood pressure and osmotic pressure cancel each other, oxygen and nutrients diffuse out of, and wastes diffuse into, the capillary. Because the red blood cells and plasma proteins are so large, they remain in the capillary.

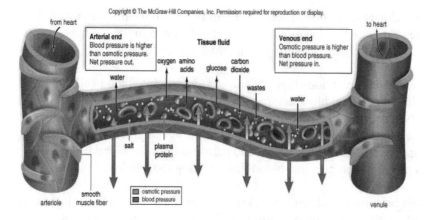

Copyright © The McGraw-Hill Companies, Inc. Permission required for reproduction or display.

Capillary exchange

In the ABO system there are two possible antigens on the red blood cells (A and B) and two possible antibodies in the plasma, which are anti-A and anti-B. The various combinations of these antigens and antibodies in a person's blood determine their blood type.

Antigen in Red Blood Cells	Plasma Antibodies	Blood Type
A	Anti-B	A
B	Anti-A	B
Both A and B	Neither Anti-A nor Anti B	AB
Neither A nor B	Both Anti-A and Anti-B	O

Type A blood cannot be given to a person with type B blood because the recipient's blood contains anti-A antibodies and **agglutination** will occur. Likewise, a person with Type A blood cannot be given type B blood.

Persons with Type AB blood can receive any blood type because there are no antibodies to cause agglutination. On the other hand, anyone can receive Type O blood because there are no antigens on the red blood cells to react with the antibodies in the plasma.

Take Note: *Homeostasis is an important concept for the AP Biology exam. As you study the next few chapters, identify the ways that each organ system contributes to the maintenance of homeostasis. Give clear examples of negative or positive feedback mechanisms in each organ system. Be sure to relate the cellular mechanisms you studied earlier to the functioning of the entire organ system you are studying.*

Multiple Choice Questions

Questions 1–4 refer to the following five choices. For each question, select the choice that is most closely related. Each choice may be used once, more than once, or not at all.

   A. frog (amphibian)
   B. jellyfish (cnidarian)
   C. crocodile (reptile)
   D. zebra (mammal)
   E. tuna (fish)

1. Use diffusion and a gastrovascular activity for oxygen exchange

2. One circuit circulatory pathway

3. Possess a three-chambered heart with two atria and one ventricle

4. Blood returns to the atrium via an enlarged chamber called the sinus venosus.

5. Which of the following pathways correctly describes blood flow in a mammal?
   A. right ventricle → pulmonary arteries → lungs → pulmonary veins → left atrium → left ventricle → aorta → body
   B. aorta → pulmonary arteries → lungs → pulmonary veins → right atrium → right ventricle → vena cava → body
   C. pulmonary veins → left atrium → left ventricle → vena cava → right atrium → right ventricle → lungs → pulmonary veins → body
   D. vena cava → right atrium → right ventricle → lungs → pulmonary arteries → left ventricle → left atrium → aorta → body
   E. pulmonary arteries → lungs → right atrium → pulmonary veins → right ventricle → vena cava → left atrium → left ventricle

6. Lymphocytes
   A. are involved in the development of immunity.
   B. are important in allergic reactions.
   C. prevent blood from clotting too quickly.
   D. are active in the inflammatory response.
   E. convert fibrinogen to fibrin.

7. Blood pressure is NOT maintained by
   A. skeletal muscle contraction in the legs.
   B. contraction of the ventricles of the heart.
   C. resistance of the blood vessels to blood flow.
   D. valves in veins to prevent backflow.
   E. decrease in insulin secretion.

8. The sinoatrial node
   A. causes contraction of the ventricles.
   B. is located at the apex of the heart.
   C. causes all chambers to rest.
   D. stimulates the atrioventricular node.
   E. sends blood through the semilunar valves.

9. An individual with type A blood
   A. cannot give blood to a person with B blood because the recipient's blood contains anti-A antibodies.
   B. can give blood to a type B recipient because there are no anti-A antibodies.
   C. can receive blood from type A and type AB donors.
   D. cannot receive blood from a person with O blood because it has anti-A antibodies.
   E. cannot give blood to anyone other than type A recipients.

10. A student is told that he should collect an organism containing hemolymph. Which of the below organisms would satisfy this requirement?
    A. starfish
    B. bird
    C. sponge
    D. insect
    E. earthworm

Free Response Question

An athlete is working hard while running. There are numerous physiological differences under these conditions compared to sitting on a couch watching television.

    A. **Trace** the path of a red blood cell from the lungs to a leg muscle and back to the heart.

    B. **Describe** how and why the pH of the blood is different while exercising and how this aids the athlete.

    C. **Explain** where carbon dioxide is produced in the athlete's body and how it is carried back to the lungs.

Annotated Answer Key (MC)

1. **B**

2. **E**

3. **A**

4. **E**

For Questions 1–4: Fish have a one-circuit circulatory pathway (heart, gills, body, heart) whereas the other organisms utilize a two-circuit pathway (heart, lungs, heart, body). They also return blood to the atrium via an enlarged chamber called the sinus venosus and have a two-chambered heart. Amphibians have a three-chambered heart and birds, reptiles and mammals have a four-chambered heart

5. **A**; Blood flows through the heart in the following manner: vena cavae ➜ right atrium ➜ right ventricle ➜ pulmonary arteries ➜ lungs ➜ pulmonary veins ➜ left atrium ➜ left ventricle ➜ aorta ➜ body. Blood that enters the right atrium is deoxygenated, while blood that enters the left atrium is oxygenated. Oxygenated and deoxygenated blood never comes in contact due to the presence of the septum.

6. **A**; Eosinophils are important in allergic reaction response, basophils prevent blood from clotting too quickly, and neutrophils are active in the inflammatory response. Within this response, thrombin converts fibrinogen to fibrin.

7. **E**; Insulin functions to regulate blood glucose levels, not blood pressure. A–D all function in maintaining blood pressure.

8. **D**; The SA node is found in the upper dorsal wall of the right atrium and initiates the heart beat, causing the atria to contract.

9. **A**

Antigen in Red Blood Cells	Plasma Antibodies	Blood Type
A	Anti-B	A
B	Anti-A	B
Both A and B	Neither Anti-A nor Anti B	AB
Neither A nor B	Both Anti-A and Anti-B	O

Type A blood cannot be given to a person with type B blood because the recipient's blood contains anti-A antibodies and **agglutination** will occur. Likewise, a person with Type A blood cannot be given type B blood. Persons with Type AB blood can receive any blood type because there are no antibodies to cause agglutination. On the other hand, anyone can receive Type O blood because there are no antigens on the red blood cells to react with the antibodies in the plasma.

10. **D**; Hemolymph is found in animals with open circulatory systems. Starfish have a water vascular system, birds and earthworms have closed systems, and sponges do not have a circulatory system.

Answer to FRQ

PART A (MAX 4)

- Lungs back to heart, pulmonary vein to left atrium to left ventricle
- Left ventricle to aorta to capillary bed in muscle
- Muscle to vein to vena cava to right atrium
- Elaboration for specific names of arteries like the femoral artery
- Elaboration: dual circulation discussed, pulmonary vs. systemic and its benefit

PART B (MAX 4)

- Description of Bohr effect
- Exercise releases CO_2 in large quantity
- CO_2 dissolves in blood stream as carbonic acid and bicarbonate ions
- Point for showing complete reaction of CO_2 dissolving into water :
    - $H_2O + CO_2 <-> H_2CO_3 <-> HCO_3^- + H^+$
    - Must show as equilibrium with double arrows (reversibility)
- Discussion of above reaction as an equilibrium reaction
- Discussion of H_2CO_3 (or bicarbonate ion) as weak acid and partially disassociating

PART C (MAX 4)

- CO_2 produced by cellular respiration
- Site of production is the mitochondria in cells
- Diffuses out of cells into blood stream, constant production means concentration gradient is always maintained
- Most carried as bicarbonate ion in blood, some also as carbonate ion
- Some binds to hemoglobin and carried in the RBC

CHAPTER 33
LYMPH TRANSPORT AND IMMUNITY

> **Take Note:** *In order to successfully prepare for the AP Biology exam, you should understand how the lymphatic system functions in the maintenance of the overall homeostasis of humans.*

33.1 The Lymphatic System

The lymphatic system consists of **lymphatic vessels** and organs. The lymphatic vessels (1) receive glycerol and fatty acids packaged as lipoproteins at intestinal villi, (2) receive excess tissue fluid collected by lymphatic capillaries, and (3) carry these to the bloodstream. The fluid that flows through the lymphatic vessels is called lymph. Lymph is cleansed of pathogens in lymph nodes, and blood is cleansed of pathogens in the spleen.

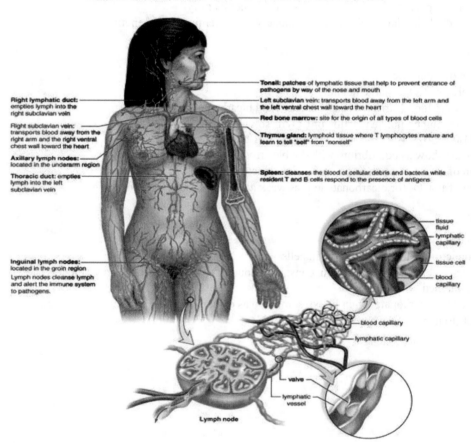

Lymphatic system

Lymphocytes, the white blood cells responsible for immunity, are produced and accumulated in the lymphatic organs (lymph nodes, tonsils, spleen, thymus gland, and red bone marrow). **T lymphocytes** mature in the thymus, while **B lymphocytes** mature in the red bone marrow where all blood cells are produced. White blood cells such as these lymphocytes are necessary for nonspecific and specific defenses.

33.2 Nonspecific Defense Against Disease

Immunity involves nonspecific and specific defenses. **Nonspecific defenses** do not require recognition of pathogens, and include barriers to entry, the inflammatory response, phagocytes and natural killer cells, and protective proteins.

Skin and mucus membranes are barriers to entry, as are secretions from oil glands and the hydrochloric acid produced in the stomach. They simply prevent foreign invaders from entering the body.

The **inflammatory response** occurs in response to tissue damage. Signs that the inflammatory response (shown in the accompanying figure) is active are redness, heat, swelling, and pain.

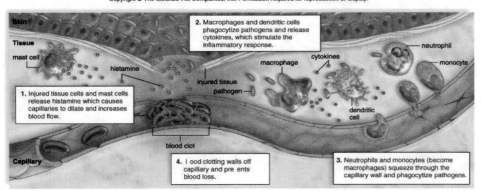

Inflammatory response

Phagocytes such as **macrophages**, found in the blood, and **dendritic cells**, found on the surface of the skin, engulf pathogens. The pathogens are destroyed by the hydrolytic enzymes in the lysosomes of the phagocytes, which then stimulate lymphocytes to initiate specific immunity. **Natural killer** (NK) cells are stimulated by dendritic cells and function to destroy cancer cells and cells infected by viruses.

Protective proteins include complement proteins and interferons. **Complement proteins** destroy pathogens in three ways: 1) by amplifying the inflammatory response, 2) binding to pathogens coated with antibodies, and 3) forming attack complexes in viruses or bacteria that cause lysis. **Interferons** are proteins released by virus-infected cells that prevent neighboring cells from also becoming infected.

33.3 Specific Defense Against Disease

Specific defenses require B lymphocytes and T lymphocytes, also called B cells and T cells, and involve recognition of specific antigens. B cells are responsible for **antibody-mediated immunity**. After their B-cell receptor combines with a specific antigen, they undergo **clonal selection**, producing **plasma cells** and **memory B cells**. Plasma cells secrete **antibodies** into body fluids; thus, antibody-mediated immunity is also called **humoral immunity**. Some progeny of activated B cells become memory B cells, which remain in the body and produce antibodies if the same antigen enters the body at a later date.

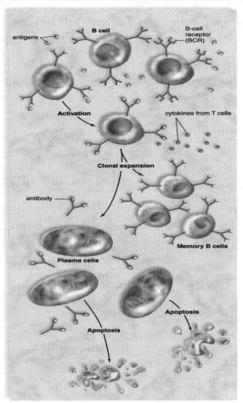

Clonal selection model as it applies to B cells

Active (long-lived) immunity occurs when a person produces antibodies in response to an illness or the administration of vaccines when they are well and in no immediate danger of contracting an infectious disease. Passive immunity is needed when an individual is in immediate danger of succumbing to an infectious disease. Passive immunity is short-lived because prepared antibodies are administered to and not made by the individual.

An antibody, or immunoglobulin, is usually a Y-shaped protein that has two antigen binding sites. Each arm of the antibody has a heavy chain and a light chain, each with a constant region and a variable region. Monoclonal antibodies, which are produced by the same plasma cell, have various functions, from detecting infections to treating cancer.

T cells have T-cell receptors and are responsible for **cell-mediated immunity**. For a T cell to recognize an antigen, the antigen must be presented by an **antigen-presenting cell (APC)**, a macrophage, along with an **MHC (major histocompatibility complex) protein**. Thereafter, the activated T cell undergoes clonal expansion until the illness has been stemmed. Then, most of the activated T cells undergo apoptosis. A few cells remain, however, as memory T cells which provide protection against a later exposure to the same antigen.

The two main types of T cells are cytotoxic T cells and helper T cells. **Cytotoxic T cells** kill virus-infected or cancer cells on contact by releasing **perforin** proteins and **granzymes**. **Helper T cells** produce cytokines and stimulate other immune cells such as NK cells, macrophages, B cells, and other T cells. Helper T cells are the primary target of human immunodeficiency virus (HIV), which causes AIDS.

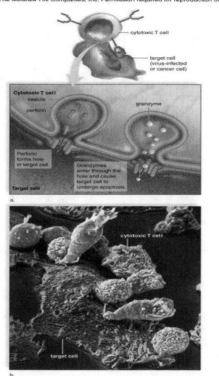

Cell mediated immunity

Take Note: *In preparation for the AP Biology exam you should be able to describe how the immune system provides nonspecific and specific responses to invading pathogens. You should also be able to explain how the body distinguishes self from nonself and how a later exposure to a pathogen differs from the initial exposure.*

33.4 Immunity Side Effects

Immune side effects include **tissue rejection** and autoimmune diseases. Tissue rejection occurs when the body recognizes self from nonself, causing antibodies and cytotoxic T cells to destroy foreign tissues. This can be controlled by using drugs to suppress the immune system and transplanting an organ from a donor that has the same MHC antigens as the recipient.

Autoimmune diseases occur when the body fails to recognize self from nonself and cytotoxic T cells attack the body's own cells. Examples of such diseases include rheumatoid arthritis, myasthenia gravis, and lupus.

Allergic responses occur when the immune system reacts vigorously to allergens, substances not normally recognized as foreign. Immediate allergic responses, usually consisting of cold-like symptoms, are due to the activity of IgE antibodies, which cause mast cells to release histamine. Delayed allergic responses, such as contact dermatitis, are due to the activity of T cells.

Multiple Choice Questions

1. Natural killer cells have the potential to attack many different cells because they
   A. recognize cells bound to MHC markers.
   B. do not kill on contact.
   C. recognize cells lacking MHC markers.
   D. have no need for an activating signal.
   E. contain interferons.

2. Which of the following is NOT a lymphatic organ?
   A. tonsils
   B. spleen
   C. thymus gland
   D. appendix
   E. red bone marrow

3. Which of the following will NOT occur shortly after a splinter punctures the skin?
   A. mast cells release histamine
   B. vasoconstriction of the blood vessels at the site of injury
   C. complement proteins attract phagocytes to the infected area
   D. neutrophils engulf the bacteria
   E. inflammation and redness at the site of injury

4. When a B cell encounters an antigen, it
   A. produces antibodies.
   B. stimulates the release of T cells.
   C. produces more antigen.
   D. stimulates the release of interferon.
   E. triggers the nonspecific defense mechanism.

5. For a T cell to recognize an antigen,
   A. natural killer cells must label the antigen with a marker.
   B. the antigen must be presented by a macrophage and an MHC protein.
   C. interferon must secrete the appropriate antibodies.
   D. memory B cells must undergo apoptosis.
   E. cytotoxic T cells must secrete perforin.

6. Delayed allergic responses are due to the activity of
   A. T cells.
   B. B cells.
   C. interferon.
   D. macrophages.
   E. interleukins.

Questions 7–10 refer to the following five choices. For each question, select the choice that is most closely related. Each choice may be used once, more than once, or not at all.
   A. B lymphocytes
   B. cytotoxic T cells
   C. macrophages
   D. dendritic cells
   E. natural killer cells

7. Kill cancer cells by releasing perforin

8. Undergo clonal selection to produce memory B cells

9. Found in the blood and engulf pathogens

10. Stimulated by dendritic cells

Free Response Question

The immune system functions at many different levels to protect the body.

A. **Describe** and give an **example** for each of how the immune system provides nonspecific and specific responses to invading pathogen.

B. **Explain** how the body distinguishes self from non-self.

Annotated Answer Key (MC)

1. **C**; Natural killer cells (NK) cells arise from stem cells in the bone marrow and appear to be a type of lumphocyte, but not identical to B or T lymphocytes. They can immediately kill tumor and virus infected cells and do not require MHC-antigen markers as do other types of cells. They originally were thought to have not required an activation signal, thus the term "natural."

2. **D**; Your pituitary gland is not a lymphatic organ. Lymphocytes are housed in many locations of the respiratory, digestive, and reproductive systems including the tonsils, spleen, thymus gland, and bone marrow.

3. **B**; Vasodilation of the blood vessels at the site of injury will actually occur along with histamine release, complement protein release, neutrophils engulfing the bacteria, and inflammation and redness.

4. **A**; When a B cell encounters an antigen, a cascade of events occurs. Once the B cell encounters the antigen and receives an activation signal from a helper T cell, it will differentiate into a plasma B cell or a memory T cell. Antibodies are produced by the B cell.

5. **B**; After digesting a pathogen, macrophages present the antigens to the correct helper T cells by attaching it to an MHC marker and integrating it into the cell membrane to ensure that the macrophage is not recognized as a pathogen.

6. **A**; Hypersensitivity to an antigen (allergy) causes specific types of T cells to interact with B cells and produce the antibody IgE. This antibody binds to mast cells and basophils, which are integral in the inflammatory response. This is the acute phase. However, as other white blood cells (neutrophils, lymphocytes, eosinophils, and macrophages) reach the initial site there can be a late phase response as well.

7. **B**

8. **A**

9. **C**

10. **E**

For Questions 7–10: Natural killer cells are stimulated by dendritic cells. Cytotoxic T cells kill cancer cells and other diseased cells, by specifically secreting the protein perforin. B lymphocytes can produce memory B cells or plasma B cells. Macrophages phagocytize pathogens and present them to lymphocytes to respond to the particular pathogen.

Answer to FRQ

PART A (MAX 6)

Nonspecific Defenses

One point for definition, one point for mechanism, one point for example (MAX 3)

- First line of defense, surface barriers (include mechanical, chemical, and biological barriers; not antigen specific; no memory)
- Time: always present
    - Protect against infection (i.e., antibacterials, lysozyme, saliva)
    - Mechanically eject pathogens (i.e., cough, sneeze)
    - Flushing (i.e., tears, urine)
    - Compete with pathogenic bacteria for food and space
    - Changing conditions of the environment (i.e., HCl, vaginal secretions, proteases)
    - Examples can include: waxy cuticle of leaves, exoskeleton, skin, tears
- Can include macrophages, neutrophils, basophils, eosinophils

Specific Defenses

One point for definition, one point for mechanism, one point for example (MAX 3)

- Second line of defense (protects against re-exposure, antigen-specific, immunological memory)
- Time: reaction rate varies with exposure
    - Humoral immunity (includes complement, coagulation, interferons, lysozyme)
    - Adaptive immunity (includes lymphocytes, killer T cells, helper T cells, B lymphocytes)

PART B (MAX 6)

- Antigens trigger an immune response.
- Antigens could be viruses, bacteria, fungi, or parasites, or a product of them.
- Must be a surface marker for identification
- Specific marker on an antigen is called an epitope (can have many epitopes).
- Antibodies read the surface proteins.
- MHC (major histocompatibility complex) markers are coded for by genes.
- MHC markers determine which antigens can be identified.

CHAPTER 34
DIGESTIVE SYSTEMS AND NUTRITION

> **Take Note:** *To successfully prepare for the AP Biology exam, you should understand how the digestive system functions in the maintenance of the overall homeostasis of an animal.*

34.1   Digestive Tracts

Most animals have a digestive tract that is responsible for the ingestion, digestion, and absorption of nutrients as well as the elimination of undigested material from the body. Some animals have an **incomplete digestive tract** that has only one opening, such as a mouth, and little specialization of parts. Planarians are one example of animals with incomplete digestive tracts. They bring food into their bodies via the pharynx, which extends out of the mouth. Undigested material leaves through the mouth. Other animals, such as earthworms, have a **complete digestive tract** with both a mouth and an anus. A complete tract tends to have specialized organs, such as a pharynx or gizzard, that have unique functions in the digestive process. Food enters via the mouth, and undigested material leaves through the anus.

Some animals are **continuous feeders**, meaning that the animal obtains food as the water continually flows through its body. Clams, for instance, are filter feeders that obtain food in this manner. Other animals are **discontinuous feeders** (e.g., squid) that need a storage area such as a crop or a stomach for food.

Animals have adaptations that correspond to their diet. Echinoderms use their tube feet to open the shells of the bivalves it eats. Most mammals have teeth, and the structure of the teeth (i.e., **dentition**) differs depending on diet. Herbivores need teeth that can clip off plant material and grind it up, so they have sharp incisors, reduced canines, and large, flat molars. Also, the stomach of an herbivore contains bacteria that can digest cellulose. Carnivores, on the other hand, need teeth that can tear and rip meat into pieces. They have pointed incisors, large canines, and jagged molars. Meat is easier to digest than plant material, so the digestive system of carnivores has fewer specialized regions and the intestine is shorter than that of herbivores. Because omnivores eat a variety of food, they exhibit a variety of tooth structures and digestive system adaptations.

> **Take Note:** *For the AP Biology exam, you should be able to compare and contrast the digestive systems of different phyla of animals.*

34.2   Human Digestive Tract

Humans have a complete digestive tract that begins with a mouth and ends with an anus. The digestive tract is a long tube—the **alimentary canal**—containing several organs, including the **mouth, pharynx, esophagus, stomach, small intestine, large intestine, rectum** and **anus**. The digestive system also has accessory organs: the **salivary glands, liver, gall bladder,** and **pancreas**. These organs are called **accessory organs** because they do not directly participate in the digestive process, but supply enzymes and other secretions that are important for digestion. There are two kinds of digestion. The first is **mechanical digestion**, which is the physical breakdown of food to make it smaller and easier for the second type, chemical digestion. **Chemical digestion** is the enzymatic breakdown of nutrients into their monomers so that they can be absorbed into the blood and transported to the cells of the body, where they are used for various cellular processes.

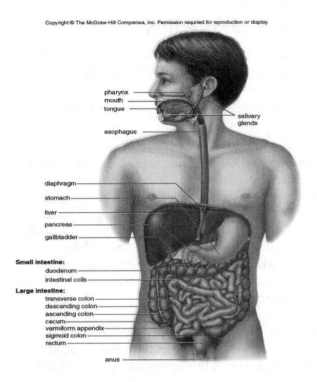

pharynx
mouth
tongue
salivary glands
esophagus

diaphragm
stomach
liver
pancreas
gallbladder

Small intestine:
duodenum
intestinal coils

Large intestine:
transverse colon
descending colon
ascending colon
cecum
vermiform appendix
sigmoid colon
rectum

anus

The human digestive tract

In the human digestive tract, food is chewed and manipulated in the mouth, where salivary glands secrete **saliva**. Chewing the food makes it smaller, increasing the surface area so that it is easier to digest later in the digestive system. Saliva contains **salivary amylase**, which begins carbohydrate digestion by breaking down starch into the disaccharide maltose. Once the food is chewed and mixed with saliva, it is formed into a **bolus**, which is swallowed.

The bolus passes to the pharynx and down the esophagus to the stomach. The esophagus is a long tube that has layers of smooth muscle. These smooth muscles contract to move the bolus down to the stomach. This is called **peristalsis**. When the bolus enters the stomach, the gastroesophogeal sphincter closes to prevent it from coming back up into the esophagus. Heartburn occurs because of acid reflex, when some of the stomach contents escape the stomach and come into the esophagus.

The stomach, a thick-walled organ in the shape of a J, stores and mixes food with mucus and gastric juices to produce **chyme**. The wall of the stomach has many folds that increase the surface area of the stomach wall for the secretion of HCl and pepsin. The pH of the stomach is very acidic because of HCl, whose role it is to kill any bacteria that may be in the bolus. Additionally, the high acidity of the stomach inhibits the salivary amylase in the bolus. **Pepsin** is the enzyme that begins protein digestion in the stomach; polypeptides are broken down into smaller peptides, which will be further digested in the small intestine.

The small intestine is where digestion is completed and absorption of nutrients occurs. The **duodenum**—the first 25 cm of the small intestine—receives bile from the liver and pancreatic juice from the pancreas. **Bile**, which is made in the liver and stored in the gall bladder, emulsifies fat and readies it for digestion by lipases produced by the pancreas. The pancreas also produces amylases and peptidases that digest starch and protein, respectively. Lipases, proteases, and nucleases are also secreted by the cells of the small intestine. The intestinal enzymes finish the process of chemical digestion.

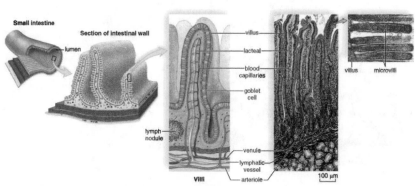

Anatomy of the small intestine

The walls of the small intestine have finger-like projections called **villi** where small nutrient molecules are absorbed. The cells of the villi have folded membranes called **microvilli**. The purpose of the villi and microvilli is to increase the surface area of the small intestine so that the absorption of nutrients into the blood can be more efficient. The nutrients must pass through the epithelial cells of the small intestine in order to get into the blood. Amino acids and glucose enter the blood vessels of a villus. Glycerol and fatty acids are joined and packaged as lipoproteins before entering lymphatic vessels called **lacteals** in a villus.

The large intestine consists of the cecum, the colon, and the rectum, and ends at the anus. The large intestine does not produce digestive enzymes; rather, it absorbs water, salts, and some vitamins. Reduced water absorption results in diarrhea. The intake of water and fiber help prevent constipation. Any undigested material will leave the body via the anus as feces.

Three accessory organs of digestion—the pancreas, liver, and gallbladder—send secretions to the duodenum via ducts. The pancreas produces pancreatic juice, which contains digestive enzymes for carbohydrates, protein, and fat. These enzymes are carbohydrases, peptidases, and lipases, respectively. The pancreas also secretes bicarbonate into the duodenum, which neutralizes the acidic chyme from the stomach. The liver produces bile, which is stored in the gallbladder. When the chyme arrives in the small intestine, the gall bladder releases bile to the duodenum through the common bile duct. The liver receives blood from the small intestine by way of the hepatic portal vein. It has numerous important functions, and any malfunction of the liver is a matter of considerable concern.

34.3   Digestive Enzymes

The basic nutrients that humans ingest are carbohydrates, lipids, and proteins. Digestive enzymes are present in digestive juices and break down these nutrients into glucose, fatty acids and glycerol, and amino acids. Salivary amylase and pancreatic amylase are responsible for the digestion of starch. Salivary amylase begins starch digestion in the mouth, and pancreatic amylase completes starch digestion in the small intestine. Carbohydrate digestion does not occur in the stomach. Pepsin and trypsin digest proteins to peptides in the stomach and duodenum, respectively. Following emulsification by bile in the duodenum, lipase digests fat to glycerol and fatty acids. Intestinal enzymes finish the digestion of starch and protein. Each digestive enzyme is present in a particular part of the digestive tract. Salivary amylase functions in the mouth; pepsin functions in the stomach; trypsin, lipase, and pancreatic amylase occur in the intestine along with the various enzymes that digest disaccharides and peptides.

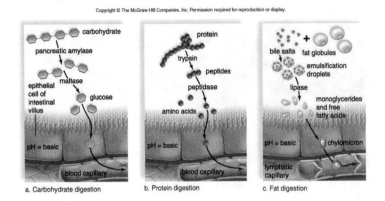

Digestion and absorption of nutrients.

> **Take Note:** *For the AP Biology exam, you should be able to describe how each nutrient ingested by humans is digested and absorbed. What adaptations of the digestive system facilitate the digestion and absorption of nutrients?*

34.4   Nutrition

Carbohydrates from fruits, vegetables, milk, and honey are necessary in the diet. Simple sugars and refined starches are not helpful because they provide calories but no fiber, vitamins, or minerals. Proteins supply us with essential amino acids, but it is wise to avoid meats that are fatty because fats from animal sources are saturated. While unsaturated fatty acids, particularly the omega-3 fatty acids, are protective against cardiovascular disease, the saturated fatty acids lead to plaque, which blocks blood vessels. Obesity is to be avoided particularly because it is now known to be associated with the development of type 2 diabetes and cardiovascular disease.

Multiple Choice Questions

1.   Which of the following characteristics is NOT seen in organisms with a two-way digestive tract?
    A.   Specialized enzymes are secreted in specific regions of the digestive system.
    B.   The surface area for absorption is increased by an intestinal fold called the typhlosole.
    C.   Different regions of the gut have specialized functions.
    D.   The digestive system is composed of a tube with a mouth and an anus.
    E.   Undigested remains are pushed back out the mouth.

2.   Which of the following adaptations would you expect to see in an herbivore?
    A.   sharp incisors
    B.   large, flat premolars
    C.   shorter digestive tract
    D.   few molars
    E.   incisors, canines, and molars

3.   Hydrochloric acid in the stomach does all of the following EXCEPT
    A.   kills bacteria and other microorganisms.
    B.   stops the action of salivary amylase in carbohydrate metabolism.
    C.   promotes the activity of pepsin for protein catabolism.
    D.   reduces the pH of the stomach to approximately 2.0.
    E.   produces mucus to protect the wall of the stomach.

4. In the mammalian digestive system, lacteals
   A. aid in the absorption of fats.
   B. are projections on the surface of villi in the small intestine.
   C. emulsify fats in the small intestine.
   D. secrete amylase to metabolize carbohydrates.
   E. are found in the gastric pits and secrete pepsin.

Questions 5–8 refer to the following five choices. For each question, select the choice that is most closely related. Each choice may be used once, more than once, or not at all.
   A. liver
   B. duodenum
   C. mouth
   D. pancreas
   E. stomach

5. Secretes bile to emulsify fats

6. Site of lipid digestion

7. Site of urea production

8. Secretes bicarbonate to neutralize acid chyme

9. Which of the following events will decrease the production of pepsin in the stomach?
   A. secretion of gastrin
   B. ingestion of food
   C. secretion of pepsin
   D. secretion of HCl
   E. secretion of bile

10. Large quantities in your diet of which of the following are most likely to cause increased risk of cardiovascular disease?
    A. monounsaturated fats
    B. omega-3 fatty acids
    C. unsaturated fats
    D. trans fatty acids
    E. polyunsaturated fats

Free Response Question

As a piece of bread is eaten, the nutrients need to be broken down and absorbed before they can be used.

   A. Trace the carbohydrates and **describe** what happens to them as they move from the mouth to the blood stream and then to the liver.

   B. **Explain** how the human body responds to a large quantity of glucose being ingested.

Annotated Answer Key (MC)

1. **E**; Organisms with a two-way digestive tract have an advantage over those with only one opening. Undigested remains are pushed back in those with only one opening, limiting the specificity of digestion, increasing the time for digestion, and decreasing the efficiency of nutrient absorption.

2. **B**; Herbivores eat primarily plant material; therefore, large, flat premolars are able to grind the plant material and begin the mechanical breakdown of cellulose. They also have a long digestive tract to increase the efficiency of plant material breakdown. Carnivores often have sharp incisors for ripping and tearing of meat materials; omnivores would have all types of teeth to accommodate their diverse diets.

3. **E**; Hydrochloric acid (HCl) kills bacteria and microorganisms that may make it through the nose and mouth, inhibits salivary amylase (which works best at a pH of 7), stimulates the cleaving of pepsinogen to produce pepsin for protein digestion, and reduces the pH of the stomach to 2, an optimum temperature for pepsin activity. HCl actually erodes the walls of the stomach; the mucus is not produced by HCl.

4. **A**; Lacteals are lymphatic capillaries on the walls of the intestine that aid in fat absorption. Microvilli are projects on the surface of villi, bile emulsifies fats, and saliva contains amylase to metabolize carbohydrates.

5. **A**; Bile is produced by the liver and secreted by the gall bladder and liver into the intestine.

6. **B**; Lipid digestion takes place in the duodenum of the small intestine.

7. **A**; Urea is produced in the liver and secreted by the kidney.

8. **D**; One of the secretions of the pancreas is bicarbonate, to neutralize the acid chyme of the stomach upon entry to the duodenum.

9. **E**; As bile is secreted into the small intestine, and chyme leaves, the cleaving of pepsinogen (by HCl and pepsin) to produce pepsin will cease. This ceases the positive feedback loop.

10. **D**; Trans fatty acids are likely to cause atherosclerosis more so than unsaturated fatty acids.

Answer to FRQ

PART A (MAX 6)
- Mouth: mechanical and chemical digestion
- Mechanical digestion (teeth grinding, stomach squeezing) increases surface area.
- Stomach: mechanical and chemical digestion
- SI: mostly chemical digestion
- Absorption in the small intestine across the brush border
- Elaboration point, description of a specific molecule during digestion as it passes across the brush border
- All capillaries lead from the capillary bed in SI to a second capillary bed in liver
- Liver processes nutrients, absorbs and stores extra nutrients
- Elaboration of idea that this is the only place in the body the blood passes through two capillary beds before returning to heart

PART B (MAX 5)
- Homeostatic mechanism: negative feedback
- Glucose is absorbed into blood and pancreas releases insulin.
- Insulin stimulates glucose uptake by cells.
- Cells convert glucose to glucose-6-phosphate.
- This traps glucose in cytoplasm where it can be used for Glycolysis.
- Elaboration for sensory mechanism in brain of high glucose levels

CHAPTER 35
RESPIRATORY SYTEMS

Take Note: *In order to successfully prepare for the AP Biology exam, you should understand the role of the respiratory system in the maintenance of the overall homeostasis of an animal.*

35.1   Gas Exchange Surfaces

The exchange of gases between the cells of an organism and the environment is called respiration, and it occurs by diffusion. In order for respiration to be efficient, the surface or membrane across which exchange occurs must be thin, moist, and have a large surface area. The circulatory system plays an important role in gas exchange because the abundance of capillaries increases the rate of gas transport to and from body tissues. This allows oxygen to reach the mitochondria much faster, so that it can be used in cellular respiration for the production of ATP.

Because they live in water, some aquatic animals, such as hydras and planarians, use their entire body surface for gas exchange because of its large surface area. Most animals, however, have a specialized gas exchange area. Though they are terrestrial, earthworms can also conduct gas exchange across their body surfaces because capillaries are close to the surface. Large aquatic animals usually pass water through **gills** in order to respire. Gills are very thin outgrowths of the pharynx or body surface, with capillaries flowing through them. In bony fishes, the blood in the capillaries flows in the direction opposite that of the water. Blood takes up almost all of the oxygen in the water as a result of this **countercurrent flow** because there is always a higher concentration of oxygen in water than there is in blood.

On land, insects use tracheal systems, and vertebrates have lungs. In insects, air enters the **tracheae** at openings called **spiracles**. From there, the air moves to ever smaller **tracheoles** until it reaches the cells, where gas exchange takes place. In vertebrates, **lungs** are found inside the body, where water loss is reduced and the membranes of the respiratory surface can remain moist. To ventilate the lungs, some vertebrates use positive pressure, but most inhale, using muscular contraction to produce a negative pressure that causes air to rush into the lungs. When the breathing muscles relax, air is exhaled. Birds have a series of **air sacs** attached to the lungs. When a bird inhales, air enters the posterior air sacs, and when a bird exhales, air moves though the lungs to the anterior air sacs before exiting the respiratory tract. The one-way flow of air through the lungs allows more fresh air to be present in the lungs with each breath, and this leads to greater uptake of oxygen from one breath of air.

The respiratory system of humans consists of the **nasal cavities, pharynx, trachea, bronchi,** and **lungs**. Within the lungs, bronchi branch into smaller tubes called **bronchioles** which end in air sacs called **alveoli** that are surrounded by capillaries. As air enters the respiratory system of humans, it is filtered and warmed as it travels toward the lungs. The epithelial cells of the trachea release mucus and are lined with cilia. The mucus traps any bacteria and debris in inhaled air, and as the cilia beat the mucus is moved out of the respiratory tract. Gas exchange occurs when oxygen and carbon dioxide diffuse across the membranes of the alveoli and the capillaries.

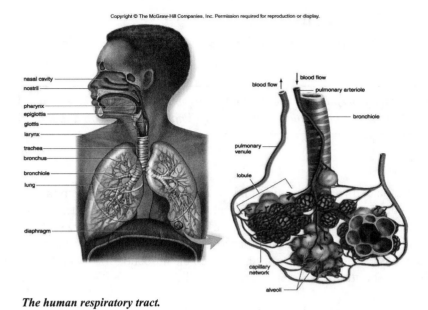

The human respiratory tract.

Take Note: *For the AP Biology exam, you should be able to compare and contrast the respiratory systems of different phyla of animals.*

35.2 Breathing and Transport of Gases

There are three steps in respiration: 1) ventilation (inhalation and exhalation), 2) external respiration (gas exchange between the air and the blood), and 3) internal respiration (gas exchange between the blood and the tissues). During inspiration, air enters the body at nasal cavities and then passes from the pharynx through the glottis, larynx, trachea, bronchi, and bronchioles to the alveoli of the lungs, where exchange occurs. During expiration, air passes in the opposite direction. Humans breathe by negative pressure, as do other mammals. During inspiration, the rib cage goes up and out, and the **diaphragm** lowers. The lungs expand and air comes rushing in. During expiration, the rib cage goes down and in, and the diaphragm rises. Therefore, air rushes out.

The rate of breathing is controlled by the **respiratory center** located in the medulla oblongata of the brain. The respiratory center sends signals to the diaphragm and the intercostals muscles that control the rate at which they contract, directly affecting the rate of breathing. The rate of breathing is influenced when chemoreceptors, such as the aortic and carotid bodies, detect that the amount of H$^+$ and carbon dioxide in the blood has changed. A decrease in pH causes the rate and depth of breathing to increase, while an increase in pH does the opposite.

Gas exchange in the lungs and tissues occurs by diffusion. At the alveoli, oxygen diffuses into the blood, where it binds to **hemoglobin**. Hemoglobin transports oxygen in the blood to the body tissues then releases it to diffuse into the cells, where it can be used for cellular respiration. Hemoglobin is a protein that consists of four polypeptide chains. Each chain has an iron-containing **heme** group that carries oxygen. Oxygen easily diffuses from hemoglobin to the tissues because the partial pressure of oxygen is lower in the tissues than it is in the blood.

Though a small amount is carried on hemoglobin, carbon dioxide is mainly transported in plasma as the **bicarbonate ion**. When carbon dioxide diffuses from the tissues into the blood, it is combined with water to form carbonic acid by the enzyme **carbonic anhydrase**. Carbonic acid then dissociates into H+ and bicarbonate. Excess hydrogen ions are transported by hemoglobin in the red blood cell while the bicarbonate ion is transported in the plasma. Once the blood reaches the lungs, the reverse reaction occurs and carbon dioxide diffuses out of the blood because there is a lower partial pressure of carbon dioxide in the air than in the blood.

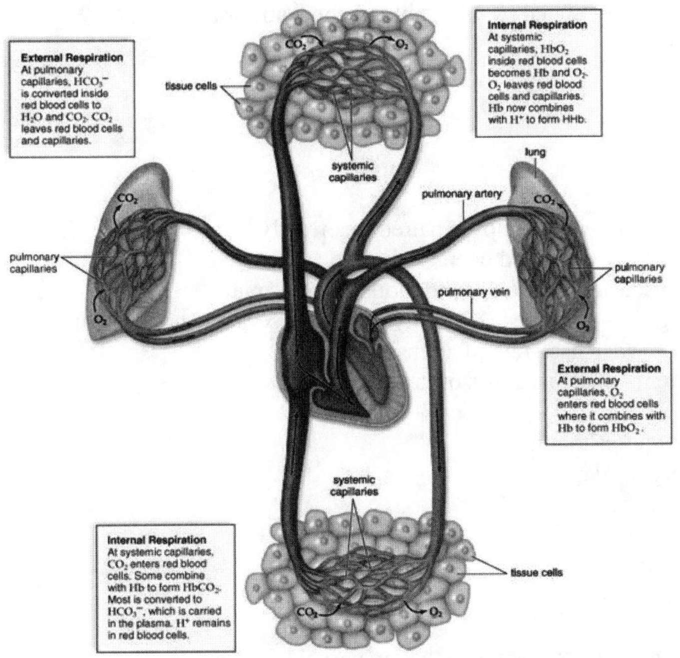

External and internal respiration

Take Note: *For the AP Biology exam, you should be able to describe how gas exchange occurs in humans. What adaptations facilitate the transport of gases within the respiratory system? How are the respiratory system and cellular respiration related?*

35.3 Respiration and Health

The respiratory tract is subject to infections such as pneumonia and pulmonary tuberculosis. New strains of tuberculosis are resistant to the usual antibiotic therapy. Major lung disorders are usually due to cigarette smoking. In chronic bronchitis the air passages are inflamed, mucus is common, and the cilia that line the respiratory tract are gone. Emphysema and lung cancer are two of the most serious consequences of smoking cigarettes. When the lungs of these patients are removed upon death, they are blackened by smoke.

Multiple Choice Questions

1. Which of the following organisms use a one-way ventilation mechanism to maximize gas exchange?
    A. reptiles
    B. amphibians
    C. fish
    D. mammals
    E. birds

2. Contraction of the diaphragm will result in a(n)
   A. decrease in temperature of the lungs, causing blood to drop off higher amounts of CO_2.
   B. increased lung volume and drawing in of outside air.
   C. collapse of the alveoli, forcing air up the trachea.
   D. decrease in rib cage size and an exhalation.
   E. increased pressure on the rib cage, causing the heart rate to increase.

3. An increase in breathing rate is primarily stimulated by which blood concentration change?
   A. decrease in CO_2 concentration
   B. decrease in O_2 concentration
   C. increase in H^+ concentration
   D. decrease in H^+ concentration
   E. increase in O_2 concentration

4. Which of the following structure – function relationships is paired incorrectly?
   A. cartilaginous rings of the trachea – reinforced open trachea
   B. air sacs located near major muscles in insects – continuous air movement
   C. finely divided outgrowths of gills – increased capability to obtain oxygen from the water
   D. highly vascularized parapodia – increased rate of diffusion
   E. countercurrent exchange system – increased loss of heat

5. The respiratory system of an annelid requires that it
   A. remains in a dry habitat during the day.
   B. conserves water via excretory pores.
   C. maintains tracheal systems at the spiracles.
   D. expends energy to secrete mucus.
   E. utilizes the clitellum to produce water.

6. Fetal hemoglobin has a higher affinity towards oxygen (O_2) than maternal hemoglobin. Therefore,
   A. fetal blood is able to release O_2 to maternal blood in the placenta.
   B. oxygen (O_2) will bind more tightly to hemoglobin to avoid rapid transference in the placenta.
   C. production of fetal hemoglobin late into life can cause respiratory difficulties.
   D. oxygen is transferred readily to fetal hemoglobin.
   E. if it remains after birth it can complicate the effects of sickle cell anemia.

7. Premature infants are often treated with synthetic surfactants, phospholipids that allow the lungs to expand and contract properly. The mechanism by which they accomplish this is by
   A. decreasing the cohesive force of H_2O molecules.
   B. increasing the surface area of the alveoli.
   C. increasing the surface tension in the bronchi.
   D. decreasing the attractive forces of O_2 in the alveoli.
   E. decreasing the surface area of the bronchioles.

8. The correct passage of air from the environment to the cells of the alveolus is
   A. nasal cavity – bronchiole – trachea – pharynx – alveolus
   B. pharynx – trachea – bronchi – bronchioles – glottis – alveolus
   C. nasal cavity – pharynx –trachea – bronchioles – alveolus
   D. glottis – bronchioles – trachea – bronchi – alveolus
   E. nasal cavity – trachea – glottis – pharynx – bronchi – alveolus

9. Each molecule of hemoglobin can bond with up to
   A. three molecules of CO_2.
   B. four molecules of O_2.
   C. five molecules of NO.
   D. two molecules of $H_2 0$.
   E. three molecules of CO.

10. Which of the following effects is NOT likely during strenuous exercise?
   A.   increased respiratory rate
   B.   decreased blood pH
   C.   increased saliva production
   D.   increased urine production
   E.   increased oxygen diffusion

Free Response Question

Organisms accomplish gas exchange in many unique ways.

   A.   **Discuss** how breathing rate is regulated.

   B.   **Explain** how the countercurrent mechanism of gills is different than the negative pressure breathing mechanism of mammals.

Annotated Answer Key (MC)

1.   **E**; A one-way ventilation mechanism provides for constant oxygen intake in the bird's unique atmospheric environment.

2.   **B**; When the diaphragm contracts, its fibers shorten and it creates a vacuum in the thoracic cavity, forcing outside air in to the space. It will not force air up the trachea and will actually increase the rib cage size (although causing inhalation, not exhalation). It does not impact temperature or directly influence heart rate.

3.   **C**; An increase in H^+ concentration indicates a decrease in pH, usually stimulated by an increased concentration of CO_2. As CO_2 combines with H_2O in the bloodstream, H_2CO_3 is formed and quickly dissociates into HCO_3^- and H^+, decreasing the pH. This stimulates an increase in breaking rate to reverse this reaction and increase blood pH back to acceptable limits.

4.   **E**; Countercurrent exchange systems provide for a decreased loss of heat by keeping the "warm" blood at the core of the organism and the "cool" blood at the extremities. The constant gradient created by a countercurrent exchange system utilizes antiparallel vessels.

5.   **D**; Due to the fact that earthworms (annelids) breathe through their skin, they must remain in a moist environment. By secreting mucus they ensure that there will always be a moist covering on their skin to aid in diffusion of oxygen.

6.   **D**; Fetal hemoglobin's affinity towards oxygen ensures that the fetus will readily receive oxygen via the maternal blood supply.

7.   **A**; Surfactants allow the lungs to expand and contract properly by not allowing the wet surfaces (to facilitate diffusion) of the lungs to stick together due to the cohesion of water. Surfactants are not made until late in gestation, as fetuses do not require the contraction and expansion of their lungs until after birth.

8.   **C**; The nasal cavity is the start of the pathway of air from the environment to the alveolus. The pharynx is the region in the back of the throat leading to the trachea, then the bronchi, bronchioles, and alveoli.

9.   **B**; There are FOUR nitrogen atoms within the hemoglobin molecule, surrounding the iron atom, providing bonds with four molecules of oxygen, carbon dioxide, or carbon monoxide.

10. **C**; During strenuous exercise, the body will increase its respiratory rate in response to a higher oxygen demand, increase urine production due to an increased blood flow, and increase its rate of oxygen diffusion in the alveoli. Blood pH will decrease as CO_2 concentration in the blood increases, as well. However, the body will most likely not produce saliva to digest food during strenuous exercise and stress.

Answer to FRQ

PART A (MAX 7)

- Breathing center (chemoreceptors) in the brain: controls the effector (lungs)
- Elaboration for any of the following structures (MAX 1)
    - Medulla oblongata
    - Pneumotaxic center
    - Apneustic center
- Motor nerves from the spinal cords send messages.
- Cyclical contraction of the diaphragm
- Concentration of CO_2 in the blood
- May address O_2 concentration during hypoxia
- CO_2 stored as HCO_3^-
- Enzymatic role: by carbonic anhydrase
- H_2CO_3 dissociation to H^+ and $HCO_3^- \rightarrow$ decrease of pH
- Lactic acid production lowers pH and increases breathing.
- Drugs, pregnancy (progesterone)

PART B (MAX 6)

- Must maintain continuous flow of water over the gills (elaboration includes continuous movement or double pump system)
- Blood flow is in the opposite direction from the water passing over the gills.
- Blood flow in the same direction $\rightarrow$ decreased efficiency of exchange (50%) $\rightarrow$ stops passive diffusion
- Creates a concentration gradient
- Water has more available oxygen than the blood, and oxygen diffusion continues.
- MOST fish cannot survive out of water due to gill collapse and a decrease in surface area.

CHAPTER 36
BODY FLUID REGULATION AND EXCRETORY SYSTEMS

Take Note: *To successfully prepare for the AP Biology exam, you should understand how the excretory system functions in the maintenance of the overall homeostasis of an animal.*

36.1   Excretion and the Environment

Animals excrete nitrogenous wastes obtained from the breakdown of proteins and nucleic acids. The amount of water and energy required to excrete nitrogenous wastes differs according to an animal's environment. Aquatic animals usually excrete **ammonia**, and land animals excrete either **urea** or **uric acid**. Ammonia is toxic, so it requires a great deal of water for excretion. Terrestrial animals need to conserve water, and have developed adaptations to excrete their nitrogen wastes (uric acid and urea) while conserving water.

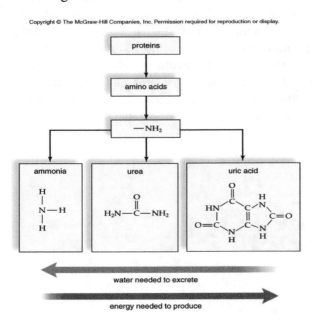

Nitrogenous wastes

Most animals have excretory organs. The **flame cells** of planarians rid the body of excess water. Earthworms have **nephridia** that exchange molecules with the blood in a manner similar to that of vertebrate kidneys. **Malpighian tubules** in insects take up metabolic wastes and water from the hemolymph. Later, the water is absorbed by the gut. Arthropods have a variety of adaptations for the removal of nitrogen wastes from the body.

Osmoregulation is important for animals, as most have to balance their water and salt intake and excretion to maintain normal solute and water concentration in their body fluids. This balance is essential for homeostasis so that all organ systems can work together for the efficient functioning of the animal.

Marine fishes live in an environment that is hypertonic to their blood, resulting in the loss of water to the environment. In order to avoid this loss, they constantly drink water, excrete salts at the gills, and pass an isotonic urine. Freshwater fishes live in a hypotonic environment, so they are constantly gaining water. In order to osmoregulate, they never drink water; they take in salts at the gills and excrete a hypotonic urine.

Terrestrial animals that live in extreme environments also have adaptations. For example, the kangaroo rat, which lives in the desert, can survive on metabolic water because of its many ways of conserving water. For instance, its fur reduces evaporation, and during the day it stays in burrows that are dark and cool. As well, their nasal passages have many folds so that the mucus membranes can capture moisture from the air as they breathe. Although they drink salt water, marine birds and reptiles have glands that extrude salt from their blood, allowing them to maintain a healthy balance of solutes and water.

> **Take Note:** *For the AP Biology exam, you should be able to compare and contrast the excretory systems of different phyla of animals.*

36.2 Urinary System in Humans

The human urinary system contains four organs: the **kidneys**, the **ureters**, the **urinary bladder**, and the **urethra**. The kidneys are the excretory organs of the urinary system because they are responsible for osmoregulation in the body. The kidney is divided into the **cortex**, the outer portion, and the **medulla**, the inner portion. The functional units of the kidneys are the **nephrons**, each of which has several parts and a blood supply. The glomerular capsule consists of the **glomerulus** (made of capillaries) surrounded by **Bowman's capsule**. Following the glomerular capsule is the **proximal convoluted tubule**, the **loop of Henle**, the **distal convoluted tubule**, and the **collecting duct**. Blood enters the nephrons from arterioles that branch off the **renal artery**, passes through the **peritubular capillaries**, and exits via venules that join to form the **renal vein**. Some nephrons are in the cortex, while others also span the medulla.

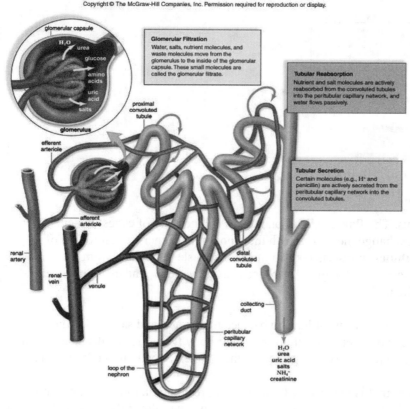

Processes in urine formation

Urine formation by a nephron requires three steps: **glomerular filtration**, when nutrients, water, and wastes enter the nephron's glomerular capsule; **tubular reabsorption**, when nutrients and most water are reabsorbed into the peritubular capillary network; and **tubular secretion**, when additional wastes are added to the convoluted tubules. When blood enters the glomerulus, water, glucose, amino acids, salts, and uric acid are filtered out of the blood and enter Bowman's capsule. This substance is called the glomerular filtrate. Proteins and cells are too large to be filtered and remain in the blood. The filtrate moves from Bowman's capsule into the proximal tubule, where salts, glucose, amino acids, and urea are reabsorbed back into the bloodstream according to the body's needs. Hydrogen ions, ammonia, and drugs are secreted into the filtrate as it passes through the proximal tubule. As the filtrate passes through the loop of Henle, water and salt are reabsorbed. The descending limb of the loop of Henle is permeable to water, but not to salts due to the abundance of aquaporins in the membranes of the epithelial cells that line the nephron. It is here that water is reabsorbed because of the high concentration of solutes in the medulla. As the filtrate moves in to the ascending limb of the loop of Henle, NaCl is passively reabsorbed in the thin portion, and actively reabsorbed in the thick portion. The filtrate then moves into the distal convoluted tubule, where more reabsorption and secretion occurs. In the collecting duct, water is reabsorbed, as is some urea. The urine that leaves the collecting duct and enters the ureters contains water, urea, uric acid, salts, ammonia, and creatinine.

Humans maintain homeostasis and conserve water by excreting a hypertonic urine. The ascending limb of the loop of the nephron actively extrudes salt so that the renal medulla is increasingly hypertonic relative to the contents of the descending limb and the collecting duct. Since urea leaks from the lower end of the collecting duct, the inner renal medulla has the highest concentration of solute. The gradient that is formed in the kidney as a result of salt and urea is what aids in the reabsorption of water via a countercurrent mechanism that ensures water will diffuse out of the descending limb and the collecting duct back into the peritubular capillaries.

Three hormones are involved in maintaining the water content of the blood. **Antidiuretic hormone (ADH)** is secreted by the posterior pituitary in response to an increase in the osmotic pressure of the blood. It makes the collecting duct more permeable to water, and it is reabsorbed in order to maintain the proper balance of water and solutes in the blood. Effective glomerular filtration occurs because of sufficient blood pressure. The hormone **aldosterone** is secreted by the adrenal cortex after low blood pressure has caused the kidneys to release **renin**. The presence of renin leads to the formation of **angiotensin II**, which causes the adrenal cortex to release aldosterone. Aldosterone acts on the kidneys to retain Na$^+$; therefore, water is reabsorbed and blood pressure rises. **Atrial natriuretic hormone**, which is secreted from the heart when blood pressure increases, prevents the secretion of renin and aldosterone. This leads to the excretion of more Na$^+$. Water is also excreted, lowering blood volume and blood pressure.

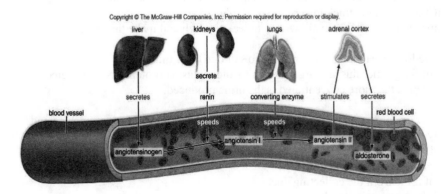

The rennin-angiotensin-aldosterone system

Another way the kidneys help the body maintain homeostasis is by working with the respiratory system to keep blood pH within normal limits. They reabsorb the HCO_3 formed as a result of decreased pH in the blood and excrete H^+ as needed to maintain pH at about 7.4.

> **Take Note:** *For the AP Biology exam, you should be able to describe how urine is formed in the nephron. What adaptations of the excretory system facilitate each process of urine formation? How does the endocrine system assist the excretory system in osmoregulation?*

Multiple Choice Questions

1. The primary source of nitrogenous waste molecules is from the breakdown of
    A. red blood cells.
    B. cholesterol.
    C. white blood cells.
    D. starch.
    E. proteins.

2. Which of the following organisms utilizes Malpighian tubules for excretion of wastes?
    A. mosquito
    B. earthworm
    C. planaria
    D. bird
    E. turtle

3. It is advantageous for aquatic invertebrates to secrete ammonia because
    A. it is not very toxic and is insoluble in water.
    B. it is synthesized by enzymatic reactions and uses less ATP than urea synthesis.
    C. it is readily concentrated for water conservation.
    D. it does not require a large ATP expenditure even though it is extremely toxic.
    E. it can be excreted in very small amounts.

4. Cells of the proximal convoluted tubule have numerous mitochondria to
    A. increase the surface area for absorption of H_2O.
    B. supply sufficient ATP needed for active transport.
    C. remove urea from the filtrate.
    D. produce more O_2 to fuel cellular respiration.
    E. increase the rate of facilitated diffusion of water.

5. Which of the following structural characteristics of the nephron helps to maintain the salt-water balance?
    A. Active transport of NaCl occurs in the thick lower loop of Henle.
    B. The bends in the proximal convoluted tubule allow urea to leak through and make the urine more concentrated.
    C. The ascending limb of the loop of Henle is impermeable to salt due to its thickness.
    D. The transport channels of the glomerulus are large enough to allow plasma proteins to be excreted.
    E. The longer the loop of Henle, the more concentrated is the urine produced.

6. Which of the following occurs in response to release of antidiuretic hormone (ADH) from the pituitary gland?
    A. decrease in permeability of the loop of Henle to H_2O
    B. reabsorption of Na^+ and K^+ by the distal convoluted tubule
    C. increase in permeability of the collecting duct to H_2O
    D. increase in production of urine from the collecting duct
    E. increase in filtration of plasma proteins by the glomerulus

7. When blood pH becomes too acidic, H^+ ions are excreted with ammonia and Na^+ and HCO_3^- are reabsorbed. This can occur when
    A. blood CO_2 levels become too low.
    B. there is too much bicarbonate in the blood.
    C. the concentration of O_2 decreases due to emphysema.
    D. hyperventilation increases blood CO_2 levels.
    E. O_2 concentration decreases at high altitudes.

8. Consumption of a large amount of caffeine, as in energy drinks, will cause
    A. replacement of electrolytes from sweating.
    B. decrease in urine production.
    C. a decrease in blood pressure.
    D. increased thirst.
    E. an increase in the reabsorption of water and NaCl.

9. Abrupt kidney failure can occur from the ingestion of antifreeze. Which of the following effects could cause this to be fatal?
    A. The ethylene-glycol based antifreeze forms crystals in the nephrons.
    B. The antifreeze increases the permeability of the collecting duct to water.
    C. The glomerulus filters the antifreeze and it is excreted in the urine.
    D. The loop of Henle is not permeable to the ethylene and it accumulates in the collecting duct.
    E. The ethylene is too large to pass through the capillary network of Bowman's capsule and it accumulates in the blood stream.

10. Glucose is primarily reabsorbed in
    A. Bowman's capsule.
    B. glomerulus.
    C. proximal convoluted tubule.
    D. distal convoluted tubule.
    E. loop of Henle.

Free Response Question

The human body uses several control mechanisms to maintain water balance.

    A. **Explain** how TWO of the following affect this water balance.
      • blood thinner medication
      • sport drinks
      • alcohol
      • caffeine

    B. **Discuss** TWO physiological causes for the renin-angiotensin-aldosterone system (RAAS) to be stimulated and how the system works to restore homeostasis.

    C. **Explain** how the principles of diffusion and osmosis influence the action of diuretics.

Annotated Answer Key (MC)

1. **E**; Proteins are the main source of nitrogenous wastes due the nitrogen component of the amino acids. Nucleic acids are also a component of this waste.

2. **A**; Insects use Malpighian tubules for excretion, not flame cells (planaria), nephridia (earthworms), or kidneys (birds and turtles).

3. **D**; Ammonia does not require a large amount of energy to be produced, whereas uric acid does (as secreted by birds and reptiles). Because aquatic invertebrates live in water, they can secrete a relatively toxic substance as it is immediately diluted and therefore can be secreted in very large amounts.

4. **B**; The proximal (first) convoluted tubule of the nephron has many mitochondria for use in active transport, which is an energy-requiring process. This helps to maintain the homeostatic water-solute balance of the body as regulated by the kidneys.

5. **E**; The longer the loop of Henle, the more water reabsorption is possible. Desert animals have very long loops of Henle to survive in their environment.

6. **C**; ADH increases the permeability of the nephron to H_2, it does not decrease it. Decreasing the permeability would allow more water to be excreted, which is not the purpose of an ADH secretion. ADH is secreted during times of water stress where the body needs to retain as much water as possible.

7. **D**; An increase in blood CO_2 will decrease pH. In the blood, CO_2 combines with H_2O to form H_2CO_3 which then dissociates into HCO_3^- and H^+ causing a drop in pH. Hyperventilation will increase CO_2 levels in the blood.

8. **D**; Caffeine is a diuretic and causes ADH to not be secreted. This will in turn cause an increase in urine production, increase in blood pressure due to an increased heart rate, decrease in the reabsorption of water and NaCl, and lack of replacement of electrolytes.

9. **A**; None of the effects other than crystal formation explain the acute nature of antifreeze poisoning, where only a small amount can be fatal. An accumulation would not be as acute.

10. **C**; Glucose is a relatively large molecule and is reabsorbed immediately in the nephron.

Answer to FRQ

PART A (MAX 5)
- Blood thinners, alcohol, and caffeine are all diuretics.
- Possible effects: increased urination, decreased blood pressure, etc.
- Inhibit Na^+ reabsorption at the ascending loop of Henle, therefore no reabsorption of H_2O follows OR inhibits H_2O reabsorption at the distal tubule
- Sport drinks will restore electrolytes.
- Possible effects

PART B (MAX 4)
- Loss of blood volume AND decrease of blood pressure (TWO points possible)
- When blood pressure is low, the body secretes renin.
- Renin stimulates the release of angiotensin.
- This causes blood vessels to constrict → increasing blood pressure
- Kidneys retain Na^+ and H_2O.

PART C (MAX 5)
- Definition of osmosis (high water concentration to low water concentration, also water potential, etc.)
- Creation of concentration gradient
- Size of molecules
- Solute potential
- Selectively permeable

CHAPTER 37
NEURONS AND NERVOUS SYSTEMS

> **Take Note:** *In order to successfully prepare for the AP Biology exam, you should understand how the nervous system functions in the maintenance of the overall homeostasis of an animal.*

37.1   Evolution of the Nervous System

A comparative study of the invertebrates shows a gradual increase in the complexity of the nervous system throughout evolution. Simple animals like sponges respond to stimuli at the cellular level. Hydras and cnidarians have **nerve nets** made of nerve cells. Planarians have two **ventral nerve cords** that extend from the **cerebral ganglia** (a cluster of neurons), as well as eyespots. This concentration of neurons and sensory receptors at the head end of the animal is known as **cephalization**. Annelids, arthropods, and mollusks all show cephalization in that they have true nervous systems that include a well-defined brain and a ventral nerve cord. The vertebrate nervous system, like that of the earthworm, is divided into the central and peripheral nervous systems. The **central nervous system** (CNS) consists of the brain and spinal cord, while the **peripheral nervous system** (PNS) consists of the nerves and ganglia in the rest of the body. The vertebrate brain is divided into three regions, (1) the **forebrain**: the olfactory bulb, cerebrum, thalamus, hypothalamus, and pituitary; (2) the **midbrain**: the optic lobe; and (3) the **hindbrain**: the cerebellum and medulla oblongata.

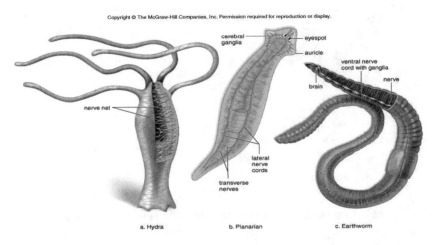

Evolution of the nervous system

> **Take Note:** *For the AP Biology exam, you should be able to explain why cephalization played a significant role in adaptations that fueled animal evolution. You should also be able to explain the embryonic development of the nervous system in vertebrates.*

Like other vertebrates, humans have a central nervous system and a peripheral nervous system. It receives sensory input from the environment, integrates the information, and generates a response that is carried out by the muscles and glands.

37.2   Nervous Tissue

The functional unit of the nervous system is the **neuron**, of which there are three types: sensory, motor, and interneuron. **Sensory neurons** carry signals from sensory receptors to the CNS, while **motor neurons** carry signals from the CNS to **effector** organs (muscles and glands). **Interneurons** are located only in the CNS and allow the different parts of the CNS to communicate with each other. Each of these neurons is made up of dendrites, a cell body, and an axon. **Dendrites** receive signals from sensory receptors or other neurons. The **cell body** contains the nucleus and other organelles that allow the neuron to function normally. The **axon** carries signals away from the cell body toward other neurons. Axons are often insulated by a **myelin sheath** made of Schwann cells that allows signals to pass through gaps called the **nodes of Ranvier**, causing faster conduction of a signal.

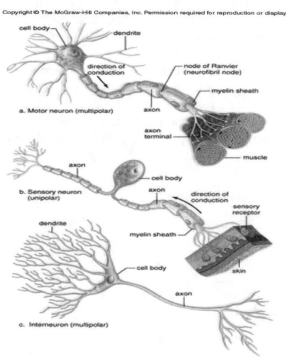

Neuron anatomy

When an axon is not conducting an action potential (nerve impulse), the **resting potential** indicates that the inside of the neuron is negative compared to the outside. There is a high concentration of sodium outside of the cell, and a high concentration of potassium inside the cell. Both of these concentrations are maintained by the **sodium-potassium pump**. When the axon is conducting a nerve impulse, an **action potential** (i.e., a change in membrane potential) travels along the fiber. **Depolarization** occurs (the inside becomes positive) due to the movement of Na^+ inside the neuron when voltage-gated sodium channels are stimulated. **Repolarization** occurs (inside becomes negative again) due to the movement of K^+ to the outside of the neuron when voltage-gated potassium channels are stimulated. After repolarization, the sodium-potassium pump restores the neuron back to resting potential, and this portion of the neuron goes through a **refractory period**, a short period during which an action potential cannot occur. Action potentials travel faster in myelinated neurons due to **salutatory conduction**, where depolarization occurs only in the nodes of Ranvier.

Transmission of the nerve impulse from one neuron to another takes place across a **synapse**. In humans, synaptic vesicles release a **neurotransmitter** into the synaptic cleft via exocytosis. This occurs when voltage-gated calcium channels open, allowing Ca^{2+} to enter the axon terminal of the presynaptic cell and stimulate the release of neurotransmitters such as acetylcholine or norepinephrine. The binding of neurotransmitters to receptors (ligand-gated channels) in the postsynaptic membrane can either increase the chance of an action potential (stimulation) or decrease the chance of an action potential (inhibition) in the next neuron. A neuron usually transmits several nerve impulses, one after the other. Once a neurotransmitter has initiated the response, it diffuses away from the synapse, is taken back into the presynaptic neuron by endocytosis, or degraded by enzymes such as acetylcholinesterase, which hydrolyzes acetylcholine.

> **Take Note:** *For the AP Biology exam, you should be prepared to describe in detail how the human nervous system responds to an external stimulus.*

37.3 Central Nervous System: Brain and Spinal Cord

The CNS consists of the spinal cord and brain, which are both protected by bone. The CNS receives and integrates sensory input and formulates motor output. The gray matter of the spinal cord contains neuron cell bodies; the white matter consists of myelinated axons that occur in bundles called **tracts**. The spinal cord sends sensory information to the brain, receives motor output from the brain, and carries out reflex actions.

In the brain, the **cerebrum** has two hemispheres connected by the **corpus callosum**. Sensation, reasoning, learning and memory, and language and speech take place in the cerebrum. The cerebral cortex is a thin layer of gray matter covering the cerebrum. The **cerebral cortex** of each cerebral hemisphere has four lobes: frontal, parietal, occipital, and temporal. The primary motor area in the frontal lobe sends out motor commands to lower brain centers, which pass them on to motor neurons. The primary somatosensory area in the parietal lobe receives sensory information from lower brain centers in communication with sensory neurons. Association areas for vision are in the occipital lobe, and those for hearing are in the temporal lobe.

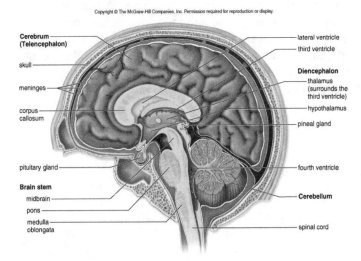

The human brain

The brain has a number of other regions. The **hypothalamus** controls homeostasis, and the **thalamus** specializes in sending sensory input on to the cerebrum. The **cerebellum** primarily coordinates skeletal muscle contractions. The **medulla oblongata** and the **pons** have centers for vital functions such as breathing and the heartbeat.

37.4   Peripheral Nervous System

The PNS lies outside of the central nervous system, and it includes the **somatic system** and the **autonomic system**. It contains **nerves**, which are bundles of axons. **Cranial nerves** are attached to the brain, and **spinal nerves** are attached to the spinal cord. Reflexes in the somatic system are automatic, and some do not require involvement of the brain. A simple reflex requires the use of neurons that make up a **reflex arc**. In response to a stimulus, a sensory neuron conducts nerve impulses from a sensory receptor to an interneuron, which in turn transmits impulses to a motor neuron, which stimulates an effector to react.

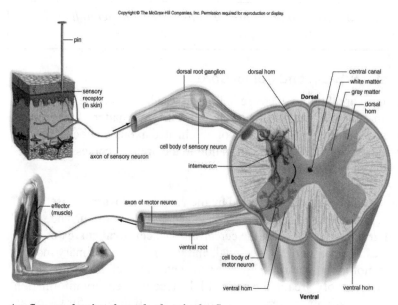

A reflex arc showing the path of a spinal reflex

While the motor portion of the somatic system of the PNS controls skeletal muscle, the motor portion of the autonomic system controls cardiac muscle and smooth muscle of the internal organs and glands. The sympathetic division and the parasympathetic division are both parts of the autonomic system. The **sympathetic division** is often associated with the reactions that occur with the "fight or flight" response during times of stress. The **parasympathetic division** is associated with the internal activities that occur during times of relaxation.

Multiple Choice Questions

1. Organisms in which of the following phyla contain the least advanced nervous systems?
    A. Cnidarians
    B. Platyhelminthes
    C. Annelid
    D. Mollusca
    E. Arthropoda

2. A particular human disorder is characterized by the poor breathing, swallowing, blood circulation, and muscle tone. Which area of the nervous system is affected in this disorder?
    A. cerebellum
    B. spinal cord
    C. medulla oblongata
    D. pons
    E. frontal lobe

Questions 3–5 refer to the following five choices. For each question, select the choice that is most closely related. Each choice may be used once, more than once, or not at all.
    A. dendrites
    B. Aaon
    C. neuroglia
    D. myelin sheath
    E. nodes of Ranvier

3. Receive information and conduct impulses toward the cell body

4. Propagate electrical impulses

5. Support and nourish neurons

6. Some of the effects of the drug methylenedioxymethamphetamine, or ecstasy, are thirst, hunger, and hyperthermia. This is due to impairment of the
    A. hypothalamus.
    B. amygdale.
    C. basal ganglia.
    D. hippocampus.
    E. cerebellum.

7. The autonomic nervous system does NOT
    A. innervate all internal organs.
    B. function in an involuntary manner.
    C. regulate control of skeletal muscle.
    D. control breathing rate and blood pressure.
    E. utilize two neurons and one ganglia for each response.

Questions 8–10 refer to the following five choices. For each question, select the choice that is most closely related. Each choice may be used once, more than once, or not at all.
    A. porifera
    B. cnidaria
    C. platyhelminthes
    D. chordata
    E. echinodermata

8. Two ventral nerve cords connected by transverse nerves

9. Radial nerves connected to a central nerve ring

10. Simplest phyla exhibiting a centralized nervous system

Free Response Question

The development of a centralized brain and nervous system were crucial in the evolution of animals.

     A. **Explain** the nervous systems of organisms in the following phyla:
- Echinoderms
- Vertebrates
- Annelida

     B. **Explain** how cephalization played a role in this evolution.

Annotated Answer Key (MC)

1. **A**; Coelenterates, cnidarians, and echinoderms have their neurons organized into a nerve net. Although they lack a brain, are radially symmetrical, and don't have division of the nervous system (CNS/PNS), they are capable of some complex behaviors. Development of bilateral summetry is associated with cephalization. Having a head with the accumulation of sensory organisms at the front end of the organism is a major advancement.

2. **C**; The medulla controls unconscious functions. The pons receives information from visual areas and controls sleep and arousal. The cerebellum coordinates muscle movements and maintains equilibrium.

3. **A**; Dendrites are fingerlike extensions off of the cell body where the neuron receives information.

4. **B**; Ions move across the plasma membrane of a nuron, specifically at the axon, to send electrical messages.

5. **C**; Glial cells provide support and nutrition and form myelin.

6. **A**; The hypothalamus regulates temperature and hunger and is very impaired by the use of ecstasy.

7. **C**; This division of the peripheral nervous system functions to maintain homeostasis in the body without conscious control (i.e., heart rate, digestion, respiration, salivation).

8. **C**

9. **E**

10. **C**

For Questions 8–10: Sponges (porifera) do not contain a nervous system. Cnidaria and Echinodermata possess differing type of nerve nets, and platyhelminthes are the first bilaterally symmetrical organisms and possess cephalization. They contain ganglionic structures, with each ganglion consisting of many axons and dendrites, and contain many longitudinal nerve cords.

Answer to FRQ

PART A (MAX 6 – Two each)

Organism	Characteristics
Echinoderms	Contain central nervous system or Central nerve ring
	Radial nerves
	No brain; OR does contain ganglia along radial nerves
	No cephalization
	No specialized sense organs
	Sensory neurons in the ectoderm
Annelida	Ventral nerve cord
	Segmented ganglia
	Cephalization
	Eyes, taste buds, tactile tentacles, statocysts
Vertebrates	Well-marked cephalization
	Large amounts of nervous tissue
	Brain
	Divided into CNS/PNS
	Dorsal nerve cord

PART B (MAX 6)

- Defined cephalization: the concentration of the nervous system towards the anterior (front) of the organism
- Tied to bilateral symmetry
- Originated with platyhelmintes (flatworms)
- Encounter a stimulus right away with sensory organs
- Quick response to a stimulus

CHAPTER 38
SPECIAL SENSES

Take Note: *In order to successfully prepare for the AP Biology exam, you should understand how the senses function in the maintenance of the overall homeostasis of an animal.*

38.1 Chemical Senses

Chemoreception is found universally in animals and is, therefore, believed to be the most primitive sense. Human taste buds and olfactory cells are **chemoreceptors**. They are sensitive to chemicals in water and air. Taste buds are primarily located in the walls of the **papillae** on the tongue. When molecules bind to receptors on the taste buds, nerve impulses are generated in the sensory nerves associated with the taste buds. These impulses are sent to the brain, which interprets the signals as one of the five types of taste: sweet, sour, salty, bitter, and umami.

Olfactory cells are modified neurons located in the olfactory epithelium of the nasal cavity that are responsible for the sense of smell. Each olfactory cell ends in five olfactory cilia that have receptors for odor molecules. Signals are sent to the **olfactory bulb** in the brain, where it is interpreted as smell. The sense of smell and the sense of taste usually work together.

38.2 Sense of Vision

Vision in humans is dependent on the eye, the optic nerves, and the visual areas of the cerebral cortex. The eye has three layers. The outer layer, the **sclera**, can be seen as the white layer that covers most of the eye; it also becomes the transparent bulge in the front of the eye called the **cornea**. The surface of the sclera is covered by a mucus membrane called the **conjunctiva** that keeps the eyes moist. The **choroid** is the middle pigmented layer that absorbs stray light rays. The colored part of the eye, known as the **iris**, is located in this layer, and it functions to regulate the size of the **pupil**. **Rod cells** (sensory receptors for dim light) and the **cone cells** (sensory receptors for bright light and color) are located in the **retina**, the inner layer of the eye. The eye has two liquid portions: the **aqueous humor**, located in between the cornea and the lens; and the **vitreous humor**, the gel-like fluid located behind the lens. The cornea, the humors, and especially the lens bring the light rays to focus on the retina. When light strikes the pigment **rhodopsin** within the membranous disks of rod cells, it splits into opsin and retinal. A cascade of reactions leads to the closing of ion channels in a rod cell's plasma membrane. Inhibitory transmitter molecules are no longer released, and nerve impulses are carried through the optic nerve to the brain. Integration occurs in the retina, which is composed of three layers of cells: the rod and cone layer, the bipolar cell layer, and the ganglion cell layer.

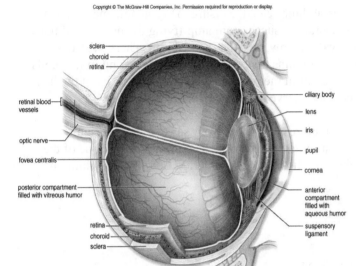

Anatomy of the human eye

Take Note: *How does vision in humans differ from vision in other animals?*

38.3 Senses of Hearing and Balance

Hearing in humans is dependent on the ear, the cochlear nerve, and the auditory areas of the cerebral cortex. The ear is divided into three parts. The **outer ear** consists of the **pinna** and the **auditory canal**, which direct sound waves to the middle ear. This part of the ear is lined with fine hairs and **ceruminous glands** that secrete the earwax that prevents the entry of foreign substances into the ear. The **middle ear** begins with the **tympanic membrane** (also called the eardrum) and contains the **ossicles** (malleus, incus, and stapes). The malleus is attached to the tympanic membrane, and the stapes is attached to the **oval window**, which is covered by membrane. The auditory (eustachian) tube, which allows air pressure to equalize in the ear, extends from the middle ear to the nasopharynx. The **inner ear** contains the **cochlea** and the **semicircular canals**, plus the **utricle** and **saccule**.

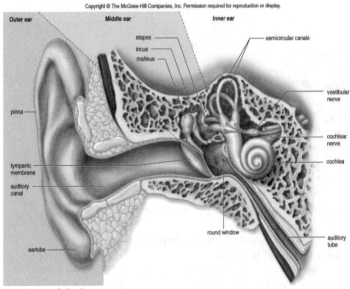

Anatomy of the human ear

Hearing begins when sound waves enter the auditory canal, causing the tympanic membrane to vibrate. In turn, the bones of the inner ear vibrate in succession, amplifying the pressure of the sound waves. The stapes then strikes the oval window, causing it to vibrate. These vibrations set up pressure waves within the cochlea, which contains the organ of Corti, consisting of hair cells whose stereocilia are embedded within the tectorial membrane. When the stereocilia of the hair cells bend, nerve impulses begin in the cochlear nerve and are carried to the brain.

The ear also contains mechanoreceptors that are responsible for our sense of equilibrium. Rotational equilibrium is dependent on the stimulation of hair cells within the ampullae of the semicircular canals. Gravitational equilibrium relies on the stimulation of hair cells within the utricle and the saccule.

Multiple Choice Questions

1. The sense of smell is a very direct sense because
   A. olfactory cells are modified neurons and lead to the same neuron in the olfactory bulb of the brain.
   B. the sense of smell is dependent upon only hundreds of olfactory cells located in the pharynx and in the brain.
   C. it actually improves with age as the number of olfactory cells increases.
   D. the sense of smell cannot be lost as the result of head trauma or respiratory infection.
   E. its receptors are nonspecific and can bind to many types of odor molecules.

Questions 2–5 refer to the following five choices. For each question, select the choice that is most closely related. Each choice may be used once, more than once, or not at all.
   A. sclera
   B. pupil
   C. optic nerve
   D. rods
   E. cones

2. Protects and supports the eyeball

3. Makes color vision possible

4. Transmits impulses to the brain

5. A fibrous layer that covers most of the eye

6. An individual who is said to be *nearsighted*
   A. wears convex lenses so that they can see close objects.
   B. has an uneven lens and sees fuzzy images.
   C. has a short eyeball, and rays focus behind the retina when viewing close objects.
   D. has difficulty focusing at a distant object.
   E. cannot undergo laser surgery to correct the abnormality.

7. Which of the following statements is true about the ear and hearing?
   A. There are 20 bones in the ear responsible for taking the pressure from the surface of the tympanic membrane.
   B. Each part of the organ of Corti is sensitive to different wave frequencies, or pitch.
   C. Even a single sound wave carries a great deal of energy.
   D. Hair cells contain chemoreceptors that are sensitive to mechanical stimulation.
   E. When nerve impulses reach the optic nerve, they are interpreted as sound.

8. When light strikes a molecule of rhodopsin,
    A. inhibitory molecules are released from the rod's synaptic vessels, causing the eyes to close.
    B. ion channels open in a rod cell's plasma membrane.
    C. the rhodopsin splits into opsin and retinal.
    D. a nerve impulse is sent to the cerebellum.
    E. vitamin A is produced.

9. In color blindness, an individual
    A. lacks certain visual pigments.
    B. does not possess rods.
    C. does not produce rhodopsin.
    D. has an uneven cornea.
    E. does not produce vitamin A.

10. Which of the following is NOT a primary type of taste?
    A. bitter
    B. sweet
    C. spicy
    D. salty
    E. umami

Free Response Question

Animals rely on stimuli and senses adaptive to their particular environment.

A. **Describe** ONE of the following:
    - the conduction of a sound wave from the ear to the central nervous system
    - the mechanism of color vision
    - the coordination of taste and smell

B. **Explain** how the special senses are not equally developed in all animals.

Annotated Answer Key (MC)

1. **A**; The olfactory epithelium in humans is located high in the nasal cavity. Between 10 and 20 million olfactory cells end in cilia that bear receptor proteins for specific odor molecules. Nerve fibers from these olfactory cells lead to the same neuron in the olfactory bulb.

2. **A**

3. **D**

4. **C**

5. **A**; The human eye has three coats, with the outermost being the sclera that functions in protection. The retina contains rods, for black and white vision, and cones, for color vision, and contains sensory receptors for sight. Messages from these receptors are transmitted by the optic nerve.

6. **D**; Nearsighted individuals have an elongated eyeball and when they attempt to look at a distant object, the image is brought to focus in front of the retina. They wear concave lenses, which diverge the light rays so that the image can be focused on the retina.

7. **B**; Three bones in the ear take the pressure from the surface of the tympanic membrane, with the stapes striking the membrane of the voal window, causing it to vibrate. The organ of Corti synapses with nerve fibers and is very sensitive to different pitches. The sensory receptors consist of hair cells that are sensitive to mechanical stimulation called mechanoreceptors.

8. **C**; Rhodopsin is a complex molecule that is a derivative of Vitamin A. When a rod absorbs light, rhodopsin splits and causes the closure of ion channels in the rod cell's plasma membrane. Inhibitory transmitter molecules go to the visual areas of the cerebral cortex.

9. **A**; Color blindness is an inherited disorder. Color vision depends on three different kinds of cones which contain pigments called the Blue, Green, and Red pigments. Each is made up of retinal and opsin, but contain a slight difference in opsin. Color blind individuals produce rhodopsin, possess rods, and have a perfectly normal cornea. However, they do lack certain visual pigments.

10. **C**; The five primary types of taste are sweet, sour, salty, bitter, and umami. Umami is Japanese savory. A particular food can stimulate more than one of these types of taste buds.

Answer to FRQ

PART A (MAX 6 – Two each)

Sound wave conduction from ear to CNS
- Outer ear: channels sound to middle ear (sound amplification)
- Middle ear: transforms energy of a sound wave → vibrations
- Middle ear: consists of three bones
- Vibrations → bones of middle ear → wave in the inner ear
- Inner ear: cochlea, semicircular canals, auditory nerve
- Inner ear: fluid-filled to transmit wave

Mechanism of color vision
- Cone cells are photoreceptors.
- Located in the retina
- Respond to a photon
- Contain rhodopsin (light-sensitive receptor in the membranes of the rod cells)
- G-protein cascade
- Optic nerve sends message to the brain.

The coordination of taste and smell
- Based upon location of olfactory receptors
- Olfactory receptors are chemoreceptors in nasal cavity
- Cilia inside the nose dissolve odor molecules.
- Taste buds on papillae
- Taste buds contain chemoreceptors.
- Sensory transduction → brain interpretation
- Many neurons are involved.

PART B (MAX 6)
- Echolocation
    - High frequency sounds
    - Special ear and brain adaptations
    - Three dimensional pictures of surroundings
- Infared vision
    - Infared sensory organs
    - Heat sensitive retina

- Electric sense
    - Modified muscle cells to produce electric charges
    - Lateral line (sensory pores in the skin) detects electrical current
- Magnetic sense
    - Deposits of magnetite in the nervous system

CHAPTER 39
LOCOMOTION AND SUPPORT SYSTEMS

> **Take Note:** *In order to successfully prepare for the AP Biology exam, you should understand how the skeletal and muscular systems function in the maintenance of the overall homeostasis of an animal.*

39.1   Diversity of Skeletons

Three types of skeletons are found in the animal kingdom: hydrostatic skeleton (cnidarians, flatworms, and segmented worms); exoskeleton (certain molluscs and arthropods); and endoskeleton (vertebrates). A **hydrostatic skeleton** is a fluid-filled gastrovascular cavity or coelom that allows support and resistance to muscle contractions so that movement can occur. **Exoskeletons** provide a place for muscle attachment and prevent desiccation. In mollusks, the exoskeleton is made of calcium carbonate, and it can grow as the animal grows. In arthropods, the exoskeleton is made of chitin. In order for an arthropod to grow, it must molt, or shed its exoskeleton and develop a new one. Echinoderms and vertebrates have an **endoskeleton** that grows with the animal and does not require molting. It also protects the internal organs and provides a place for muscles to attach. It also supports the weight of a large animal and allows flexible movements. The rigid but jointed skeleton of arthropods and vertebrates helped them colonize the terrestrial environment.

> **Take Note:** *For the AP Biology exam, you should be able to compare and contrast the structure and function of the skeletons of different animals. How did having a jointed skeleton help arthropods and vertebrates colonize the terrestrial environment?*

39.2   The Human Skeletal System

The human skeleton gives support to the body, helps protect internal organs, provides sites for muscle attachment, and is a storage area for calcium and phosphorus salts, as well as a site for blood cell formation.

Most bones are cartilaginous in the fetus but are converted to bone during development. A long bone undergoes **endochondral ossification** in which a cartilaginous growth plate remains between the **primary ossification center** in the middle and the secondary centers at the ends of the bones. Growth of the bone is possible as long as the **growth plates** are present, but eventually they too are converted to bone. Bone is constantly being renewed; **osteoclasts** break down bone, and **osteoblasts** build new bone. A long bone has a shaft made of compact bone and two ends that contain spongy bone. The shaft contains a **medullary cavity** with yellow marrow, and the ends contain red marrow. **Osteocytes** are mature bone cells that are located in the **lacunae** of the **osteons** of compact bone. The formation of red blood cells occurs in the red bone marrow.

The human skeleton is divided into two parts: (1) the **axial skeleton**, which is made up of the skull, the vertebral column, the sternum, and the ribs; and (2) the **appendicular skeleton**, which is composed of the pectoral and pelvic girdles and their appendages. Joints are classified as immovable, like those of the cranium; slightly movable, like those between the vertebrae; and freely movable (synovial joints), like those in the knee and hip. In synovial joints, ligaments bind the two bones together, forming a capsule containing synovial fluid.

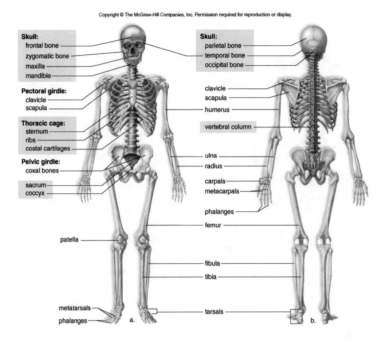

Skull:
frontal bone
zygomatic bone
maxilla
mandible

Pectoral girdle:
clavicle
scapula

Thoracic cage:
sternum
ribs
costal cartilages

Pelvic girdle:
coxal bones

sacrum
coccyx

patella

metatarsals
phalanges

Skull:
parietal bone
temporal bone
occipital bone

clavicle
scapula
humerus

vertebral column

ulna
radius

carpals
metacarpals

phalanges

femur

fibula
tibia

tarsals

a.

b.

The human skeleton

39.3 The Human Muscular System

There are three types of muscles in the human body: 1) **smooth muscle**, which lines hollow organs; 2) **cardiac muscle**, which is found in the heart; and 3) **skeletal muscle**, which is attached to the skeleton and is responsible for movement. Whole skeletal muscles can only shorten when they contract, and they work in **antagonistic pairs**. For example, if one muscle flexes the joint and brings the limb toward the body, the other one extends the joint and straightens the limb. A muscle at rest exhibits **tone**, in which some muscle fibers are contracted. If tone were not present, a person would collapse.

A whole skeletal muscle is composed of muscle fibers. Each muscle fiber is a cell that contains **myofibrils** in addition to the usual cellular components. The fibers are surrounded by a plasma membrane called the **sarcolemma**. Muscles also contain a modified endoplasmic reticulum called the **sarcoplasmic reticulum** that stores calcium. Longitudinally, myofibrils are divided into contractile units called **sarcomeres**, which are made of **actin** and **myosin** filaments that alternate and overlap to give skeletal muscle its striated appearance.

The **sliding filament model** of muscle contraction says that myosin filaments have cross-bridges, which attach to and detach from actin filaments, causing actin filaments to slide and the sarcomere to shorten. (The H zone disappears as actin filaments approach one another.) Myosin hydrolyzes ATP, and this supplies the energy for muscle contraction. Anaerobic creatine phosphate breakdown and fermentation quickly generate ATP. Sustained exercise requires cellular respiration for the generation of ATP.

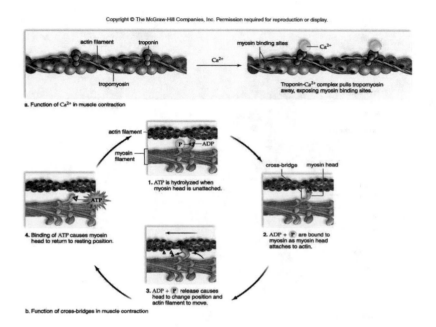

a. Function of Ca²⁺ in muscle contraction

1. ATP is hydrolyzed when myosin head is unattached.

2. ADP + P are bound to myosin as myosin head attaches to actin.

3. ADP + P release causes head to change position and actin filament to move.

4. Binding of ATP causes myosin head to return to resting position.

b. Function of cross-bridges in muscle contraction

The role of calcium and myosin in muscle contraction

Nerves innervate muscles. Nerve impulses traveling down motor neurons to neuromuscular junctions cause the release of ACh, which binds to receptors on the sarcolemma (plasma membrane of a muscle fiber). Impulses begin and move down T tubules that approach the sarcoplasmic reticulum, and calcium ions are released and bind to **troponin**. The troponin-Ca²⁺ complex causes the **tropomyosin** threads winding around actin filaments to shift their position, revealing myosin binding sites. Myosin filaments are composed of many myosin molecules with double globular heads. When myosin heads break down ATP, they are ready to attach to actin. The release of ADP + P causes myosin heads to change their position. This is the power stroke that causes the actin filament to slide toward the center of a sarcomere. When more ATP molecules bind to myosin, the heads detach from actin, and the cycle begins again.

Multiple Choice Questions

1. During muscle contraction, Ca²⁺
   A. allows myosin to bind to actin.
   B. combines with actin.
   C. is a cofactor in the production of ATPase enzymes.
   D. is released during muscular relaxation.
   E. binds with ATP to the myosin heads.

2. In athletes who train at a high level
   A. the amount of endoplasmic reticulum increases to accommodate the additional Ca²⁺.
   B. there is a greater reliance on fermentation to produce ATP.
   C. an increased amount of lactate is produced.
   D. oxygen debt still occurs.
   E. lactic acid is used as a source of fuel, preventing lactate build up.

3. Compact bone
    A. contains lacunae in concentric circles for strength.
    B. is filled with red bone marrow to produce blood cells.
    C. is separated by irregular spaces.
    D. does not contain a blood supply.
    E. is located beneath the spongy bone.

4. Which of the following organisms contains an endoskeleton?
    A. flatworm
    B. mollusk
    C. starfish
    D. insect
    E. jellyfish

5. Which of the following is NOT an advantage of a jointed endoskeleton?
    A. ability to grow with the animal
    B. supports the weight of large animals
    C. protects vital internal organs
    D. prevents drying out on land
    E. allows swimming, jumping and running.

6. The human skeletal system does NOT
    A. protect vital internal organs.
    B. provide sites for muscular attachment.
    C. serve as an important storage reservoir for minerals.
    D. produce blood cells.
    E. synthesize myelin.

Questions 7–10 refer to the following five choices. For each question, select the choice that is most closely related. Each choice may be used once, more than once, or not at all.
    A. hydra
    B. snail
    C. insect
    D. sea cucumber
    E. dog

7. Fluid-filled gastrovascular cavity as a hydrostatic skeleton

8. Calcium carbonate shells

9. Skeleton composed of chitin

10. Contains an exoskeleton consisting of spicules

Free Response Question

Skeletons serve as support systems for animals.

    A. **Discuss** THREE functions of skeletons.

    B. **Explain** the difference in types of skeletons in the following phyla
        • Annelida (earthworm)
        • Arthropoda (arthropods)
        • Vertebrata (vertebrates)

Annotated Answer Key (MC)

1. **A**; Ca^{2+} is released from the sarcoplasmic reticulum and combines with troponin. After this reaction, the tropomyosin threads shift and the myosin binding sites are exposed. The myosin heads function as ATPase enzymes and split ATP into ADP and P. The heads then bind to actin. Calcium is a cell-signaling molecule, and not usually considered as a cofactor of the enzymes it regulates.

2. **E**; There is a tendency to see an increased number of mitochondria in endurance athletes. These individuals rely less on fermentation and produce less lactate. While oxygen debt still occurs, lactic acid is used as a source of fuel, preventing lactate build up and muscle fatigue.

3. **A**; Compact bone does not have spaces. It contains many mineral deposits. Spongy bone has spaces, is lighter than compact bone, and is located on the inside of a long bone. The spaces are often filled with red bone marrow to produce blood cells.

4. **C**; Cnidarians (jellyfish), flatworms, roundworms and annelids (earthworms) have a hydrostatic skeleton. Molluscs and arthropods have an exoskeleton and sponges, echinoderms (starfish) and vertebrates possess an endoskeleton.

5. **D**; Exoskeletons of arthropods protect them against wear and tear, predators, and drying out.

6. **E**; Myelin is produced by the glial cells of the nervous system.

7. **A**

8. **B**

9. **C**

10. **D**; There are many different forms of skeletons within the animal kingdom. Hydrostatic skeletons utilize water to maintain turgor pressure and structure redistributing internal body fluids within the organism. Other components, such as chitin and calcium carbonate, compose rigid body parts to receive the applied force of muscle contraction. Endoskeletons have rigid internal body parts to receive the applied force.

Answer to FRQ

PART A (MAX 5)

*   Protection of vital internal organs OR less vulnerable to injury
*   Sites for muscle attachment OR coordination with muscles for movement
*   Storage reservoir for minerals
*   Production of blood cells (hematopoiesis in red marrow)
*   Rigid framework for support (against the pull of gravity)

PART B (MAX 9)
Three per phyla

Platyhelminthes

*   Hydrostatic pressure
    *   o Fluid filled cavity (coelom)
    *   o Surrounded by muscles
*   May aid in locomotion
*   Cuticle for body protection

Arthropoda
- Composed of chitin
  - Polysaccharide/carbohydrate
- Divided into regions
  - Abdomen, thorax, head (or fused cephalothorax)
  - Jointed appendages (i.e., legs, antennae, mouth parts)
    - Example
  - Number of appendages, size, shape, color have functional significance

Vertebrates
- Composed of bone
  - Calcium phosphate/calcium carbonate
- Periosteum
- Marrow
- Spongy bone
- Initially laid down as cartilage → bone
- Contain a blood supply, metabolically active, repairable
- Vertebral column, ribs, skull, appendicular skeleton, ligaments
- Number, size, and shape may vary amidst species and sexes

CHAPTER 40
HORMONES AND ENDOCRINE SYSTEMS

Take Note: *In order to successfully prepare for the AP Biology exam, you should understand how the endocrine system functions in the maintenance of the overall homeostasis of an animal. You should familiarize yourself with the organs of the endocrine system and the hormones produced and secreted by each. Also, you should be able to describe the involvement each hormone in the regulation of its homeostatic mechanism.*

40.1   Endocrine Glands

The nervous system and the endocrine system both use chemical signals to facilitate communication within the body. While communication in the nervous system depends on nerve impulses and neurotransmitters, endocrine glands secrete hormones into the bloodstream, and from there they are distributed to their target organs or tissues. The organs of the endocrine system are the **endocrine glands** that secrete their products directly into the bloodstream. These organs include the hypothalamus, pituitary gland, pineal gland, thyroid, parathyroids, thymus, adrenal glands, pancreas, and gonads. Like the nervous system, the endocrine system uses negative feedback to regulate homeostatic mechanisms.

Hormones are chemical signals that usually act at a distance between body parts. Hormones are either peptides or steroids, and have a variety of effects on cells. **Peptide hormones** do not enter cells; rather, they bind to receptors located in the plasma membrane of the cells of their target organs. This results in the formation of cAMP, leading to the activation of an enzyme cascade inside the cell that ultimately results in the response of the target organ.

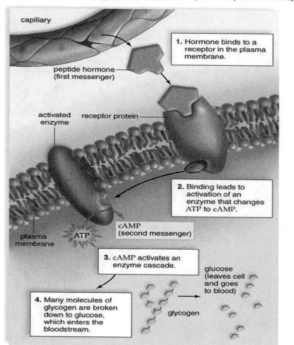

Peptide hormone

Steroid hormones, on the other hand, can pass through the plasma membrane of the cell. Once inside the cell they enter the nucleus, where they combine with a receptor, and this complex attaches to DNA, activating transcription. Protein synthesis follows.

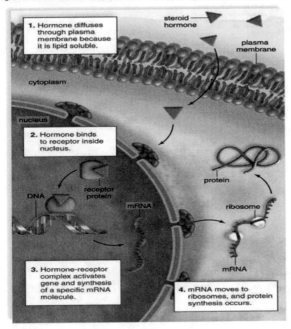

1. Hormone diffuses through plasma membrane because it is lipid soluble.

steroid hormone

plasma membrane

cytoplasm

nucleus

2. Hormone binds to receptor inside nucleus.

protein

receptor protein

DNA

mRNA

ribosome

mRNA

3. Hormone-receptor complex activates gene and synthesis of a specific mRNA molecule.

4. mRNA moves to ribosomes, and protein synthesis occurs.

Steroid hormone

> **Take Note:** *For the AP Biology exam, you should be prepared to compare and contrast the mechanism of action of peptide and steroid hormones.*

40.2   Hypothalamus and Pituitary Gland

The **hypothalamus** regulates the internal environment of the body, controlling heartbeat, body temperature, and water balance. It also controls the **pituitary gland**, a small gland that is connected to the hypothalamus. There are two parts of the pituitary gland: 1) the posterior pituitary, which stores the hormones ADH and oxytocin, made in the hypothalamus; and 2) the anterior pituitary, which makes the hormones TSH, ACTH, prolactin, growth hormone, LH, and FSH. Neurosecretory cells in the hypothalamus produce ADH and oxytocin, which are stored in axon endings in the posterior pituitary until they are released by neuronal stimulation. The hypothalamus also produces hypothalamic-releasing and hypothalamic-inhibiting hormones, which pass to the anterior pituitary by way of a portal system. These hormones stimulate the pituitary to release its hormones, some of which act on other glands to stimulate the release of other hormones. Other hormones released from the anterior pituitary (prolactin, growth hormone, and MSH) do not act on other glands, but directly on target organs.

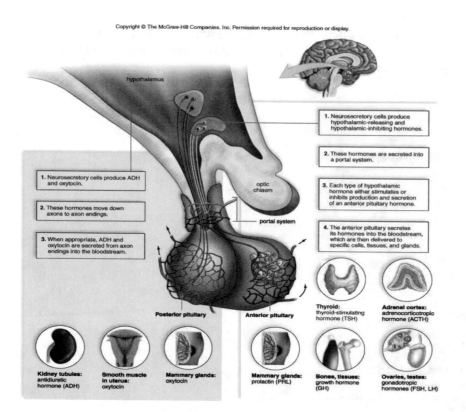

1. Neurosecretory cells produce hypothalamic-releasing and hypothalamic-inhibiting hormones.

2. These hormones are secreted into a portal system.

3. Each type of hypothalamic hormone either stimulates or inhibits production and secretion of an anterior pituitary hormone.

4. The anterior pituitary secretes its hormones into the bloodstream, which are then delivered to specific cells, tissues, and glands.

1. Neurosecretory cells produce ADH and oxytocin.

2. These hormones move down axons to axon endings.

3. When appropriate, ADH and oxytocin are secreted from axon endings into the bloodstream.

Hypothalamus and the pituitary

40.3   Other Endocrine Glands and Hormones

The **thyroid gland** is the largest endocrine gland in the body and is located in the neck below the larynx. It requires iodine to produce **thyroxine (T₄)** and **triiodothyronine (T₃)**, which increase the metabolic rate. If iodine is available in insufficient quantities, a simple goiter develops; if the thyroid is overactive, an exophthalmic **goiter** develops. The thyroid gland also produces **calcitonin**, which helps lower the blood calcium level when it is too high. The four **parathyroid glands** attached to the thyroid gland secrete **parathyroid hormone** (PTH), which raises the blood calcium and decreases the blood phosphate levels. Calcitonin and PTH are antagonistic hormones whose actions in regulating calcium homeostasis are opposed to each other.

The **adrenal glands** respond to stress: Immediately, the adrenal medulla secretes **epinephrine** and **norepinephrine**, which bring about responses we associate with emergency situations. On a long-term basis for response to stress, the adrenal cortex produces the glucocorticoids (e.g., cortisol) and the mineralocorticoids (e.g., aldosterone). **Cortisol** stimulates hydrolysis of proteins to amino acids that are converted to glucose; in this way, it raises the blood glucose level. **Aldosterone** causes the kidneys to reabsorb sodium ions (Na⁺) and to excrete potassium ions (K⁺) to increase water reabsorption and blood pressure in the body. The antagonistic hormone to aldosterone is **atrial natriuretic factor** (ANH) secreted from the heart, which cause the kidneys to excrete Na⁺ and water, thus lowering blood pressure. Addison disease develops when the adrenal cortex is underactive, and Cushing syndrome develops when the adrenal cortex is overactive.

Recall that the **pancreas** is an accessory organ of the digestive system. The pancreatic islets secrete **insulin**, which lowers the blood glucose level, and **glucagon**, which has the opposite effect. The most common illness caused by hormonal imbalance is diabetes mellitus, which occurs because of the failure of the pancreas to produce insulin or the failure of body cells to take it up.

The **gonads**—the testes and ovaries—produce the sex hormones, **testosterone**, and **estrogen and progesterone**, respectively. Among other effects, these hormones are responsible for the secondary sex characteristics that result in sexual dimorphism in humans. The **pineal gland** produces **melatonin**, which may be involved in circadian rhythms and the development of the reproductive organs; and the **thymus** secretes **thymosins**, which stimulate T-lymphocyte production and maturation.

Tissue and organs having other functions also produce hormones. Leptin is a newly described hormone secreted by adipose tissue that regulates appetite, and erythropoietin from the kidneys stimulates the production of red blood cells. Cells can produce local hormones, for example, prostaglandins are produced and act locally in various organs of the body.

> **Take Note:** *In order to be successful on the AP Biology exam, you should be able to describe the role of each hormone in the maintenance of homeostasis and the consequences of changes—both positive and negative—in hormone levels.*

Multiple Choice Questions

Questions 1–4 refer to the following five choices. For each question, select the choice that is most closely related. Each choice may be used once, more than once, or not at all.

    A. pituitary gland
    B. hypothalamus
    C. adrenal glands
    D. thyroid
    E. pancreas

1. Production of insulin and glucagon

2. Secretion of epinephrine and norepinephrine

3. Secretion of calcitonin

4. Secretion of aldosterone for the reabsorption of Na^+

5. Which of the following occurs in response to high blood pressure?
    A. Kidneys excrete Na^+ and water in urine.
    B. Adrenal cortex secretes aldosterone in the blood.
    C. Kidneys secrete rennin into the blood.
    D. Atrial natriuetic hormone is released.
    E. Aldosterone is secreted.

6. Hypothyroidism does NOT result in
    A. weight gain.
    B. decreased pulse rate.
    C. unusual sweating.
    D. lowered body temperature.
    E. lethargy.

7. Prostaglandins
    A. affect neighboring cells and are not secreted into the bloodstream.
    B. are chemical signals released by animals that influence behavior.
    C. cause the mammary glands to develop and produce milk.
    D. cause skin color changes in reptiles having melanophores.
    E. decrease pain and inflammation.

8. The main function of thyroid hormones is to
    A. cause uterine contractions.
    B. promote bone and tissue growth.
    C. stimulate development of eggs and sperm.
    D. increase metabolic rate.
    E. promote water conservation.

9. Anabolic steroids do NOT cause
    A. balding in men and women.
    B. liver dysfunction.
    C. decreased size of ovaries.
    D. severe acne.
    E. breast reduction in women.

10. In an athlete accused of blood doping, you would see
    A. decreased number of red blood cells.
    B. thinner blood.
    C. lower blood pressure.
    D. decreased maturation rate of red blood cells.
    E. decrease in cardiac output.

Free Response Question

The hypothalamus regulates the internal nervous system.

    A. **Explain** the areas of regulation and how the hypothalamus coordinates endocrine secretions.

    B. **Discuss** the function of THREE of the following hormones.
- adrenocorticotropic hormone (ACTH)
- prolactin (PRL)
- growth hormone (GH)
- melanocyte-stimulating hormone (MSH)

Annotated Answer Key (MC)

1. **E**

2. **C**

3. **D**

4. **C**; Major glands of the human endocrine secrete specific hormones. Calcitonin lowers blood calcium levels, aldosterone aids in water conservation, insulin and glucagon regulate blood glucose levels and epinephrine and norepinephrine are active in emergency situations and raise blood glucose levels.

5. **A**; The regulation of blood pressure and volume is a negative feedback loop. When the blood Na^+ level is low, the kidneys secrete renin that leads to the secretion of aldosterone from the adrenal cortex. Aldosterone causes the kidneys to reabsorb Na^+, and water follows so that blood volume and pressure return to normal. When the blood Na^+ is high, atrial natriuetic hormone is secreted, causing the kidneys to excrete Na^+, and water follows, returning the blood pressure to normal.

6. **C**; Unusual sweating is not a symptom of hypothyroidism but is a symptom of hyperthyroidism or oversecretion of thyroid hormone.

7. **A**; Prostaglandins often cause pain and inflammation in neighboring cells and are NOT carried in the bloodstream. Pheromones are signals that influence behavior, and prolactin is the hormone that causes milk production.

8. **D**; Oxytocin stimulates uterine contraction. LH and FSH stimulate development of eggs and sperm. GH promotes bone and tissue growth, and ADH promotes water conservation.

9. **C**; Steroids decrease testicular size, lower sperm count, and cause impotency in men. They also increase the size of the ovaries and cause menstruation and ovulation to cease in females.

10. **E**; Blood doping occurs when some athletes use erythropoietin (EPO) to improve performance by increasing the oxygen carrying capacity of the blood. This increases the number of red blood cells, the blood becomes thicker, blood pressure can become elevated, and the athlete is at an increased risk of heart attack or stroke.

Answer to FRQ

PART A (MAX 6)

- Autonomic nervous stem
- Controls:
  - Heartbeat
  - Body temperature
  - Water balance
- Produces hormones which are stored and secreted by the posterior pituitary
- Controls the secretions of the anterior pituitary → controls thyroid, adrenal cortex, and gonads
- Hypothalamus-releasing and inhibiting hormones
- Neurosecretory cells
- Portal system

PART B (MAX 6)

Hormone	Function
ACTH	Produces glucocorticoid
	Stimulates the adrenal cortex
TSH	Stimulates the release of T4 and T3
	Affects metabolism
Prolactin	Milk production
	Produced only after childbirth
GH	Skeletal and muscle growth/fat metabolism
	Affects rate at which amino acids enter cells → protein synthesis
MSH	Skin-color changes in fishes, amphibians, and reptiles
	Low concentrations in humans
	Cause change in melanophores

CHAPTER 41
REPRODUCTIVE SYSTEMS

> **Take Note:** *To successfully prepare for the AP Biology exam, you should understand how the reproductive system functions in the maintenance of the overall homeostasis of an animal. Be prepared to explain how the endocrine system regulates the reproductive cycle of humans.*

41.1   How Animals Reproduce

Animals reproduce in one of two ways: sexually or asexually. Ordinarily, **asexual reproduction** may quickly produce a large number of offspring genetically identical to the parent, unless a mutation occurs. Several invertebrates can reproduce asexually. Sponges produce **gemmules** that can develop into new sponges. Cnidarians (polyps) reproduce asexually by **budding**, and planarians can reproduce asexually by **regeneration**. **Parthenogenesis** is a form of asexual reproduction often seen in insects in which an offspring develops from an unfertilized egg.

Sexual reproduction involves gametes and produces offspring that are genetically different from the parents. Dioecious animals have separate sexes, while other animals are monoecious or **hermaphroditic**. Most hermaphrodites reproduce by **cross-fertilization**, where two animals exchange gametes. Some hermaphrodites can reproduce by **self-fertilization**. The **gonads** are the primary sex organs for the production of gametes, but there are also accessory organs. Eggs are produced in **ovaries** and sperm are produced in **testes**. Eggs and sperm are produced from **germ cells** by meiosis. The **accessory organs** consist of storage areas for sperm and ducts that conduct the gametes. They also contribute to formation of the semen. In sexual reproduction, fertilization can occur internal or externally. Internal fertilization occurs after **copulation**, when the male deposits sperm inside the body of the female.

Animals typically protect their eggs and embryos. The eggs of **oviparous** animals, such as birds, contain yolk, and in terrestrial animals a shelled egg prevents drying out. The amount of yolk is dependent on whether there is a larval stage.

Reptiles and birds have extraembryonic membranes that allow them to develop on land; these same membranes are modified for internal development in mammals. **Ovoviviparous** animals incubate their eggs internally until the offspring have hatched, and **viviparous** animals incubate the embryo that is not contained within an egg. Placental mammals are viviparous animals. The **placenta** is a membrane derived from the chorion that allows an embryo to exchange oxygen and nutrients with its mother during its period of gestation.

> **Take Note:** *On the AP exam, you may be asked to compare and contrast reproduction in animals with reproduction in plants. How is asexual reproduction in plants similar to and different from asexual reproduction in animals? How is gamete formation in plants similar to and different from gamete formation in animals?*
>
> *You should also be prepared to give examples of organisms that reproduce sexually, organisms that reproduce asexually, and organisms that reproduce both sexually and asexually. What adaptations do these organisms have that allow them to reproduce the way they do? How do environmental conditions affect reproduction?*

41.2 Male Reproductive System

In human males, sperm are produced in the **testes**, mature in the **epididymides**, and may be stored in the **vasa deferentia** before entering the urethra, along with **seminal fluid** (produced by seminal vesicles, the prostate gland, and bulbourethral glands). Sperm are ejaculated during male orgasm, when the penis is erect.

Spermatogenesis occurs in the **seminiferous tubules** of the testes, which also produce testosterone in interstitial cells. **Testosterone** is a hormone that brings about the maturation of the primary sex organs during puberty and promotes the secondary sex characteristics of males, such as low voice, facial hair, and increased muscle strength. Sperm have three parts at functional maturity: 1) a head, 2) a midpiece, and 3) a tail. The head contains a nucleus, and is covered by an enzyme-containing cap called an **acrosome**. These enzymes aid the sperm in penetrating the egg. The midpiece contains mitochondria that provide the energy for movement, and the tail is a flagellum made of mictotubules in the typical 9+2 arrangement of cilia and flagella.

Sexual function is controlled by the endocrine system. **Follicle-stimulating hormone** (FSH) from the anterior pituitary stimulates spermatogenesis, and **luteinizing hormone** (LH) stimulates testosterone production. A hypothalamic-releasing hormone, **gonadotropic-releasing hormone** (GnRH), controls anterior pituitary production and FSH and LH release. The level of testosterone in the blood controls the secretion of GnRH and the anterior pituitary hormones by a negative feedback system.

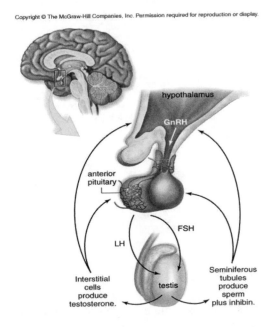

Hormonal control of testes

Take Note: *For the AP Biology exam, you should be prepared to describe the role of negative feedback in the regulation of male sexual function.*

41.3 Female Reproductive System

In females, an oocyte produced by an ovary enters an **oviduct**, which leads to the **uterus**. The oviducts have are not attached to the ovary; they have finger-like projections called **fimbriae** that extend around the ovaries to guide the oocyte toward the uterus. The uterus has a thick wall of muscle that is lined with

endometrium. The uterus is where the embryo will complete development during the gestation period. Through the cervix, the uterus opens into the **vagina**, which serves as the birth canal. The external genital area of women includes the vaginal opening, the clitoris, the labia minora, and the labia majora.

In either ovary, one follicle a month matures, produces a secondary oocyte by meiosis, and becomes a **corpus luteum** once the oocyte is released. This is called the **ovarian cycle**. The follicle and the corpus luteum produce estrogens, collectively called **estrogen**, and **progesterone**, the female sex hormones.

The **uterine cycle** occurs concurrently with the ovarian cycle. In your textbook, refer to the figure that depicts the uterine cycle. In the first half of these cycles (days 1–13, before ovulation), the anterior pituitary produces FSH and the follicle produces estrogen. Estrogen causes the endometrium to increase in thickness to prepare for the implantation of a growing embryo. In the second half of these cycles (days 15–28, after ovulation), the anterior pituitary produces LH and the follicle produces progesterone. Progesterone causes the endometrium to become secretory. It separates from the wall of the uterus and is released out of the body if a woman is not pregnant. Feedback control of the hypothalamus and anterior pituitary causes the levels of estrogen and progesterone to fluctuate. When they are at a low level, fertilization has not occurred and menstruation begins.

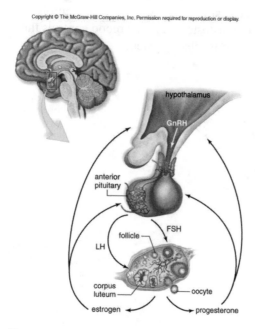

Hormonal control of ovaries

If fertilization occurs, a zygote is formed and development begins. The resulting embryo travels down the oviduct and implants itself in the prepared endometrium. A placenta, which is the region of exchange between the fetal blood and the mother's blood, forms. At first, the placenta produces **human chorionic gonadotropin** (HCG), which maintains the corpus luteum until the; later, it produces progesterone and estrogen until the placenta begins producing these hormones, which will preserve the pregnancy by opposing the effects of the anterior pituitary and maintaining the lining of the uterus.

The female sex hormones (estrogen and progesterone) also affect other traits of the body. Primarily, estrogen brings about the maturation of the primary sex organs during puberty and promotes the secondary sex characteristics of females, including less body hair than males, a wider pelvic girdle, a more rounded appearance, and development of breasts.

Take Note: *For the AP Biology exam, you should be prepared to describe the role of negative feedback in the control of the ovarian and uterine cycles. How does fertilization of an egg affect these cycles?*

41.4 Control of Reproduction

Numerous birth control methods and devices, such as the birth control pill, diaphragm, and condom, are available to prevent pregnancy. A morning-after pill, RU-486, is now available. Also known as Mifepristone, this drug prevents the uterine lining from thickening to accommodate the implanted embryo. Research is currently being conducted to develop other types of birth control for women, as well as hormonal birth control for men. Some couples are infertile and may use assisted reproductive technologies to have a child. **Artificial insemination** and **in vitro fertilization** have been followed by more sophisticated techniques, such as gamete intrafallopian transfer and intracytoplasmic sperm injection.

41.5 Sexually Transmitted Diseases

Sexually transmitted diseases include AIDS, an epidemic disease; genital warts, which lead to cancer of the cervix; genital herpes, which repeatedly flares up; hepatitis, especially types A and B; chlamydia and gonorrhea, which cause pelvic inflammatory disease (PID); and syphilis, which has cardiovascular and neurological complications if untreated. The causes for various sexually transmitted diseases are summarized in the table below.

Disease	Cause
AIDS	Human Immunodeficiency Virus
Genital warts	Human papillomaviruses
Genital herpes	Herpes simplex viruses
Hepatitis	Heptatitis viruses A, B, C
Chlamydia	*Chlamydia trachomatis*
Gonorrhea	*Neisseria gonorrhoeae*
Syphilis	*Treponema pallidum*

Take Note: *For the AP Biology exam, you should know which sexually transmitted diseases are caused by viruses and which by bacteria. What diseases can be treated by antibiotics? What other organisms can cause a homeostatic imbalance in the reproductive system?*

Multiple Choice Questions

Questions 1–4 refer to the following five choices. For each question, select the choice that is most closely related. Each choice may be used once, more than once, or not at all.

    A. sponge
    B. hydra
    C. honeybee
    D. earthworm
    E. tapeworm

1. Produce asexual gemmules that develop into new individuals

2. Reproduce through parthenogenesis

3. Hermaphroditic; capable of self-fertilization

4. Reproduce asexually by budding

5. Which of the following is an example of sequential hermaphroditism?
    A. Cnidarians produce only temporary gonads in the fall when sexual reproduction occurs.
    B. In wrasses, if a male fish dies, the largest female becomes a male.
    C. Earthworms contain both male and female sex organs and cross-fertilize.
    D. If a sea star is chopped up, it has the potential to regenerate into several new individuals.
    E. A species of hydra has an asexual colonial stage and a sexually reproducing medusa stage.

6. Any animal that deposits an egg in the external environment is
    A. viviparous.
    B. marsupial.
    C. a monotreme.
    D. oviparous.
    E. a placental mammal.

7. Which of the following events occurs during ovulation?
    A. breakdown of the endometrium
    B. development of the corpus luteum
    C. LH spike
    D. decreased secretion of progesterone
    E. decreased secretion of estrogen

8. Acrosomal enzymes in the sperm
    A. mature spermatids.
    B. support and nourish the sperm.
    C. allow a sperm to penetrate the egg.
    D. promote spermatogenesis.
    E. control the production of testosterone.

9. An advantage of sexual reproduction is
    A. the production of a genetically identical individual.
    B. increase in genetic variability.
    C. a decrease in mutations.
    D. increased reproduction rate.
    E. increased ability to colonize favorable environments.

10. If pregnancy does not occur,
    A. HCG is secreted.
    B. the corpus luteum degenerates.
    C. the endometrium thickens.
    D. progesterone levels increase.
    E. estrogen is secreted.

Free Response Question

Most animals reproduce sexually, however some are capable of asexual reproduction.

A. **Discuss** the advantages of sexual reproduction in animals.

B. **Explain** each of the following strategies:
- placental mammals
- marsupials
- monotremes
- egg-laying reptiles

Annotated Answer Key (MC)

1. **A**

2. **C**

3. **E**

4. **B**

For Questions 1–4: Although the majority of animals reproduce sexually, a few are also capable of asexual reproduction. Pathenogenesis is a modification of sexual reproduction in which an unfertilized egg develops into a complete individual. The queen honeybee makes and stores sperm to fertilize eggs. Any unfertilized eggs become haploid males. In budding, a new individual arises as an outgrowth of the parent.

5. **B**; While all of these statements are true, sequential hermaphroditism is a form of sex reversal, with the organisms expressing only one gender at a time.

6. **D**; Most aquatic animals deposit their eggs in the water, where they undergo development. These animals are oviparous. Many have a larval stage. Marsupials and placental mammals do not lay eggs. Placental mammals are termed viviparous because they do not lay eggs and development occurs inside the female's body until offspring can live independently.

7. **C**; During the follicular phase, FSH is released by the anterior pituitary gland. The ovarian follicle produces increasing levels of estrogen, which causes the endometrium to thicken during the proliferative phase prior to ovulation. After ovulation, the corpus luteum also develops.

8. **D**; The acrosome is the cap on the head of a sperm. It contains enzymes to assist in penetration. LH controls the production of testosterone, which is necessary for the maturation of sperm.

9. **B**; In sexual reproduction, the parents unite to form a genetically unique individual. In asexual reproduction, a single parent gives rise to offspring that are identical to the parent, unless mutations have occurred. The adaptive advantage of asexual reproduction is that organisms can reproduce rapidly and colonize favorable environments quickly.

10. **B**; If fertilization occurs, human chorionic gonadotropin is secreted, which maintains the corpus luteum until the placenta begins its own production of progesterone and estrogen. The endometrium also thickens. However, if fertilization does not occur, estrogen and progesterone are not secreted, the lining does not thicken, and the corpus luteum disintegrates.

Answer to FRQ

PART A (MAX 4)
- Production of a genetically unique individual
- Increased SPECIES resistance to pathogens
- Genetic variability of the species

- Variability to guard against predation (coevolution)

PART B (MAX 8)

Reptiles
- Leathery shelled eggs → extraembryonic membranes
- Prevents dessication
- Yolk → nutrient-rich material

Marsupials
- Yolk sac membrane → supplies nutrients
- Development in a pouch → milk production

Placental
- Internal exchange of materials
- Continues to supply after birth

Monotremes
- Egg-laying mammals

CHAPTER 42
ANIMAL DEVELOPMENT

42.1 Early Developmental Stages

Development begins at **fertilization**, the fusion of a sperm and an egg to form a zygote. Only one sperm actually enters the egg, though several attempt to penetrate the **corona radiata** and the **zona pellucida**, the protective covering of the egg and extracelluar matrix of the egg, respectively. Once a sperm has attached to the plasma membrane of the egg, several things happen to prevent **polyspermy**, including the depolarization of the oocyte membrane and formation of the zygote, which block other sperm from penetrating the egg. Both sperm and egg contribute chromosomes to the diploid zygote.

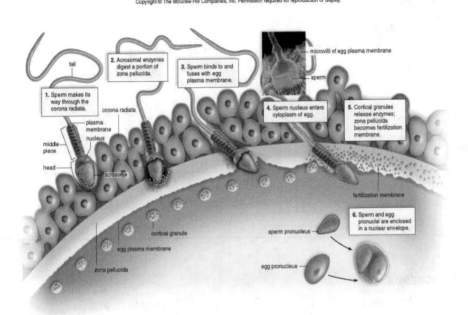

Fertilization

The early developmental stages in animals proceed from cellular stages to tissue stages to organ stages. During the cellular stage, **cleavage** (cell division) occurs, but there is no overall growth. The result is a **morula**, a ball of 16 cells, which becomes the hollow **blastula** when an internal, fluid-filled cavity (the **blastocoel**) appears. Although all animals undergo this process during development, the cleavage stage in vertebrates is not always equal due to the presence of the yolk. As well, the blastocoel in vertebrates is the space that separates the blastula from the yolk as the blastula spreads out over the yolk.

During the tissue stage, **gastrulation** involves invagination of cells into the blastocoels, and migration of cells, and results in formation of the germ layers: ectoderm, mesoderm, and endoderm. The **ectoderm** is the outer layer of cells that will eventually become the nervous and integumentary systems, the mouth and the rectum. The **mesoderm** becomes the musculoskeletal, cardiovascular, urinary, lymphatic, and reproductive systems. The **endoderm** and the cavity called the **archenteron** that it surrounds develop into the digestive and respiratory systems.

Organ formation can be related to germ layers. For example, during **neurulation**, the nervous system develops. A **neural tube** is formed from midline ectoderm, just above the notochord. The anterior end of the neural tube will become the brain, and the rest will become the spinal cord. At this point, it is possible to draw a typical cross-section of a vertebrate embryo in which the notochord has not been replaced by the vertebral column

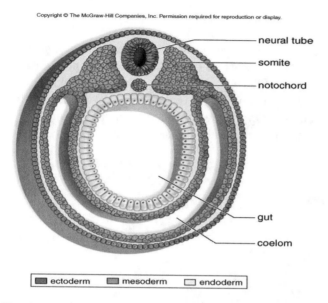

- neural tube
- somite
- notochord
- gut
- coelom

☐ ectoderm   ☐ mesoderm   ☐ endoderm

Vertebrate embryo, cross-section

Take Note: *Using a vertebrate of your choice as an example, you should be able to explain the processes of cleavage, gastrulation, and neurulation. What is accomplished at each stage of development?*

42.2   Developmental Processes

Embryonic development requires three processes: 1) growth, 2) cellular differentiation, and 3) morphogenesis. **Cellular differentiation**, the specialization of cells in structure and function, begins with cytoplasmic segregation in the egg. After the first cleavage of a frog embryo, only a daughter cell that receives a portion of the **gray crescent** is able to develop into a complete embryo. Therefore, cytoplasmic segregation of maternal determinants occurs during early development of a frog. **Induction**, the ability of embryonic tissues to influence the development of other tissues, is also part of cellular differentiation. For example, the notochord induces the formation of the neural tube in frog embryos. The reciprocal induction that occurs between the lens and the optic vesicle is another good example of induction. In *C. elegans,* investigators have shown that induction is an ongoing process in which one tissue after the other regulates the development of another, through chemical signals coded for by particular genes.

The shape and form of the body results from **morphogenesis**, which involves cell movement, pattern formation, and apoptosis. In **pattern formation**, cells divide and differentiate, organizing themselves into tissues and organs. **Morphogens** are transcription factors that bind to DNA. Some morphogen genes determine the axes of the body, and others regulate the development of segments. During development, sequential sets of master genes code for morphogen gradients that activate the next set of master genes in sequential order. Apoptosis is necessary for the formation of hands and feet.

Homeotic genes control pattern formation such as the presence of antennae, wings, and limbs on the segments of *Drosophila*. Homeotic genes code for proteins that contain a **homeodomain**, a particular sequence of 60 amino acids. These proteins are also transcription factors, and the homeodomain is the portion of the protein that binds to DNA. Homologous homeotic genes have been found in a wide variety of organisms, and therefore they must have arisen early in the history of life and been conserved.

> **Take Note:** *For the AP Biology exam, you should be able to explain the role of induction in embryonic development. Can you describe the experimental evidence that demonstrated how induction occurs?*

42.3 Human Embryonic and Fetal Development

Human development can be divided into **embryonic development** (months 1 and 2) and **fetal development** (months 3–9). The early stages in human development resemble those of the chick. The similarities are probably due to their evolutionary relationship, not to the amount of yolk the eggs contain, because the human egg has little yolk.

The extraembryonic membranes appear early in human development. The **trophoblast** of the **blastocyst** is the first sign of the chorion, which goes on to become the fetal part of the placenta. Exchange between fetal and maternal blood occurs at the placenta. The amnion contains **amniotic fluid**, which cushions and protects the embryo. The yolk sac and allantois are also present. The yolk sac is the location where blood cell formation first occurs. The blood vessels of the allantois become the umbilical cord. The allantois, which will eventually become the urinary bladder, collects the urine produced by the fetal kidneys.

Fertilization occurs in the oviduct, and cleavage occurs as the embryo moves toward the uterus. The morula becomes the blastocyst before implanting in the endometrium of the uterus. Organ development begins with neural tube and heart formation. There follows a steady progression of organ formation during embryonic development. During fetal development, refinement of organ systems occurs, and the fetus adds weight. When the brain of the fetus is mature, the amnion bursts, and the fetus is delivered, followed by the placenta.

Multiple Choice Questions

Questions 1–4 refer to the following five choices. For each question, select the choice that is most closely related. Each choice may be used once, more than once, or not at all.
  A. ectoderm
  B. mesoderm
  C. endoderm
  D. neural plate
  E. archenteron

1. Gives rise to the musculoskeletal system

2. Dorsal surface of the embryo

3. Gives rise to the epithelial lining of the digestive and respiratory tracts

4. A split of this layer becomes the coelom.

5. The function of the acrosome is to
    A. produce semen for transport through the oviduct.
    B. provide enzymes for digestion of the zona pellucida.
    C. provide ATP for locomotion by the flagella.
    D. fuse with the fertilization membrane to prevent polyspermy.
    E. contain DNA for fusion with the oocyte.

6. The placenta
    A. functions in gas, nutrient, and waste exchange.
    B. begins to develop upon implantation of the blastocyst.
    C. secretes molecules to avoid detection by the immune system.
    D. receives blood from the umbilical arteries.
    E. contains a barrier that prevents toxic substances from crossing to the fetus.

7. A mother gives birth to a child with spina bifida, a neural tube defect. She may not have consumed adequate amounts of
    A. vitamin D.
    B. citric acid.
    C. folic acid.
    D. vitamin K.
    E. lactose.

8. Apoptosis is NOT critical in
    A. normal functioning of the immune system.
    B. preventing the occurrence of cancer.
    C. morphogenesis.
    D. preventing webbing of hands and feet.
    E. cleavage to produce a morula.

9. The second trimester is characterized by
    A. development of organs and organ systems.
    B. embryonic and fetal development.
    C. major growth of functional organ systems.
    D. cleavage from a zygote to a blastocyst.
    E. implantation of the zygote.

10. Homeotic genes do NOT
    A. code for transcription factors.
    B. directly interfere with the cell cycle.
    C. direct the activities of morphogenesis.
    D. regulate cell to cell adhesion.
    E. remain highly conserved and present in many organisms.

Free Response Question

In vertebrates, specific events occur at precise times during embryonic development.

   A. **Describe** the placement of the extraembryonic membranes.

   B. **Discuss** how the nervous system of the vertebrate embryo develops.

Annotated Answer Key (MC)

1. **B**; The mesoderm gives rise to the musculoskeletal system and skin, cardiovascular system, urinary system, and more. It is the middle embryonic germ layer.

2. **D**; Neural folds develop on either side of the neural groove, which becomes the neural tube when these folds fuse.

3. **C**; The endoderm is the inner layer of the embryo. Gastrulation is not complete until three layers of cells that will develop into adult organs are produced.

4. **B**; A splitting of the mesoderm produces a coelom, which is completely lined by mesoderm.

5. **B**; Human sperm have three distinct parts: a tail, which is flagellated and allows the sperm to swim toward the egg; a middle piece, which contains mitochondria; and a nucleus containing the DNA. This nucleus is capped by a membrane-bound acrosome that contains enzymes for digestive purposes.

6. **E**; Harmful substances can cross the placenta. Serious birth defects can result if substances are taken at critical times of embryonic development.

7. **C**; Folic acid is present in many breads and cereals, green leafy vegetables, nuts and citrus fruits. Neural tube defects occur just a few weeks after conception.

8. **E**; As the zygote moves along the oviduct, it undergoes cleavage to produce a morula. Multiple cell divisions, not cell death, have occurred.

9. **A**; Development is divided into trimesters characterized by specific developmental accomplishments.

10. **B**; They do not interfere directly with the cell cycle but they appear to determine which cell cycle schedule is carried out.

Answer to FRQ

PART A (MAX 8)
One point for identification of membrane, one point for description of function

- Allantois: collects nitrogenous waste
- Amnion: contains the protective fluid
- Yolk sac: nutrition/nourishment
- Chorion: gas exchange

PART B (MAX 6)
- Ectoderm identified as tissue of origin
- Neural plate formation → neural tube when the folds fuse
- Anterior end → brain
- Remaining portions → spinal cord
- Cells migrate to various locations → skin/muscles/adrenal medulla/ganglia of PNS
- Spinal cord protects the CNS
- Notochord signals development
- Homeotic genes cause developmental signals (Hox)

CHAPTER 43
BEHAVIORAL ECOLOGY

43.1   Inheritance Influences Behavior

Behavior is the observable responses that organisms have to environmental stimuli. Investigators have long been interested in the degree to which nature (genetics) or nurture (environment) influences behavior. Studies with birds, snakes, humans, snails, and mice have been done, among many others. Hybrid studies with lovebirds produce results consistent with the hypothesis that behavior has a genetic basis. Garter snake experiments indicate that the nervous system controls behavior. Twin studies in humans show that certain types of behavior are apparently inherited. *Aplysia* DNA studies indicate that the endocrine system also controls behavior.

43.2   The Environment Influences Behavior

Though there is a genetic basis to behavior, it is also affected by environment influences. Even behaviors formerly thought to be **fixed action patterns** (FAPs), or otherwise inflexible, sometimes can be modified by learning. **Learning** is the modification of behavior based on experience. For instance, the red bill of laughing gulls initiates chick begging behavior. However, with experience, chick begging behavior improves and they demonstrate an increased ability of chicks to recognize parents.

Other studies suggest that learning is involved in behaviors. **Imprinting** in birds, during a sensitive period, causes them to follow the first moving object they see. Song learning in birds involves including the existence of a sensitive period during which an animal is primed to learn, and the positive benefit of social interactions with other birds.

Associative learning is a change in behavior that occurs in response to the association of two events with each other. Classical conditioning and operant conditioning are types of associative learning. In **classical conditioning**, the pairing of two different types of stimuli causes an animal to form an association between them. In this way, dogs will salivate at the sound of a bell because they have learned to associate the bell with food. In **operant conditioning**, animals learn behaviors because they are rewarded when they perform them.

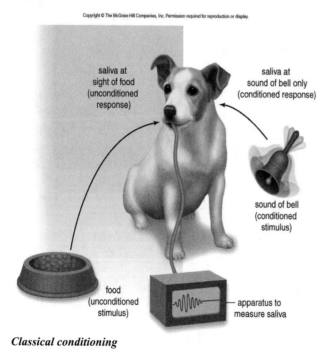

saliva at sight of food (unconditioned response)

saliva at sound of bell only (conditioned response)

sound of bell (conditioned stimulus)

food (unconditioned stimulus)

apparatus to measure saliva

Classical conditioning

Animals migrate when they travel long distances from one location to another. Orientation and **migratory behavior** occur in several groups of animals. **Orientation** is the ability to move in a certain direction, such as when birds migrate from north to south in the winter. Migration can also involve **navigation**, a learned ability to change direction in response to environmental cues that indicate that an animal is traveling in the wrong direction. Animals use the sun, stars, and the earth's magnetic field in order to migrate.

Imitation and insight learning—**cognitive learning**—also occurs in animals. Some animals learn certain behaviors by imitating others. **Insight learning** has occurred when an animal can solve a new and different problem without prior experience.

> **Take Note:** *For the AP Biology exam, you should be able to describe several different types of behavior and how they are affected by environmental influences. What adaptive advantages do these behaviors confer on the animals that display them?*

43.3   Animal Communication

Communication is an action by a sender that affects the behavior of a receiver. Communication is a vital aspect of social behavior in animals, because it allows them to work cooperatively in groups. Chemical, auditory, visual, and tactile signals are forms of communication fostering cooperation that benefits both the sender and the receiver.

Pheromones are chemical signals that are passed between members of the same species. Pheromones are important in attracting a mate for reproduction and for territory marking, and are as effective at night as during the day.

Auditory communication includes language, which may occur between other types of animals and not just humans. Auditory communication can be modified by volume, pattern, duration, and repetition, and includes calls, songs, and language. It is also equally effective during the day and the night.

Visual communication allows animals to signal others without the need of auditory or chemical messages. Because it is used by animals that are active during the day, that is the time when it is more effective. Courtship and defense behavior are types of visual communication.

Tactile communication occurs when animals touch each other, and is especially associated with sexual behavior. It also fosters bonding between members of the same social or family group. Tactile communication is also used by animals to communicate the location of food relative to the sun, as seen in the waggle dance done by honeybees.

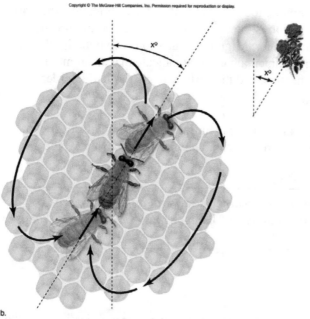

b.
Communication among bees

Take Note: *For the AP Biology exam, you should be able to describe several different types of communication and how they are affected by environmental influences. What adaptive advantages does communication confer on the animals that display them?*

43.4   Behaviors That Increase Fitness

Traits that promote reproductive success are expected to be advantageous overall, despite any possible disadvantage. Since there is a genetic aspect to behavior, it stands to reason that behavior is affected by natural selection; this is the study of **behavioral ecology**. Some animals are **territorial** and defend a territory where they have food resources and can reproduce. When animals choose those foods that return the most net energy, they have more energy left over for reproduction. This **optimal foraging model** suggests that animals will forage for food that provides more energy than is spent in obtaining it.

Reproductive strategies include monogamy, polygamy, and polyandry. Which strategy is employed depends on the animal and its environment. **Monogamy** occurs when there are limited mating opportunities for males or when territoriality exists. **Polygamy** is seen in animals where the females expend time and energy caring for their young, and thus may not be available for mating. **Polyandy** is when a female mates with several males. This occurs in situations where multiple offspring cannot be supported by the environment at the same time. **Sexual selection** is a form of natural selection that selects for traits that increase an animal's fitness. The result of sexual selection is often female choice and male competition. Males produce many sperm and are expected to compete to inseminate females. By producing as many offspring as possible, males increase their own fitness. Females produce few eggs and can be selective about their mates, thereby increasing their own fitness.

Living in a social group can have its advantages (e.g., ability to avoid predators, raise young, and find food). It also has disadvantages (e.g., tension between members, spread of illness and parasites, and reduced reproductive potential). When animals live in groups, the benefits must outweigh the costs or the behavior would not exist.

In most instances, the individuals of a society act to increase their own fitness (ability to produce surviving offspring). In this context, it is necessary to consider **inclusive fitness**, which includes personal reproductive success as well as the reproductive success of relatives. Sometimes, animals perform **altruistic** acts, as when individuals help their parents rear siblings. Social insects help their mother reproduce, but this behavior seems reasonable when we consider that siblings share 75% of their genes. Among mammals, a parental helper may be likely to inherit the parent's territory. In **reciprocal altruism**, animals aid one another for future benefits.

Multiple Choice Questions

Questions 1–4 refer to the following five choices. For each question, select the choice that is most closely related. Each choice may be used once, more than once, or not at all.
    A. fixed action pattern
    B. imprinting
    C. classical conditioning
    D. operant conditioning
    E. habituation

1. You become nauseous when you think of a food that once caused you to be sick to your stomach.

2. Once you begin a yawn, it is difficult to stop and it must run its course.

3. A baby recognizes its mother and father.

4. A child associates a pleasant feeling with reading books while sitting on her mother's lap.

5. Communication via sound is advantageous for animals because it
    A. does not attract predators.
    B. cements social bonds within a group.
    C. uses body language to send messages.
    D. requires organisms to look at one another.
    E. can be modified by pattern, duration, and repetition.

6. Monogamy does NOT occur
    A. when territoriality exists.
    B. frequently in mammals.
    C. when one female mates with more than one male.
    D. when males have limited mating opportunities.
    E. when the male is certain the offspring are his.

7. Living in groups
    A. prevents individuals from contracting parasites.
    B. assists with reducing disputes over food and shelter.
    C. helps animals avoid predators and rear offspring.
    D. eliminates dominance hierarchies.
    E. decreases reproductive benefits.

8. Which of the following is NOT a factor in sexual selection?
    A. possession of antlers in deer
    B. bright coloration of birds
    C. decreased size of female spiders
    D. symmetrical feathers of peacocks
    E. burrowing to avoid prey in prairie dogs

9. Migration
    A. often occurs due to environmental stimuli.
    B. does not require orientation.
    C. decreases survival and reproductive success.
    D. is in response to a fixed action pattern.
    E. is tied with circadian rhythms.

10. When sowbugs (pillbugs, squash bugs, potato bugs, rolly pollies) are exposed to either dry or moist habitats, they are more active in the moist habitat. This is an example of
    A. migration.
    B. kinesis.
    C. taxis.
    D. fixed action pattern.
    E. behavioral rhythm.

Free Response Question

Animals exhibit a wide diversity of social behaviors. This requires that they develop communication strategies.

    A. **Discuss** the different communication strategies animals have developed.

    B. **Explain** evidence showing that communication affects the behavior of the receiver.

Annotated Answer Key (MC)

1. **C**; This is classical conditioning because there was an automatic response (nausea) from a particular stimulus (food).

2. **A**; Fixed action patterns are behaviors that are elicited by a stimulus and are unable to start once finished. It is hypothesized that once we see a person yawn, we are unable to stop our response to that particular stimulus.

3. **B**; Imprinting is considered a form of learning and often happens during a sensitive or critical time of development. Social interactions between parent and offspring during this time seem key to normal imprinting.

4. **C**; The child associates an automatic response (pleasure) with a stimulus (book reading) and sitting on her mother's lap.

5. **E**; Auditory communication is faster than chemical communication and is effective during both night and day. It can be modified in so many ways that vocalizations can have so many meanings.

6. **C**; Polygamy occurs when males monopolize multiple females. Polyandry occurs when one female mates with more than one male. Monogamy occurs when both male and female help with the rearing of the young and pair bond. Monogamy occurs when males have limited mating opportunities, territoriality exists, and the male is fairly certain the offspring are his.

7. **C**; Living in societies has greater reproductive benefits than costs. Animals can avoid predators and increase their reproductive success by raising more young to reproductive maturity. However, there are many disputes over food and shelter, which often requires dominance hierarchies, and often living in close quarters exposes individuals to illness and parasites that can easily pass from one animal to another. However, this increases social behaviors such as grooming and, in the case of humans, an investment in medical care.

8. **E**; Sexual dimorphism often influences sexual selection. Males compete for females and therefore characteristics that are appealing to females often proliferate in reproductively successful males.

9. **A**; Migration is long distance travel from one location to another. It requires the ability to travel in a particular direction, navigation, and has a proximate cause and ultimate cause. Often environmental stimuli such as temperature or day length tell the animals it is time to travel.

10. **B**; Kinesis is a change in an activity in response to a stimulus such as humidity; it is neither an attraction nor a repulsion. If the pillbugs were attracted to the humidity, it would be taxis. Pillbugs exhibit a negative phototaxis when they are moved away from light towards darker areas.

Answer to FRQ

Part A (MAX 8)
Two points each type

Tactile Communication
- One animal touches another animal.
- Cements social bonds
- Can occur during night/day
- Does not alert predators
- Often uses environmental guides: magnetic field, biological clock

Visual Communication
- Does not require chemical or auditory messages
- Active during the day
- May result in reduced fitness/alerts predators
- Increases sexual selection by females
- Affects dominance hierarchies

Chemical Communication
- Pheromones: chemical signals in low concentration between MEMBERS OF THE SAME SPECIES
- Many pheromones → different meanings
- Do not alert predators
- Receptors may detect pheromones →hypothalamus release → hormones

Auditory Communication
- Fast type of communication
- Effective night/day
- Many modifications
- May alert predators
- Impacts language

PART B (MAX 4)
Two points each example with explanation

- Examples are numerous. May include any example dictating a specific behavioral response due to one of the above types of communication.

CHAPTER 44
POPULATION ECOLOGY

44.1 Scope of Ecology

Ecology is the study of the interactions of organisms with other organisms and with the physical environment. Ecology and evolution are intertwined; ecological interactions are the selection pressures that lead to natural selection, which affects the ecological interactions. Ecology encompasses several levels of study: organism, population, community, ecosystem, and finally the biosphere. A **population** consists of all the organisms of the same species that live in a particular area at the same time, while a **community** is made of several populations of multiple species. An **ecosystem** refers to a community and the abiotic environment, and the **biosphere** consists of the zones of soil, air, and water where life exists. Ecologists are particularly interested in how interactions affect the distribution, abundance, and life history strategies of organisms.

44.2 Demographics of Populations

Demographics include statistical data about a population, such as population density, distribution of individuals, and growth rate. **Population density** is the number of individuals per unit area or volume. **Distribution** of individuals is the dispersal pattern of individuals within an area, and can be uniform, random, or clumped. A **uniform** distribution occurs in a population when all of the individuals are evenly spaced apart. A **random** distribution has no apparent pattern to the spacing of individuals in a population. In a **clumped** distribution, some individuals are spaced close to each other, but far from others. A population's distribution is often determined by the availability of resources, that is, abiotic factors such as water, temperature, and the availability of nutrients and mates.

Population growth is dependent on number of births and number of deaths, immigration, and emigration. The number of births minus the number of deaths results in the rate of natural increase (growth rate). For instance, if in a population of 100 individuals there are 15 births and and 5 deaths, the **growth rate** can be calculated as:

$$(15 - 5)/100 = 0.01 = 1.0\%$$

Mortality (deaths per capita) within a population can be recorded in a life table and illustrated by a survivorship curve, which predicts the probability of individuals in a cohort (all of the members of a population born at the same time) surviving to particular ages. There are three types of **survivorship curves**, shown below:

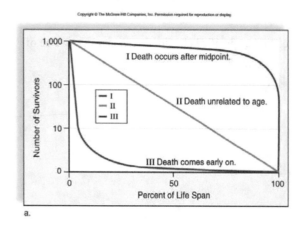

Survivorship curves

In a type I curve, most individuals in a population survive well past 50% of their expected life span. In a type II curve, there is a constant rate of death within the population. In a type III curve, the rates of survival at young ages are low.

The pattern of population growth is reflected in the age distribution of a population, which consists of pre-reproductive, reproductive, and post-reproductive segments. **Age structure diagrams** can be used to predict the growth rate of a population. Populations that are growing exponentially have a pyramid-shaped age distribution pattern.

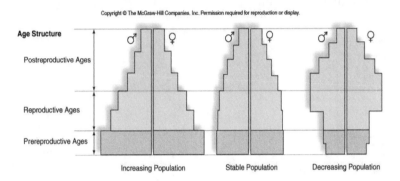

Age structure diagram

> **Take Note:** *On AP Biology exam, you should be able to analyze an age structure diagram and identify each type of survivorship curve. Can you give examples of organism that exhibit each type of curve?*

44.3 Population Growth Models

There are two models for population growth: semelparity and iteroparity. In **semelparity**, the individuals in a population have one reproductive event in their lifetime. In **iteroparity**, members of a population have multiple reproductive events during their lifetime. Scientists have developed mathematical models of population growth based on these two reproductive patterns. One model for population growth assumes that the environment offers unlimited resources. In this example, the members of the population have discrete reproductive events, and therefore the size of next year's population is given by the equation $N_t + 1 = RN_t$. Under these conditions, **exponential growth** results in a J-shaped curve containing a lag phase where growth is slow due to a small population, as well as a growth phase, where growth is increasing.

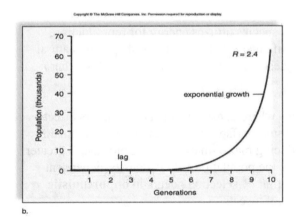

Model for exponential growth

Most environments, however, restrict growth, and exponential growth cannot continue indefinitely. Under these circumstances, an S-shaped, or **logistic growth** curve results. In this graph, there is also a lag phase and a growth phase, but there is also a deceleration phase, where growth slows down before reaching a stable equilibrium phase, where growth stabilizes. The growth of the population is given by the equation $dN/dt = rN (K - N)/K$ for populations in which individuals have repeated reproductive events. The term $(K - N)/K$ represents the unused portion of the **carrying capacity** (K), the maximum number of individuals of a given species the environment can support. When the population reaches carrying capacity, the population stops growing because environmental conditions oppose **biotic potential**, the maximum rate of natural increase (growth rate) for a population.

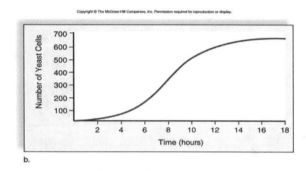

b.

Model for logistic growth

Take Note: *In preparation for the AP Biology exam, you should be able to draw and explain the two models of population growth. Can you give examples of organisms that demonstrate each model?*

44.4   Regulation of Population Size

Population growth is limited by density-independent factors and density-dependent factors. **Density-independent factors** include abiotic factors such as weather. A natural disaster could lead to a drastic decrease in population size, however the effect of the event is not dependent on the density of the population. **Density-dependent factors** include biotic factors such as predation and competition. In this case, the percentage of the population affected does depend on the population density. Other means of regulating population growth exist in some populations. For example, territoriality and dominance hierarchies are means of population regulation, as are recruitment and migration. Other populations seem to not be regulated, and their population size fluctuates widely.

44.5   Life History Patterns

The **life history** of a species considers the ways in which a species utilizes energy for reproduction, parental care, and daily activities during its life span. The life history of a species is shaped by natural selection. The logistic growth model has been used to suggest that the environment promotes either *r*-selection or *K*-selection.

So-called *r*-selection occurs in unpredictable environments where density-independent factors affect population size. Energy is allocated to producing as many small offspring as possible. Adults remain small and do not invest in parental care of offspring. Producing many small offspring provides a greater chance that some of them will survive. Since the density of the population is low, density-dependent factors play little role in determining the size of the population. R-selection occurs in **opportunistic species** such as bacteria, annual plants, and several species of insects.

K-selection occurs in environments that remain relatively stable, where density-dependent factors affect population size. Energy is allocated to survival and repeated reproductive events. The adults are large and invest in parental care of offspring. These species, known as **equilibrium species**, are larger species that mature late and have a longer life span. They are strong competitors, but cannot easily adapt to environmental changes and can quickly become extinct if their way of life changes. Examples of organisms that are k-selected include large mammals, birds of prey, and long-lived plants.

Actual life histories contain trade-offs between these two life history patterns, and such trade-offs are subjected to natural selection. R-strategists and k-strategists are in fact at two ends of a continuum, and most populations have a life history strategy that lies somewhere in between.

> **Take Note:** *In preparation for the AP Biology exam, you should be able to explain r-selected and k-selected strategies. How does each strategy affect the size of a population over time?*

44.6   Human Population Growth

The human population is still expanding, but deceleration has begun. It is unknown when the population size will level off, but it may occur by 2050. Substantial increases are expected in certain LDCs (less-developed countries) of Asia and also Africa. Support for family planning, human development, and delayed childbearing could help prevent the increase.

Multiple Choice Questions

1. Which of the following does NOT describe a limiting factor that determines where an organism lives?
    A. Trout only live in cool mountain streams where the oxygen content is high.
    B. Trees cannot grow above the timberline because of low temperatures and a frozen water source.
    C. The red kangaroo does not live outside arid inland areas because it is adapted to feeding on the grasses that grow there.
    D. Carp and catfish are found in rivers near the coast where they can tolerate warm temperatures.
    E. Male peacocks with longer tails living in the forest have higher rates of predation.

2. Which of the following is a density-independent factor regulating population growth?
    A. competition for resources
    B. territoriality
    C. disease
    D. predation
    E. desertification

3. Which of the following is a characteristic of an opportunistic life history pattern, as seen in dandelions?
    A. large individuals
    B. long life span
    C. much care of offspring
    D. early reproductive age
    E. few and large offspring

4. Genetic diversity comprises not only individual genetic variation within a population but also between populations. Species diversity is the number of species. Ecosystem diversity refers to the network of community interactions within an ecosystem.

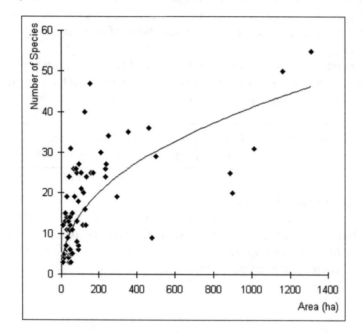

The graph above shows the number of bird species found in areas of different sizes. Which statement below is supported by the data shown in the graph?
   A. Cow birds, an edge species, are more prevalent in the US now than 30 years ago.
   B. The number of bird species in the US is increasing over time.
   C. The numbers of songbirds in the US is decreasing due to habitat loss.
   D. The larger a geographic area of a community is, the more species it contains.
   E. There are fewer large areas of habitat in the US than middle-sized areas.

5. The closer population size nears the carrying capacity,
   A. the less likely resources will become scarce.
   B. the population rate stabilizes.
   C. competition and predation will decrease.
   D. birthrate increases.
   E. death rate decreases.

6. In order to calculate population size from year to year, you do NOT need to know
   A. number of females present.
   B. birth rate.
   C. death rate.
   D. reproductive age of females.
   E. number of males in the population.

Questions 7–10 refer to the following five choices. For each question, select the choice that is most closely related. Each choice may be used once, more than once, or not at all.
   A. population
   B. community
   C. habitat
   D. ecotone
   E. biosphere

7. Boundary area between a forest and grassy field comprised of deciduous trees and grasses

8. Contains interdependent populations and their abiotic environments

9. A specific number of dandelions in a particular field

10. All of the organisms in a coral reef

Free Response Question

Populations tend to exhibit specific survivorship curves based upon the dynamics of the environment and the genetic basis of the organisms in the population.

    A. **Explain** the different types of survivorship curves and give an example of each.

    B. **Discuss** how density influences population growth.

Annotated Answer Key (MC)

1. **E**; Limiting factors are those environmental factors that determine where an organism lives.

2. **E**; All other choices are density dependent because the percentage of the population affected does increase as the density of the population increases.

3. **D**; The early reproductive age ensures that there will be many offspring produced (as many die before reproducing themselves).

4. **D**; The graph shows habitat size (x-axis) in relation to number of species (y-axis). It does not show the number of species in relation to time. The larger the habitat, the more species it can maintain due to an increase in biotic and abiotic factors.

5. **B**; Carrying capacity is the maximum number of individuals of a given species the environment can support. It varies throughout time depending on fluctuating conditions.

6. **E**; Ecologists focus on the number of females in sexually reproducing populations and how many female offspring they produce.

7. **D**; Energy flow and chemical cycling are significant aspects of how an ecosystem works. They do not have distinct boundaries and contain transitional zones called ecotones to allow interactions.

8. **B**; Multiple species in a given location constitute a community.

9. **A**; populations are all organisms belonging to the same species within an area at the same time.

10. **B**; Communities consist of all the various populations of multiple species at one location.

Answer to FRQ

PART A (MAX 8)
One point description, one point example.
- Natural populations may fit combination of curves.
- Impacted by age distribution (mortality patterns are affected)

- I: population where most individuals survive well past the midpoint of the life span/death near the end of the lifespan
- Examples: large mammals, humans, others
- II: death is unrelated to age/survivorship decreases at a constant rate
- Examples: small mammals, songbirds, others
- III: most individuals die very young
- Examples: invertebrates, fish, others

PART B (MAX 6)
- Density dependent factors:
  - Competition
  - Availability of resources – dispersal patterns
  - Reproductive success – finding mates
  - Age distribution – reproductive maturity
  - Habitat
  - Light (for plants growing close together)
  - Reproductive strategy (polyandry, polygamy, monogamy)
  - Examples are numerous.

CHAPTER 45
COMMUNITY AND ECOSYSTEM ECOLOGY

45.1 Ecology of Communities

A **community** is an assemblage of populations interacting with one another within the same environment. Communities differ in their composition (species found there) and their diversity (species richness and relative abundance). The **species composition**, or species richness, of a community consists of all the species found in that community. Different communities contain different species. **Species diversity** includes species richness and the relative abundance of each species. The greater the species richness and the more even the distribution of species within a community, the greater the diversity.

An organism's **habitat** is where it lives in the community. An **ecological niche** is defined by the role an organism plays in its community, including its habitat and how it interacts with other species in the community. A species's niche is determined by both biotic (living) and abiotic (nonliving) factors. The **fundamental niche** of a species is comprised of all abiotic conditions under which a species could survive if unfavorable biotic conditions were not present. The **realized niche** of a species, then, is determined by the presence of adverse biotic interactions, such as competition and predation. A species's fundamental niche is larger than its realized niche. Competition, predator-prey, parasite-host, commensalistic, and mutualistic relationships help organize populations into an intricate dynamic system.

The **competitive exclusion principle** states that no two species can indefinitely occupy the same niche at the same time. **Resource partitioning** decreases the competition between two species. When resources are partitioned between two or more species, increased **niche specialization** occurs. But the difference between species can be more subtle, as when warblers feed at different parts of the tree canopy. Barnacles competing on the Scottish coast may be an example of present ongoing competition.

Predator-prey interactions between two species are especially influenced by amount of predation and the amount of food for the prey. A cycling of population densities may occur, depending on the prevalence of either the predator or the prey. Prey defenses take many forms: camouflage, use of fright, and warning coloration are three possible mechanisms. **Batesian mimicry** occurs when one species has the warning coloration but lacks the defense. **Müllerian mimicry** occurs when two species with the same warning coloration have the same defense. **Coevolution** can occur within a community. For example, the better the predator becomes at catching prey, the better the prey becomes at escaping the predator.

Like predators, parasites take nourishment from their host. Whether parasites are aggressive (kill their host) or benign probably depends on which results in the highest fitness. Symbiotic relationships occur when two organisms live in close association with each other, and are classified as commensalistic, parasitic, or mutualistic. **Mutualistic** relationships, as when Clark's nuthatches feed on but disperse whitebark pine seeds, are critical to the cohesiveness of a community because both organisms involved in the relationship benefit from it. **Commensalism** occurs when one organism benefits from the symbiotic relationship and the other is unaffected. In a **parasitic** relationship, one organism benefits and the other is harmed.

> **Take Note:** *For the AP Biology exam, you should be prepared to discuss the interactions that occur between organisms in a community and how these interactions affect the population densities of the organisms involved. What kinds of relationships exist in a community? In what ways do organisms adapt and evolve due to interactions with other organisms?*

45.2   Community Development

Ecological succession involves a series of species replacements in a community. **Primary succession** occurs where there is no soil present, such as after a volcanic eruption or the retreat of a glacier. Soil develops from rock exposed to wind and water. **Secondary succession** occurs where soil is present and certain plant species can begin to grow. Pioneer species—plants that invade the area—are the first species to begin secondary succession. Eventually, long-lived trees will grow. A **climax community** forms when the stages of succession lead to a particular type of community.

Take Note: *On the AP Biology exam, you may be asked to describe how ecological succession occurs. How do biomass and energy flow change as a climax community develops?*

45.3   Dynamics of an Ecosystem

Populations interact among themselves and the environment in **ecosystems**, which have biotic and abiotic components. The biotic components are autotrophs, heterotrophs, detritus feeders, and decomposers. **Autotrophs** are organisms that make their own food, and are the producers of the ecosystem. **Heterotrophs** obtain their organic nutrients from preformed food, and are the consumers of an ecosystem. Depending on their diet, heterotrophs can be classified as herbivores, carnivores, omnivores, detritivores, or decomposers. The abiotic components of an ecosystem are resources, such as nutrients, and conditions, such as type of soil and temperature.

Ecosystems are characterized by energy flow and chemical cycling. All energy comes from the sun. Energy flows because as food passes from one population to the next, each population makes energy conversions that result in a loss of usable energy. Ecosystems contain **food webs**, interactions of food chains, in which the various organisms are connected by **trophic relationships**. In grazing food webs, food chains begin with a producer. In a detrital food web, food chains begin with detritus. **Ecological pyramids** are graphic representations of the number of organisms, biomass, or energy content of trophic levels.

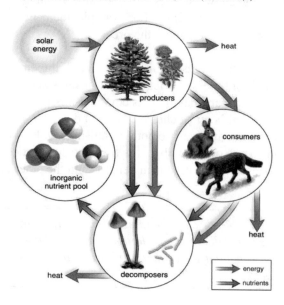

Nature of an ecosystem

Chemicals cycle because they pass from one population to the next until decomposers return them once more to the producers. **Biogeochemical cycles** may be sedimentary (phosphorus cycle) or gaseous (carbon and nitrogen cycles). Chemical cycling involves a reservoir, an exchange pool, and a biotic community.

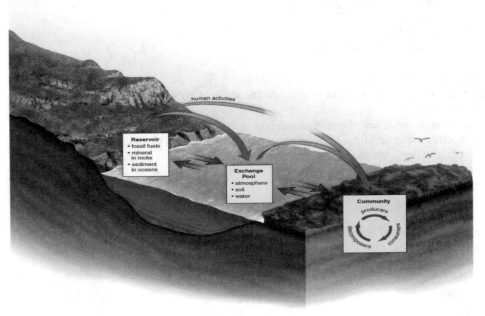

Model for chemical cycling

In the **water cycle**, evaporation over the ocean exceeds that provided by precipitation. On land, transpiration contributes to evaporation. Precipitation over land results in bodies of fresh water plus groundwater, including aquifers. Eventually, all water returns to the oceans.

In the **carbon cycle**, carbon dioxide in the atmosphere is an exchange pool; both terrestrial and aquatic plants and animals exchange carbon dioxide with the atmosphere. Living and dead organisms serve as reservoirs for the carbon cycle because they contain organic carbon. Human activities increase the level of CO_2 and other greenhouse gases contributing to global warming.

In the **phosphorus cycle**, geological upheavals move phosphorus from the ocean to land. Slow weathering of rocks returns phosphorus to the soil. Most phosphorus is recycled within a community, and phosphorus is a limiting nutrient.

Plants cannot use nitrogen gas directly from the atmosphere. In the **nitrogen cycle**, **nitrogen fixation** occurs. Nitrogen-fixing bacteria convert N_2 to ammonium, making nitrogen available to plants. **Nitrification** is the production of nitrates, while **denitrification** is the conversion of nitrate back to N_2, which enters the atmosphere. Human activities increase transfer rates in the nitrogen cycle. Acid deposition occurs when nitrogen oxides enter the atmosphere, combine with water vapor, and return to Earth in precipitation.

Multiple Choice Questions

1. Resource partitioning may NOT lead to
    A. a change in migration patterns of birds.
    B. a decrease in competition between two species.
    C. increased niche specialization.
    D. each species of bird using different parts of a tree.
    E. owls and hawks both becoming nocturnal hunters.

2. Which of the following is an example of mutualism?
    A. Barnacles attach themselves to the backs of whales and the shells of horseshoe crabs are provided with a home and transportation.
    B. Spanish moss grows in the branches of trees and receives light but takes no nourishment from the tree.
    C. The acacia tree receives protection from herbivores by providing a home and a food source for ants.
    D. Leeches remain attached to the exterior of a body by means of specialized organisms, and the host is weakened over time.
    E. HIV reproduces within human lymphocytes.

3. Detritovores
    A. feed on decomposing particles of organic matter.
    B. feed on other animals.
    C. produce their own food.
    D. are chemosynthetic.
    E. are nonliving parasites.

4. The primary type of organisms responsible for fixing nitrogen are
    A. prokaryotes.
    B. fungi.
    C. protists.
    D. invertebrates.
    E. plants.

5. The principle source of nitrogen for plants is
    A. nitrates.
    B. nitrites.
    C. nitrogen gas.
    D. ammonium.
    E. amino acids.

6. The process directly releasing carbon dioxide into the atmosphere is
    A. photosynthesis.
    B. photorespiration.
    C. reduction.
    D. chemosynthesis.
    E. respiration.

7. Which of the following would NOT be a result of the death of all prokaryotic organisms in different ecosystems?
    A. decreased available nitrogen
    B. dead organisms decomposing faster
    C. decreased denitrification
    D. fewer exotoxins released
    E. less photosynthesis and oxygen released

8. Which of the below is an example of primary succession?
    A. the growth of trees on a hillside that was clear cut for lumber
    B. colonization of a sand dune by plants
    C. replacement of grasses by trees after a forest fire
    D. the filling in of a pond with silt and sediments, becoming a swamp
    E. the explosion of algae in a lake due to over-nitrification

Questions 9–10 refer to the following food chain.

Sunflower → Slug → Mouse → Snake → Hawk

9. In the food chain, the slug is classified as a
    A. primary producer.
    B. omnivore.
    C. autotroph
    D. primary consumer.
    E. decomposer.

10. A decrease in the population of mice would most likely cause
    A. a decrease in the sunflower population.
    B. a decrease in the hawk population.
    C. no change in population size of snakes and hawks.
    D. an increase in the slug population.
    E. an increase in the snake population.

Free Response Question

Symbiotic relationships frequently occur in nature. For any of the FOUR examples listed:

- viral infection
- malaria
- plant root nodules
- epiphytes
- cleaning symbiosis

A. **Explain** the nature of the relationship and which organisms specifically are involved.

B. **Discuss** how the relationships may affect co-evolution.

Annotated Answer Key (MC)

1. **E**; In resource partitioning there is less niche overlap. If both organisms became nocturnal, this would increase niche overlap and cause more competition between the species.

2. **C**; When both organisms benefit from the symbiotic relationship it is classified as mutualism.

3. **A;** Detritovores include fan worms, clams, earthworms, some beetles, termites, and ants. These organisms are important in the cycling of nutrients in the environment.

4. **A;** Prokaryotes fix nitrogen into usable forms for plants.

5. **A;** While gaseous nitrogen is abundant in the atmosphere, plants use nitrates as their principal usable form of nitrogen.

6. **E;** All living organisms respire, directly releasing carbon dioxide into the environment.

7. **B;** Bacteria are decomposers and a decrease in their population would decrease the rate that dead organisms would decompose.

8. **B;** Primary succession occurs in areas where there is no soil formation. Secondary succession occurs in areas where soil is present.

9. **D;** Primary consumers are in the second trophic level and are herbivores.

10. **B;** A decrease in the population of mice would make starvation possible for many snakes, therefore decreasing the food source of the hawks who are carnivores and could not feed on snakes or mice.

Answer to FRQ

PART A (MAX 8)
One point for organisms, one point for correct type, one point for involvement benefit/deficit.

- Viral infection: virus (genetic machinery/habitat)/host (decrease in immune function/death) → parasitism
- Malaria: *Plasmodium* (nourishment/live/reproduce)/Mosquitoes (neutral/vector) / Humans(host/live/nourishment/reproduce)→ commensalism/parasitism
- Plant Root Nodules: plants (nitrogen in usable form)/bacteria (nutrients) → mutualism
- Epiphytes: trees (neutral)/fern, moss, ivy, orchid (anchor) →commensalism
- Cleaning: Invertebrates, fish, birds (nutrition)/large fish/vertebrates (cleaning) → mutualism

PART B (MAX 4)
- Two species adapt in response to selective pressures.
- Provokes natural selection (neither species would have evolved without the influence of the other)
- Relationship to endosymbiosis (parasitism → mutualism)
- Increased tolerance → commensalism → mutualism

CHAPTER 46
MAJOR ECOSYSTEMS OF THE BIOSPHERE

46.1   Climate and the Biosphere

Climate is defined as the weather in the particular region, which is influenced by temperature and rainfall. Because the earth is a sphere, the sun's rays at the poles are distributed out over a larger area than the direct rays at the equator. The temperature at the surface of the earth therefore decreases from the equator toward each pole. Earth is tilted on its axis, and the seasons change as the earth revolves annually around the sun.

Warm air rises near the equator, loses its moisture, and then descends at about 30° north and south latitude, and so forth, to the poles. When the air descends, it absorbs moisture from the land, and therefore the great deserts of the world are formed at 30° latitudes. Because the earth rotates on its axis daily, the winds blow in opposite directions above and below the equator. **Topography** (the physical features of the land) also plays a role in the distribution of moisture. Air rising over coastal ranges loses its moisture on the windward side, making the leeward side arid.

46.2   Terrestrial Ecosystems

A **biome** is a major type of terrestrial community. Biomes are distributed according to climate, that is, temperature and rainfall influence the pattern of biomes about the world. The effect of temperature causes the same sequence of biomes when traveling to northern latitudes as when traveling up a mountain. Keep in mind that biomes are not independent of each other. Each contributes to the others surrounding it, because there are no clear lines of demarcation between them; rather, there is a gradual transition between them.

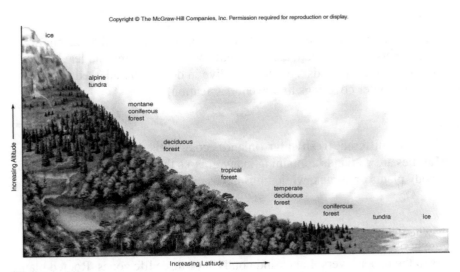

Climate and biomes

The **Arctic tundra** is the northernmost biome and consists largely of short grasses, sedges, and dwarf woody plants. Because of cold winters and short summers, most of the water in the soil is frozen year-round. This is called the **permafrost**. The plants and animals that live in the tundra are adapted for the extremely cold temperatures and short growing seasons.

The **taiga**, a coniferous forest, has less rainfall than other types of forests. **Coniferous forests** are populated with cone-bearing trees that are well-adapted to the cold because of the thick covering on their leaves and bark, as well as thin leaves that can support the weight of heavy snow. There are also lots of mosses and lichens on the floor of these forests.

The **temperate deciduous forest** has trees that gain and lose their leaves because of the alternating seasons of summer and winter. This forest is stratified in that a canopy is formed by the tallest trees, though sunlight can still penetrate to the understory and shrubs below. Mosses, lichens, and ferns populate the forest floor.

Tropical rain forests are continually warm and wet. These are the most complex and productive of all biomes, and they are home to a great diversity of species. Like the temperate deciduous forests, rainforests are stratified into the canopy, the understory, and the forest floor. Unlike the temperate forests, however, sunlight does not penetrate to the lower layers very well due to the broad leaves of the trees in the canopy. Most animals live in the canopy, though there are a few that live on the forest floor. **Epiphytes**, plants that grow on other plants but have their own root systems, are common in the rainforest.

Shrublands usually occur along coasts that have dry summers and receive most of their rainfall in the winter. Shrubs are plants that are shorter than trees, do not have a central trunk, and have thick evergreen leaves with a cuticle. These adaptations help them to withstand the arid environment and grow quickly after a fire. Shrublands are exemplified by the chaparral in California, which is highly flammable. Plants that populate this biome germinate due to the heat and scarring that occurs during a fire.

Among **grasslands**, the savanna, a tropical grassland, supports the greatest number of different types of large herbivores. Temperate grasslands, such as that found in the central United States, have a limited variety of vegetation and animal life. The rainfall in grasslands is insufficient to support many tall trees, and the grasses are adapted to withstand the constantly changing environment due to drought, fire, flooding, and grazing.

Deserts are characterized by a lack of water—they are usually found in places with less than 25 cm of precipitation per year. Some desert plants, such as cacti, are succulents with thick stems and leaves, and others are shrubs that are decidous during dry periods. Animals that live in deserts are adapted to be nocturnal or burrowing, and have thick outer coverings to prevent dessication.

46.3 Aquatic Ecosystems

Streams, rivers, lakes, and wetlands are different **freshwater** (inland) ecosystems. Wetlands include marshes, swamps, and bogs. **Marshes** are wetlands containing many grasses that are frequently iundated with water. **Swamps** contain woody plants or shrubs, while **bogs** have acidic water, peat deposits, and sphagnum moss.

In deep lakes of the temperate zone, the temperature and the concentration of nutrients and gases in the water vary with depth. The entire body of water is cycled twice a year, in fall and spring, distributing nutrients from the bottom layers to the upper layers. Lakes and ponds have three life zones. Rooted plants and clinging organisms live in the **littoral zone**, plankton and fishes live in the sunlit **limnetic zone**, and bottom-dwelling organisms such as crayfishes and molluscs live in the **profundal zone**. The **benthic zone** is the sediment at the soil-water interface.

Marine ecosystems are divided into coastal ecosystems and the oceans. The coastal ecosystems, especially estuaries, are more productive than the oceans. **Estuaries** (and associated salt marshes, mudflats, and mangrove forests) are near the mouth of a river, and are considered the nurseries of the sea.

An **ocean** is divided into the pelagic division and the benthic division. The oceanic province of the **pelagic division** (open waters) has three zones: 1) the epipelagic zone, 2) the mesopelagic zone, and 3) the bathypelagic zone. The **epipelagic zone** receives adequate sunlight and supports the most life. The **mesopelagic zone** contains organisms adapted to minimum or no light. The **bathypelagic zone** consists of the deepest water that is in complete darkness. The **benthic division** (ocean floor) includes organisms living on the continental shelf in the **sublittoral zone**, the continental slope in the bathypelagic zone, and the abyssal plain in the **abyssal zone**.

Take Note: *Be prepared to describe the characteristics of terrestrial and aquatic biomes on the AP Biology exam. What characteristics do the plants and animals that live in these biomes have that make them successful?*

Multiple Choice Questions

1. Which environmental factors are most important for classifying biomes?
   A. soil type and rainfall
   B. elevation and temperature
   C. longitude and humidity
   D. temperature and precipitation
   E. humidity and soil type

2. A biome that has an unusually high primary productivity must also have a
   A. small concentration of decomposers.
   B. large number of primary producers.
   C. low amount of precipitation.
   D. large number of primary consumers.
   E. large concentration of archaebacteria.

Questions 3–6 refer to the following five choices. For each question, select the choice that is most closely related. Each choice may be used once, more than once, or not at all.
   A. taiga
   B. deciduous forest
   C. tundra
   D. rain forest
   E. chaparral

3. Characterized by arid conditions and drought-resistant shrubs and oaks

4. Deep snow in the winter and dense evergreen forests

5. Contains trees that lose their leaves each winter and has moderate amounts of rainfall

6. Climate includes a warm growing season and cold winters spanning much of the Eastern United States.

7. Nearly all organisms in the intertidal zone
   A. dwell underground.
   B. feed during the day.
   C. are capable of withstanding changing salinity levels.
   D. are dependent on staying in water.
   E. depend on sunlight as an energy source.

8. During the winter in temperate deciduous forests, mammals do NOT forage on
   A. deciduous leaves.
   B. fruits.
   C. nuts.
   D. berries.
   E. bark.

9. Bogs differ from marshes in that
   A. marshes are nutrient poor.
   B. bogs are characterized by acidic water.
   C. marshes provide a habitat for many waterfowl.
   D. bogs do not contain mosses or other plants.
   E. marshes are not extremely productive ecosystems.

10. Temperature moderation near the equator is NOT due to
    A. the tilt of the earth.
    B. the production of low pressure areas.
    C. high rainfall.
    D. the direction of the winds.
    E. movement of water between the north and south hemispheres.

Free Response Question

Aquatic ecosystems are some of the most productive in the biosphere.

   A. **Discuss** lake stratification in a temperate region throughout the year.

   B. **Describe** the zones of a lake and the types of organisms that occupy each zone.

Annotated Answer Key (MC)

1. **D**; Biomes are primarily classified into land biomes, freshwater biomes, and marine biomes. They are characterized by climate, but primarily by temperature and precipitation.

2. **B**; High primary productivity requires large numbers of autotrophs, or organisms that are considered primary producers and synthesize their own carbohydrate source.

3. **E**; The chaparral is characterized by arid conditions and dense stands of evergreen shrubs, often found in coastal areas.

4. **A**; All biomes are characterized by temperature and moisture levels. While there is some overlap, specific characteristics define different biomes. The taiga or the northern coniferous forest biome extends across most of North America just south of the tundra. Characterized by cold winters, deep snow, and often hot summers. Characterized by dense stands of cone-bearing trees.

5. **B**; The deciduous forest or temperate broadleaf is characterized by mild winters, and often hot and humid winters, with significant amounts of rainfall during all seasons. This biome covers most of the Eastern US.

6. **B**; The deciduous forest or temperate broadleaf is characterized by mild winters, and often hot and humid winters, with significant amounts of rainfall during all seasons. This biome covers most of the Eastern US.

7. **C**; The intertidal zone is a harsh area for organisms to live in. As a result, these organisms can tolerate wide swings in light, moisture, and temperature.

8. **A**; During the winter, deciduous trees lose their leaves. Animals will forage on leftover fruits, nuts, berries, and bark during the wintertime.

9. **B**; Bogs are characterized by acidic waters, peat deposits, and sphagnum moss. They receive most of their water from precipitation and are nutrient poor. Many plants thrive in bogs, as well as other larger mammals.

10. **E**; The predominant ocean currents carry water away from the equator, not between the hemispheres.

Answer to FRQ

PART A (MAX 6)

Lake Stratification
- Fall: epilimnion cools → surface waters sink, deep water rises
- Occurs until temperature is uniform
- Wind circulates water (mixing of warm and cold).
- Even distribution of nutrients and oxygen
- Winter: water cools → ice formation
- Insulation of ice → prevents cooling of water below
- Spring: epilimnion warms → cooler water sinks below warmer water on the bottom
- Surface waters absorb solar radiation → thermal stratification (change in temperature of differing levels → impacts life forms → impacts oxygen concentration)

PART B (MAX 6)
One point example per zone.
- Littoral zone: closest to the shore
- Organisms: aquatic plants, protozoans, invertebrates, fishes and some reptiles, wading birds
- Limnetic zone: sunlit part of the lake
- Organisms: small fish, plankton, large fish
- Profundal zone: below light penetration
- Organisms: zooplankton, invertebrates, fish feed on debris
- Benthic zone: sediment at the soil-water interface
- Organisms: worms, snails, clams, crayfishes, some insects, bacteria (decomposers)

CHAPTER 47
CONSERVATION OF BIODIVERSITY

47.1   Conservation Biology and Biodiversity

Conservation biology is the scientific study of biodiversity and its management for sustainable human welfare. The unequaled present rate of extinctions has drawn together scientists and environmentalists in basic and applied fields to address the problem. Four ethical principles are supported by conservation biology in order to avoid the disruption of ecosystems:

1. Biodiversity is advantageous for the biosphere.

2. Extinctions—especially those due to human actions—are unwanted.

3. Ecological interactions are necessary for biodiversity.

4. Biodiversity that occurs as a result of evolutionary change is valuable.

It is estimated that 10–20% of all species may become extinct within 20–50 years if action is not taken. Conservation biologists use **bioinformatics** (the science of collecting and analyzing biological information and making it available) to understand and protect biodiversity as well as to educate the general public on the value of preserving biodiversity.

Biodiversity is the variety of life on Earth; the exact number of species is not known, but there are assuredly many species yet to be discovered and recognized. Biodiversity must be preserved at the genetic, community (ecosystem), and landscape levels of organization. There are 30,000 species worldwide that are **endangered species** (in danger of extinction throughout most of their range); many species are also **threatened** (close to being endangered). Both endangered and threatened species are losing their genetic diversity as their populations decrease in number.

Conservationists have discovered that biodiversity is not evenly distributed in the biosphere, and therefore saving particular areas may protect more species than saving other areas. This will help to conserve the **ecosystem diversity** that exists in communities by assuring that ecological interactions are enhanced.

47.2   Value of Biodiversity

The direct value of biodiversity is seen in the observable services of individual wild species for human beings. Wild species are our best source of new medicines to treat human ills, and they help meet other medical needs. Wild species also provide a means for studying organisms about which little is known. For example, the bacterium that causes leprosy grows naturally in armadillos, but not in laboratory conditions; horseshoe crab blood contains a bacteria-fighting substance that is used to sterilize pacemakers, surgical implants, and prosthetic devices.

Wild species have agricultural value. Domesticated plants and animals are derived from wild species, and wild species are used as a source of genes for the improvement of their phenotypes. Instead of pesticides, wild species can be used as biological controls; this not only saves the environment, but also money. Most flowering plants benefit from animal pollinators from the wild, which are more resistant to parasites than domesticated pollinators. Much of our food, particularly fish and shellfish, is still caught in the wild. Hardwood trees from natural forests supply us with lumber for various purposes, such as making furniture.

The indirect services provided by ecosystems are largely unseen and difficult to quantify but absolutely necessary to our wellbeing. These services include the workings of biogeochemical cycles, waste disposal, provision of fresh water, prevention of soil erosion, and regulation of climate. Many people enjoy vacationing in natural settings. Various studies show that ecosystems with greater diversity function better than less diverse systems.

> **Take Note:** *In preparation for the AP Biology exam, you should understand biodiversity and the methods that conservation biologists are using to preserve it. What are the direct and indirect values of biodiversity?*

47.3 Causes of Extinction

Researchers have identified the major causes of extinction. Habitat loss is the most frequent cause, followed by introduction of exotic species, pollution, overexploitation, and disease. (Pollution often leads to disease, so these were discussed at the same time.)

Habitat loss has occurred in all parts of the biosphere, but concern has now centered on tropical rain forests and coral reefs, where biodiversity is especially high. Forests have become fragmented due to the building of roads as well as human occupancy and farming. Coral reefs, mangrove forests, and wetland areas are also being degraded by human actions.

Exotic species have been introduced into foreign ecosystems through colonization, horticulture and agriculture, and accidental transport. They disrupt the food webs of the environments into which they have been introduced, affecting the ecological interactions, succession, and biodiversity of local communities.

Pollution is an environmental change that affects the lives and health of living things in an adverse manner, weakening organisms and causing disease. Various causes of pollution include fertilizer runoff, industrial emissions, and improper disposal of wastes, among others. One effect of pollution is **eutrophication** in lakes due to over-enrichment. An excess of nutrients from runoff causes algal blooms to develop. The bacteria that decompose the algae when they die use the oxygen in the lake, decreasing its availability for other organisms in that habitat. **Global warming** is an increase in the average temperature of the environment that occurs as a result of the presence of greenhouse gases in the atmosphere.

Overexploitation is exemplified by commercial fishing, which is so efficient that fisheries of the world are collapsing. The overexploitation causes the population of wild species to drastically decrease; the individuals of a small population are then more valuable, making the incentive to capture them greater.

47.4 Conservation Techniques

To preserve species, it is necessary to preserve their habitat. Some emphasize the need to preserve biodiversity hotspots because of their richness. Often today it is necessary to save **metapopulations** (populations that are divided into smaller populations) because of past habitat fragmentation. If so, it is best to determine the source populations and save those instead of the sink populations. The preservation of a **keystone species** such as the grizzly bear requires the preservation of a landscape consisting of several types of ecosystems over millions of acres of territory. Obviously, in the process, many other species will also be preserved because of the influence of the keystone species.

Since many ecosystems have been degraded, **habitat restoration** may be necessary before sustainable development is possible. Three principles of restoration are (1) start before sources of wildlife and seeds are lost; (2) use simple biological techniques that mimic natural processes; and (3) aim for sustainable development so that the ecosystem maintains itself and fulfills the needs of humans.

Take Note: *For the AP Biology exam, you should be able to explain the causes of extinction and discuss ways in which habitat restoration can help to restore biodiversity.*

Multiple Choice Questions

1. Which of the following is an example of an introduced species?
    A. thousands of fiddler crabs feeding on mudflats
    B. wide swaths of saltmarsh cord grasses growing along the entire North American seaboard
    C. fire ants migrating northward, killing the majority of other ant species in their path
    D. the predominance of the red-tailed hawk as the most common hawk in North America
    E. black bears found across most of North America, feeding on many different types of food

2. Which of the following is an example of an area with a high biodiversity?
    A. 200-acre cornfield in Iowa
    B. 200 square meters of rain forest in Central America
    C. 20 square miles of salt marsh on the Virginia coast
    D. 10-mile long sandy beach in Florida
    E. 200 square meters of tundra in Alaska

3. Which of the following is NOT a major threat to the loss of biodiversity?
    A. increased quantity of edge habitats through conversion of land to farmland
    B. introduction of nonnative species to help control erosion
    C. commercial fishing of the oceans to feed the world's population
    D. importing genetic variation from one population to another
    E. elimination of top carnivores, such as the wolf, from an area to allow others to thrive

4. What are the three primary levels on which biodiversity is studied?
    A. microscopic, individual, population
    B. bacterial, animal, plant
    C. subcellular, cellular, organismal
    D. organs, Systems, organismal
    E. genetic, species, ecosystem

5. The greatest number of species in the biosphere are
    A. insects.
    B. plants.
    C. animals.
    D. protists.
    E. bacteria.

6. Eutrophication occurs when
    A. sulfur dioxide from power plants combines with water vapor in the atmosphere.
    B. lakes receive excess nutrients due to runoff from agricultural fields.
    C. global trade brings new species from one country to another.
    D. lakes turnover in the spring and fall due to temperature changes.
    E. organic chemicals mimic the effects of hormones and impact fish reproductive rates.

Questions 7–10 refer to the following five choices. For each question, select the choice that is most closely related. Each choice may be used once, more than once, or not at all.

    A. flagship species
    B. keystone species
    C. exotic species
    D. endangered species
    E. source population

7. An organism in peril of immediate extinction

8. A population that lives in a favorable area; high birthrate and low death rate

9. Species that influence the ecosystem while population remains relatively low

10. Kudzu flourishes in the Southern United States after being transported here from Asia.

Free Response Question

Biodiversity is of immense value.

    A. **Discuss** THREE of the various ways that biodiversity is of value to the human population.

    B. **Explain** TWO of the causes of extinction.
        • habitat loss
        • exotic species
        • pollution

Annotated Answer Key (MC)

1. **C**; Fire ants migrating northward and impacting/killing other species are an introduced species. Simply large numbers of a species like fiddler crabs does not make them introduced, nor does the predominance of one top predator or one type of grass in a stressful habitat such as saltmarsh cordgrass.

2. **B**; The rainforest is one of the biosphere's most diverse biomes.

3. **D**; Kudzu is an example of a nonnative species being bought to the United States, which has since flourished in the South, and populations of other animals actually rebounded with the wolf reintroduction to Yellowstone. The Illinois prairie chicken population rebounded when variation was imported from another population, preventing the extinction vortex.

4. **E**; Biodiversity cannot occur on an organismal level because it is independent of all others.

5. **A**; There are nearly 900,000 species of insects; 240,000 species of plants; 280,000 species of insects; 55,000 species of protists; and 5,000 species of bacteria and archae combined.

6. **B**; During eutrophication, lakes are stressed due to over-enrichment. Algae begin to grow. While the decomposers break down the algae, they use up oxygen, sometimes leading to massive fish kills.

7. **D**; Threatened species are not as near extinction in the foreseeable future.

8. **E**; Saving metapopulations sometimes requires determining which of the populations are source and sink.

9. **B**; Keystone species are critical in the viability of a community. They often have integral jobs within a community, such as being pollinators or dispersing seeds and berries.

10. **C**; Exotic species often interfere with energy flow within ecosystems.

Answer to FRQ

PART A (MAX 6)

Medicinal value	Prescription drugs
	Antibiotics
	Source for bacteria (leprosy)
	Hemolymph from horseshoe crabs
Agricultural	Wheat, corn, rice →derived from wild plants
	Locating resistance genes to transfer to crops
	Biological pest controls
	Pollinators
Biogeochemical cycles	Energy flow → removal of carbon dioxide, uptake of excess nitrogen, provision of phosphate
Waste disposal	Decomposers
Fresh water provision	Drinking, irrigation, food source
Soil erosion	Intact ecosystems naturally retain soil
Climate regulation	Shade, reduce electrical needs, solar radiation, global climate control
Ecotourism	Many examples

PART B (MAX 6)
Three points per cause.

Habitat Loss
- Removal/destruction of habitats → reduction in species diversity
- Increased human populations → reduction in species diversity

Exotic Species
- Particular influence on islands
- Colonization: transported to new locations
  - Exotic species → reduction in species diversity
  - Dandelion, pigs
- Horticulture and agriculture: escape from cultivated areas
  - Exotic species → reduction in species diversity
  - Kudzu
- Accidental Transport: transported to new locations
  - Exotic species → reduction in species diversity
  - Zebra mussels

Pollution
- Acid deposition: sulfur dioxide and nitrogen oxide
- Return as acid rain/snow/dry deposition
- Occurs across geographic boundaries
- Impacts lower trophic levels first
- Eutrophication: over enrichment
- Algal bloom → decomposers → oxygen depletion → upper trophic levels impacted
- Ozone depletion: shields UV light
- Reduction of immune system of all organisms
- Chemical pollution: mimics hormones → impacts normal physiological mechanisms
- Global warming: increase in average temperature
- Burning of fossil fuels and methane

BIOLOGY PRACTICE EXAM I

1. Which of the following enzymes works best at an acidic pH?
   A. salivary amylase
   B. pepsin
   C. carbonic anhydrase
   D. lipase
   E. pancreatic amylase

2. A test cross
   A. involves two heterozygous individuals.
   B. produces all homozygous recessive offspring.
   C. can only work with sex-linked genes.
   D. mates a dominant phenotype with a homozygote recessive.
   E. is often used in human genetic testing.

3. Phototropisms occur when
   A. blue light activates a photoreceptor in the plasma membrane.
   B. plants move in response to touch.
   C. auxin causes the cells on the illuminated side of the stem to elongate.
   D. red light activates phytochrome and the shoots elongate.
   E. when the root cap is removed and the roots no longer respond to light.

4. The relatedness between a turtle and a duck is most significant in the fact that both
   A. use the same DNA triplet code.
   B. code for the same 20 amino acids.
   C. produce similar proteins.
   D. maintain a high degree of DNA base sequence similarity.
   E. contain cytochrome *c*.

5. Introns are NOT
   A. removed and exons joined together to form a functional mRNA transcript.
   B. able to regulate how exons are put together so that different proteins can result from a single gene.
   C. able to determine which genes are expressed and how they are to be spliced.
   D. located between exons on the DNA strand.
   E. found in the single circular chromosome of eubacteria.

6. An increase in secretion of epinephrine by the adrenal medulla will
   A. decrease blood glucose levels.
   B. suppress the immune system.
   C. dilate arterioles in the skin.
   D. inhibit lipid breakdown.
   E. constrict the pupils.

7. In the following food chain, which organism is a secondary consumer?

leaves → caterpillars → birds → hawks → decomposers

   A. leaves.
   B. caterpillars.
   C. birds.
   D. hawks.
   E. decomposers.

8. X-inactivation in humans may result in
   A. complete albinism in females.
   B. patches of skin lacking sweat glands in females.
   C. color blindness in males.
   D. two Barr bodies to be present in males.
   E. a dominant Y chromosome in males.

9. An open circulatory system is advantageous in grasshoppers because
   A. they only require one respiratory pigment.
   B. tracheae are found throughout the body.
   C. blood is pumped within vessels throughout the body.
   D. the hemolymph carries oxygen and nutrients throughout the body.
   E. capillaries are able to reach the appendages.

10. Investigators have found that the cancer drug ceramide forms holes in the membranes of organelles responsible for ATP synthesis in targeted cells, triggering cell death. Which of the following organelles does ceramide affect?
    A. rough endoplasmic reticulum
    B. nucleus
    C. mitochondria
    D. lysosome
    E. ribosome

11. Which of the following scenarios depicts a population being regulated by a density-independent factor?
    A. A winter freeze destroys a citrus crop in the southern United States.
    B. Tuberculosis is spread through the air when an infected individual coughs.
    C. A woodpecker population competes for nesting sites in nearby trees.
    D. Gypsy moths in New England defoliate entire forests in the summer and a massive die-off results shortly afterwards.
    E. Infection of house finches by bacteria drive populations of finches down in greater populated areas.

12. The muscular system is NOT responsible for which of the following life processes?
    A. coordinating body activities
    B. providing a means for locomotion
    C. acquiring materials and energy
    D. maintaining body shape
    E. heat production

13. An organism that exhibits bilateral symmetry and deuterostome development is a(n)
    A. cnidarian (jellyfish).
    B. mollusc (snail).
    C. flatworm (planaria).
    D. echinoderm (starfish).
    E. chordate (bird).

14. Five species of frogs are all found in the same geographic area. The species remain separate because the period of most activity, feeding, and active mating differs. This is an example of
    A. habitat isolation.
    B. temporal isolation.
    C. behavioral isolation.
    D. hybrid sterility.
    E. zygote mortality.

15. Nondisjunction of chromosomes during meiosis I of oogenesis and subsequent fertilization by haploid sperm will result in
    A. four diploid zygotes.
    B. two diploid zygotes with one additional chromosome, and two diploid zygotes with one missing chromosomes.
    C. two diploid zygotes and two diploid zygotes with one additional chromosome.
    D. four haploid zygotes.
    E. three haploid zygotes and one diploid zygote with three extra chromosomes.

16. If you put a black dot to the right of center on a piece of paper and use your right hand to move the paper slowly towards your right eye, the dot will disappear. This phenomenon illustrates that
    A. there are no rods or cones where the optic nerve exits the retina.
    B. an uneven cornea does not allow the image to be seen clearly.
    C. rhodopsin only splits when colored images are present.
    D. there are no receptors in the fovea so the image is blurred.
    E. the lens is sometimes opaque and incapable of transmitting light rays.

17. Prions are
    A. not found in healthy brains.
    B. do not cause chaperone proteins to impact the protein folds.
    C. capable of binary fission in brain tissue.
    D. able to interact with normal proteins and change their shape.
    E. contain a nucleoid region and consist of a single chromosome.

18. Red fruit (R) is dominant over yellow fruit (r) and tallness (T) is dominant over shortness (t). In a dihybrid cross involving two heterozygote individuals for both traits, the F_1 generation will
    A. all be red and tall.
    B. consist of 9/16 red and tall individuals.
    C. be 1/2 red and tall and 1/2 yellow and short.
    D. all be yellow and short.
    E. be 1/4 red and short.

19. Which of the following is NOT true regarding roots?
    A. Adventitious roots can emerge above the soil line.
    B. Rootlike projections can grow into host plants and extract water and nutrients.
    C. Mycorrhizae are associations between roots and fungi in a mutualistic relationship.
    D. Root nodules contain nitrogen-fixing bacteria.
    E. Fibrous root systems are adapted to expand and store starch.

20. Cancer cells
    A. contain normal nuclei.
    B. undergo apoptosis.
    C. exhibit contact inhibition.
    D. consist of one, organized layer.
    E. undergo angiogenesis.

21. In the following enzymes cascade, a faulty code for enzyme E_A will result in which of the following effects?

    A $\rightarrow$ E_A $\rightarrow$ B $\rightarrow$ E_B $\rightarrow$ C
    phenylalanine   tyrosine   melanin

    A. Tyrosine will build up and cause sickle cell anemia.
    B. Phenylalanine will accumulate and cause PKU.
    C. Tyrosine will not be converted to melanin and the individual will be albino.
    D. Melanin will accumulate and cause the skin to become dark and blotchy.
    E. Enzyme E_B will replace E_A and convert tyrosine to melanin.

22. Which of the following DNA sequences is complimentary to the following mRNA sequence?

    5'-CAGUCCUUCCUC-3'

    A. 5'-GTCAGGTTGGTC-3'
    B. 5'-GYCAGGAAGGAG-3'
    C. 3'-GUCAGGAAGGAG-3'
    D. 5'-UGCAAGGAAGGC-3'
    E. 3'-GTCAGGAAGGAG-5'

23. The efficiency of translation is in part due to
    A. splicesomes.
    B. delayed chaperone proteins.
    C. multiple A, P, and E sites on a single ribosome.
    D. polyribosomes.
    E. mRNA processing.

24. Which of the following plants produce flagellated sperm?
    A. bryophytes (mosses)
    B. pteridophytes (ferns)
    C. gymnosperms (conifers)
    D. angiosperms (flowering plants)
    E. dicots (trees and shrubs)

25. A eukaryotic, multicellular organism with no internal protection of the zygote and a haploid life cycle is a
    A. bacteria.
    B. protist.
    C. fungus.
    D. plant.
    E. animal.

26. Which of the following is true regarding meiosis and mitosis?
    A. Mitosis involves a pairing of homologous chromosomes in the first division.
    B. Meiosis produces two diploid daughter cells.
    C. In meiosis I, sister chromatids separate and move to the poles.
    D. In mitosis, a diploid number of duplicated chromosomes are at the metaphase plate.
    E. Genetically identical daughter cells are produced in meiosis.

27. An athlete who takes erythropoietin (EPO), which stimulates red blood cell production, will
    A. cause red blood cell production to diminish in the bone marrow.
    B. decrease the oxygen carrying capacity of the blood.
    C. decrease the viscosity of the blood.
    D. lower blood pressure.
    E. increase the risk of heart attacks and strokes.

28. In honeybees, the queen bee makes and stores sperm she uses to selectively fertilize eggs. Any unfertilized eggs become haploid males. This is an example of
    A. asexual reproduction.
    B. parthenogenesis.
    C. copulation.
    D. oogenesis.
    E. mitosis.

29. During this test, you are most likely stressed and nervous. Therefore, which of the following physiological responses is occurring in your body right now?
    A. stimulation of the gall bladder to release bile
    B. increased intestinal activity
    C. dilation of air passages
    D. stimulation of saliva
    E. constriction of pupils

30. An embryo that develops a heart defect may have been influenced by a teratogen during development of which of the following layers?
    A. ectoderm
    B. neural plate
    C. archenteron
    D. mesoderm
    E. endoderm

31. Organisms in the same class must also be in the same
    A. species.
    B. order.
    C. phylum.
    D. family.
    E. genus.

32. All of the organisms interacting in a coral reef constitute a
    A. population.
    B. community.
    C. ecosystem.
    D. biosphere.
    E. species.

33. Fungal cells differ from plant cells in that fungi
    A. contain chloroplasts.
    B. have a nucleus.
    C. possess a large central vacuole.
    D. have cell walls composed of chitin.
    E. do not contain centrioles.

34. Red algae are specialized for absorbing blue wavelengths of light for photosynthesis and therefore
    A. contain carotenoids as photosynthetic pigments.
    B. grow at greater ocean depths.
    C. cannot grow in saltwater.
    D. are not classified with the green algae in the same genus.
    E. form mutualistic relationships with coral.

35. The bond between a purine and a pyrimidine in the double helix of DNA is what type?
    A. polar covalent
    B. ionic
    C. nonpolar covalent
    D. hydrogen
    E. van der Waals

36. Why are many people concerned with the general widespread use of antibiotics in many products, such as hand creams, soaps, and sunscreens?
    A. The antibiotics will kill many symbiotic bacteria on human's skin.
    B. Drug companies are worried about being put out of business.
    C. The number of antibiotic-resistant bacteria is growing.
    D. The antibiotics can cause skin cancer.
    E. Bacteria become immune to antibiotics after being exposed to them.

37. In recombinant DNA technology, the restriction enzymes used to cut the plasmid do NOT
    A. recognize more than one recognition sequence.
    B. cut more than one time.
    C. usually create sticky ends.
    D. search for exact sequences of defined length.
    E. break bonds in the DNA backbone.

38. Which of the following is NOT an example of the edge effect?
    A. Forest edges are brighter, warmer, drier, and windier.
    B. More vines, shrubs, and weeds grow in the edge than the forest interior.
    C. Turkeys and deer are more plentiful in the edge region.
    D. Songbird populations have decreased at the edge of a forest.
    E. Mature deciduous trees inhabit the edge of a forest while grasses grow throughout the center.

39. Bacteria carry out denitrification and counter balance which of the following if not for human activities?
    A. nitrogen fixation
    B. ammonification
    C. renitrification
    D. sedimentation
    E. decomposition

40. Living in close quarters
    A. decreases time invested in grooming.
    B. exposes individuals to illness and parasites.
    C. limits social behavior.
    D. decreases protection from predators.
    E. minimizes dominance hierarchies.

41. Which of the following statements is NOT evidence for endosymbiotic theory?
    A. Present day mitochondria and chloroplasts have a size range within the range of that of bacteria.
    B. Mitochondria and chloroplasts have their own DNA and make some of their own proteins.
    C. Mitochondria and chloroplasts both do not have ribosomes.
    D. The mitochondria and chloroplasts divide by binary fission, as do bacteria.
    E. The membranes of mitochondria and chloroplasts differ—the outer membrane resembles a eukaryote and the inner membrane resembles a prokaryote.

42. The correct progression of evolution of biological processes would be
    A. aerobic respiration, photosynthesis, glycolysis.
    B. glycolysis, aerobic respiration, photosynthesis.
    C. aerobic respiration, glycolysis, photosynthesis.
    D. photosynthesis, aerobic respiration, glycolysis.
    E. glycolysis, photosynthesis, aerobic respiration.

43. A toxin that blocks photosystem I recycling electrons during photosynthesis would result in
    A. carbon dioxide being released.
    B. too much ATP being produced.
    C. photons of red light being reflected.
    D. too much NADPH being produced.
    E. water being produced.

44. The molecule 2,6-dichloroindophenol (DPIP) is frequently used in experiments studying photosynthesis. The purpose of DPIP is to
    A. act as an electron acceptor.
    B. allow photons of green light to be utilized.
    C. stabilize chlorophyll outside of the thylakoid membrane.
    D. allow NAD^+ to accept electrons from the photosystems.
    E. balance the amount of oxygen produced with carbon dioxide used.

45. Water potential as a concept in studying plants is preferable to osmosis because water potential
    A. takes pressure into account as well as solute concentration.
    B. only examines the concentration of water molecules.
    C. relates the concentrations of trace minerals to macronutrients in water.
    D. relates rates of precipitation to rates of transpiration.
    E. measures the amount of energy in water molecules.

46. Often, people will stop taking antibiotics for an infection after their symptoms subside, but before the entire prescription is taken. This may cause
    A. bacteria that are most sensitive to the antibiotic to remain in the body.
    B. the enzymes within the bacteria to renature and cause serious side effects.
    C. resistant bacteria to proliferate in the body.
    D. symbiotic bacteria in the digestive tract to be mutated.
    E. transformation of bacteria absorbing the antibiotic.

47. According to the RNA-first hypothesis,
    A. after the protocell developed it could synthesize DNA from RNA.
    B. RNA was responsible for DNA and protein formation.
    C. RNA genes could have replicated because of the presence of proteins.
    D. protein synthesis occurred before there were DNA genes.
    E. reverse transcription could not occur within the protocell.

The groups below consist of five choices followed by a list of numbered phrases. Select the choice that is most closely related. Each choice may be used once, more than once or not at all.

Questions 48–51 refer to the following species.
    A. keystone species
    B. exotic species
    C. sink population
    D. flagship species
    E. source population

48. The grizzly bear in northwestern United States; disperse berries, kill hoofed animals; move soil

49. The giant panda encourages preservation of biodiversity through being treasured as an icon.

50. Ballast water released in Oregon contained nearly 400 species from Japan.

51. An organism that impacts the viability of a community, although the population number may not be very high.

Questions 52–55 refer to the following reproductive barriers.

    A. habitat isolation

    B. hybrid sterility

    C. behavioral isolation

    D. mechanical isolation

    E. zygote mortality

52. Many animal species are restricted to a particular level of the forest canopy.

53. Male dragonflies have claspers suitable only for holding females of their species.

54. Female gypsy moths release pheromones that are detected miles away by receptors on the antennae of males.

55. A cross between a cabbage and a radish produces offspring that cannot form gametes.

Questions 56–59 refer to the following molecules.

    A. RuBP

    B. NADPH

    C. ADP

    D. $FADH_2$

    E. Acetyl CoA

56. Central to the balance between carbohydrate metabolism and lipid metabolism

57. Used to reduce carbon dioxide to carbohydrate

58. A five-carbon sugar intermediate molecule in the Calvin Cycle

59. Metabolism of this molecule causes ketosis.

Questions 60–63 refer to the following organelles.

    A. smooth endoplasmic reticulum

    B. ribosomes

    C. mitochondria

    D. chloroplasts

    E. vesicles

60. A proliferation of this organelle occurs with alcohol abuse.

61. Organelles only found in autotrophs

62. Alzheimer's disease is associated with a decreased rate and capacity for protein synthesis at this organelle.

63. After withdrawal from a drug, excess of this organelle is removed through autophagocytosis.

Questions 64–67 refer to the following diagram.

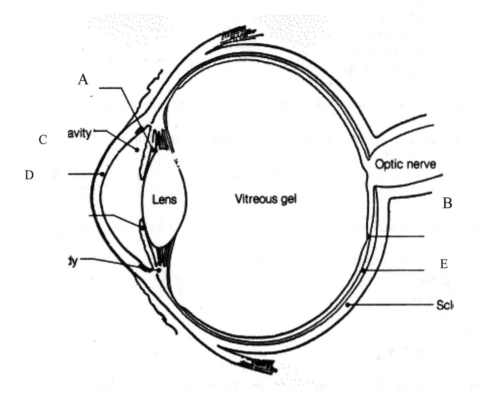

64. Rods and cones are located here.

65. Tough, white outer layer for protection

66. Allows more light in at night and less light in during the day

67. Light enters the eye here.

Questions 68–71 refer to the following groups of plants.

     A. angiosperms
     B. gymnosperms
     C. conifers
     D. bryophytes
     E. ferns

68. Possess vascular tissue, produce naked seeds and needle-like leaves

69. Ovules are completely enclosed within ovaries, which becomes a fruit.

70. Largest phyla of nonvascular plants

71. Includes broadleaved evergreen trees, grasses. and hardwood deciduous trees

Questions 72–75 refer to the following hormones.

    A. ethylene
    B. abscisic acid
    C. cytokinins
    D. gibberellins
    E. auxin

72. Synthetic forms of this hormone were used as a defoliant in the Vietnam War.

73. If you purchase "extra-large" grapes in the grocery store, they may have been treated with this hormone.

74. Tomatoes can be shipped across a country without ripening because of a lack of this hormone.

75. Plants under environmental stress will produce this hormone.

Questions 76–80 refer to the following hormones.

    A. erythropoietin
    B. testosterone
    C. melatonin
    D. prostaglandins
    E. estrogen

76. Frequently associated with circadian rhythms in humans

77. Teenagers are often tired in the morning and awake late at night due to this hormone that controls our sleep-wake cycle.

78. Causes uterine muscles to contract during childbirth

79. Regulates sexual development in males and females

80. Maintains secondary sexual characteristics such as facial hair and a lower voice during puberty in males

Questions 81–84 refer to the following pedigree.

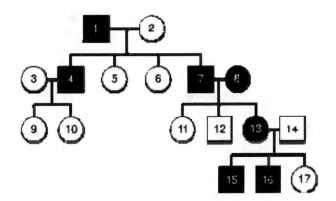

81. Which of the following inheritance patterns fits this pedigree?
    A. autosomal dominant
    B. X-linked recessive
    C. mitochondrial Inheritance
    D. autosomal recessive
    E. X-linked dominant

82. What can be concluded about the offspring of two heterozygous individuals?
    A. They must be homozygous dominant.
    B. There is a 50% chance they will be afflicted.
    C. All will be carriers.
    D. There is a 75% chance they will be genotypically normal.
    E. There is a 25% chance they could be homozygous recessive.

83. What are the chances that offspring from individual #15 and a heterozygous female will be afflicted?
    A. 0%
    B. 25%
    C. 50%
    D. 75%
    E. 100%

84. This particular disease could be
    A. sickle cell anemia.
    B. hemophilia.
    C. Down syndrome.
    D. Turner syndrome.
    E. Huntington's disease.

Questions 85–89 refer to the following.

A student submerges decalcified eggs into 4 different solutions of sucrose – 0 M, 0.5 M, 1.0 M, and 2.0 M. Below is a graph of the results:

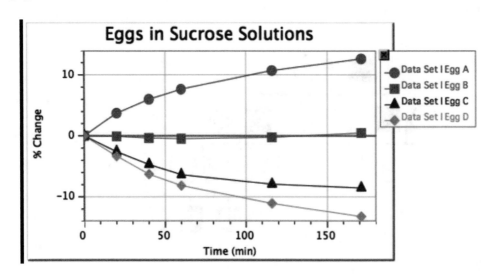

85. Based on the data above, which egg was in the distilled water?
    A. Egg A, because it gained mass.
    B. Egg B, because it remained unchanged.
    C. Egg C, because it lost mass.
    D. Egg C because it lost the least amount of mass.
    E. Egg D, because it lost the most mass.

86. What substance was most likely diffusing in or out of the eggs?
    A. sucrose
    B. proteins
    C. salts
    D. water
    E. carbohydrates

87. Why is % change used on the Y-axis instead of change in mass (grams)?
    A. Each egg has individual mass differences.
    B. Grams as a unit should not be used on the Y-axis.
    C. % change is the independent variable.
    D. % change looks better as a unit.
    E. The graph looks better with 0 in the middle.

88. What is the dependent variable in this experiment?
    A. time
    B. number of eggs
    C. sucrose molarity
    D. % change
    E. number of different solutions

89. At approximately 115 minutes Egg C is showing a flat line. Which statement below best describes what is happening?
    A. It is isotonic with its surrounding solution.
    B. Diffusion has stopped because the concentrations are equal.
    C. The water potential inside the egg is greater than the water potential outside.
    D. The egg is hypotonic to the hypertonic solution.
    E. The egg is hypertonic to the isotonic solution.

Questions 90–94 refer to the following.

A student inoculates two strains of *Sordaria* (wild-type and tan mutant) on a plate of agar. Where the mycelia of the two strains meet, fruiting bodies called perithecia develop. Meiosis occurs within the perithecia during the formation of asci. Perithecia may rupture, but the ascospores remain in the asci when pressure is applied to the slide. Using a compound microscope, the student viewed the slide and located a group of hybrid asci, with the results as follows:

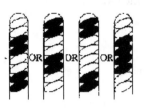 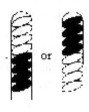

84   28

90. The percentage of crossover asci is divided by 2 because
    A. only half the spores in each ascus are the result of a crossover event.
    B. the spores cross over in groups of two.
    C. half of the asci are tan and half wild, so only half can be a variant.
    D. only half of the ascospores remain in the asci when the perithecia rupture.
    E. only one strain of *Sordaria* is capable of crossing over.

91. The gene to centromere distance is
    A. used to determine the distance between linked genes.
    B. based on many years of research and data collection.
    C. the actual position of the genes.
    D. the same in all species of fungi.
    E. determined with a monohybrid cross.

92. What is the gene to centromere distance for the cross above?
    A. 12.5
    B. 14
    C. 28
    D. 37.5
    E. 42

93. *Sordaria* is a good organism to use for genetic mapping because it
    A. is autotrophic and easy to feed.
    B. has a short lifespan and is easy to culture.
    C. is small and requires a microscope to see the perithecia.
    D. thrives on any medium.
    E. is a mold and is nonallergenic.

94. While counting the perithecia, the student also saw 36 asci that were all black. Why didn't the student include these in the counts?
    A. These are the result of a mutation and there was no crossing over.
    B. These were formed by asexual reproduction with no crossing over occurring.
    C. These are immature perithecia and it cannot be determined if there was crossing over.
    D. These were the result of a cross between two tan strains, and it is impossible to determine if there was crossing over.
    E. These are the result of parthenogenesis and show no crossing over.

Questions 95–98 refer to the following.

The graph below depicts the catalase enzyme at varying temperatures. Students collected the bubbles given off from the reaction of catalase and hydrogen peroxide (H_2O_2).

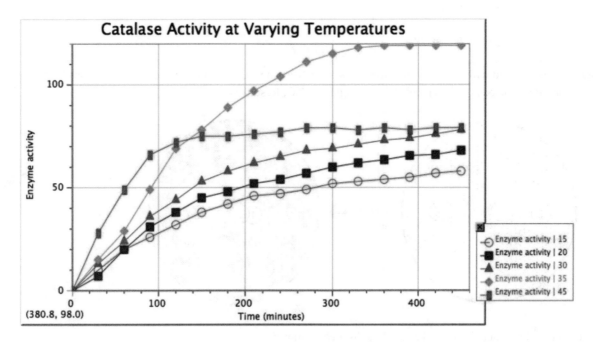

95. The most likely reason that the catalase at 45°C became a flat line at approximately 100 minutes was because the
    A. catalase used up all of the hydrogen peroxide.
    B. enzyme denatured due to the increased temperature.
    C. water bath cooled and the enzyme was not capable of working.
    D. a competitive inhibitor was activated and decreased enzymatic activity.
    E. active site was blocked by the product and stalled the reaction.

96. At approximately what temperature is the optimal temperature for catalase?
    A. 15°C
    B. 25°C
    C. 35°C
    D. 45°C
    E. 55°C

97. If the catalase at 15°C were able to continue for an unlimited time, at approximately what value would the enzyme activity flat line?
    A. 55
    B. 60
    C. 70
    D. 100
    E. 120

98. What is the gas released by the reaction between catalase and hydrogen peroxide?
    A. hydrogen (H_2)
    B. water (H_2O)
    C. ozone (O_3)
    D. oxygen (O_2)
    E. carbon dioxide (CO_2)

Questions 99–100 refer to the following.

An agar plate is streaked with bacteria, and three discs are placed on it. Disc 1 contains no antibiotic, discs 2 and 3 contain antibiotics. The petri dish is incubated overnight at 34°C. Results are as follows.

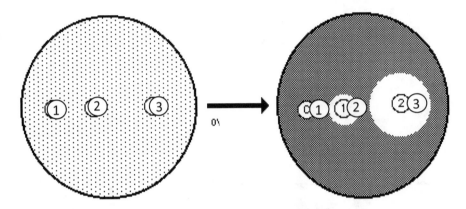

The plate is then incubated for several days. Results are as follows:

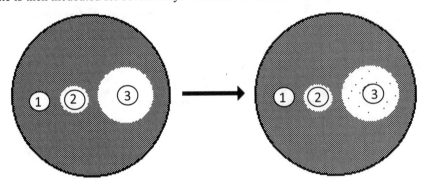

99. Which of the following can account for the lawn on the outside of the dish and the clear area near the antibiotics after one night of incubation?
    A. The bacteria were inhibited by the paper disc.
    B. The antibiotic on the paper disc inhibited bacterial growth.
    C. The paper reflected light around it, destroying bacteria.
    D. A contaminant on the dish killed the bacteria.
    E. More bacteria were initially spread on the outer edges of the dish.

100. Penicillin-resistant bacteria produce an enzyme that
   A. prevents attachment by the penicillin to the nuclear pores.
   B. increases the fluidity of the cell membrane.
   C. inhibits protein synthesis at the ribosomes.
   D. is secreted and dismantles the antibiotic.
   E. decreases cell wall connections allowing increased permeability.

Free Response Questions

1. Regulation is important at all levels of organization. For THREE of the following, **explain** the regulatory mechanism and how homeostasis or equilibrium is maintained.
   - Water movement in vascular plants
   - Carbohydrate storage in vertebrates
   - Proteins and the cell cycle
   - Temperature in ectotherms
   - Water balance in freshwater protists
   - Population size and ecosystem functioning

2. Structural relationship to function is demonstrated in many ways. Choose THREE of the following and **explain** how structure and function are related.
   - Introns in eukaryotic DNA
   - Chloroplasts in autotrophic organisms
   - Leaves of angiosperms
   - Cellulose in cell walls

3. Different colors of light have different effects on photosynthesis. The processes of photosynthesis and transpiration are intrinsically linked together.
   A. **Design** an experiment to study the effects of different colored light on transpiration.
   B. **Explain** how stomata are involved in the regulation of transpiration.

4. Without genetic variation, some of the basic mechanisms of evolution may not operate.
   A. **Discuss** THREE ways that variation can be introduced into a specific gene pool.
   B. Biotechnology has the potential to modify organisms. **Discuss** how biotechnology is used to create genetically modified organisms and their potential impact on the environment.

1. B; Enzymes have optimum pH ranges. Pepsin works in the vertebrate stomach where the pH is very low.
2. D; In a test cross, an individual with a dominant phenotype (BB or Bb) is crossed with a homozygote recessive to determine the genotype of the dominant individual. If phenotypically recessive offspring result, the dominant individual was a heterozygote.
3. A; Phototropism of stems occurs because the cells on the shady side of the stem elongate due to the presence of auxin. In the presence of blue light, a photoreceptor called phototropin is activated and becomes phosphorylated. A transduction pathway begins.
4. D; Biochemical evidence for evolution maintains that all organisms use the same DNA triplet code and therefore the same 20 amino acids in their proteins. Organisms may produce similar proteins, but have differences in their genetic codes due to the degeneracy of the code. However, when the degree of similarity in DNA base sequences is high, common descent and relatedness can be deduced.
5. E; Eubacteria do not contain introns. While introns were once thought to only be intervening sequences in the pre-mRNA transcript, scientists now know that introns are regulatory in nature and may influence gene flow.
6. B; Epinephrine plays a role in the body's reaction to short-term stressful stimuli. Blood glucose levels will increase and arterioles in skeletal muscle will dilate to receive glucose and oxygen. Nonessential physiological responses will slow or cease until the stressful stimulus has passed.
7. C; Plants are considered primary producers as autotrophs, caterpillars are primary consumers (herbivores), and birds are considered secondary consumers (carnivores).
8. B; Human females possess Barr bodies, or inactive X chromosomes, as an embryo. The inactive X chromosome does not produce gene products, and therefore female cells have a reduced amount of product from genes on the X chromosome.
9. B; The hemolymph of a grasshopper is colorless because it does not contain hemoglobin or any other pigment. It carries nutrients but no oxygen. Oxygen is taken to cells by way of tracheae found throughout the body, efficiently delivering oxygen and restricting water loss.
10. C; ATP synthesis occurs in the mitochondria.
11. A; Density-independent factors are usually abiotic. The intensity of the effect does not increase with increased population density.
12. A; The nervous and endocrine systems regulate and coordinate body activities.
13. E; In deuterostome development, the anus appears at or near the blastopore and only later does a second opening form by the mouth. Bilateral symmetry was preceded by radial symmetry. In bilaterally symmetrical animals, only one longitudinal cut yields two identical halves.
14. B; Prezygotic isolating mechanisms (habitat isolation, temporal isolation, and behavioral isolation) prevent mating attempts. Temporal isolation describes the fact that because species reproduce at different times of day or during different seasons they will not interbreed.
15. B; Nondisjunction can occur during meiosis I and results in abnormal eggs that also have one more or one less than the normal number of chromosomes. Nondisjunction during meiosis II may result in two eggs having one more or one less than the usual number of chromosomes. Meiosis normally results in four haploid eggs and, after fertilization by haploid sperm, four diploid zygotes.
16. A; Your blind spot occurs because no vision is possible where the optic nerve exits the retina. If there are no rods or cones, there is no synapse with the bipolar cells and ganglion cells that initiate nerve impulses and no stimulus is detected.
17. D; Normal prion proteins are found in healthy brains, although their function is unknown. Damage results when a normal prion protein changes shape so that the polypeptide chain is in a different configuration. Misshapen proteins interact with normal proteins inducing shape changes.
18. B; The results are 9/16 red and tall, 3/16 red and short, 3/16 yellow and tall, and 1/16 yellow and short.

19. E; Fibrous root systems often are associated with erosion control. Tap roots are able to expand and store starch.

20. E; Cancer cells disregard many cellular control mechanisms. They do not undergo apoptosis (cell suicide) and reproduce at will. They often consist of many layers and require a substantial blood supply in order to provide nutrients to the potential tumor.

21. C; The enzyme cascade is interrupted immediately after the missing enzyme; therefore, tyrosine will not be produced, nor melanin.

22. E; It is important to remember the orientation of the strands. 5' ends of strands must bond with the 3' end of the other strand. Adenine bonds with thymine and cytosine bonds with guanine. However, in RNA, uracil takes the place of thymine and bonds with adenine.

23. D; Many ribosomes are translating a eukaryotic mRNA at the same time, increasing the rate of protein synthesis.

24. A; Bryophytes produce flagellated sperm as they live in very moist environments where sperm can travel.

25. C; Bacteria are not eukaryotic or multicellular. Plants and animals do have internal protection of the zygote. Protists do not have a haploid life cycle, as fungi do.

26. D; Homologous chromosomes do not pair in mitosis. Meiosis produces haploid daughter cells. Sister chromatids migrate towards the poles in meiosis II, yielding the production of genetically unique daughter cells.

27. E; Erythropoietin stimulates red blood cell formation, thereby increasing the oxygen carrying capacity of the blood. It actually increases the viscosity of the blood and increases blood pressure, thereby increasing the risk of heart attacks and strokes.

28. B; Parthenogenesis occurs when an unfertilized egg becomes a haploid male.

29. C; The autonomic nervous system responds during times of stress in a "fight or flight" response. Your body is not expending energy for digestion (bile release, intestinal activity, saliva production) and your pupils will dilate to allow more light in to the eye.

30. D; The circulatory system is derived from the mesoderm.

31. C; The hierarchy of classification is: Kingdom, Phylum, Class, Order, Family, Genus, and Species. An organism must be classified in the same higher order division.

32. B; More than one population interacting in a given environment is a community.

33. D; While many diverse organisms possess cell walls, the component of those cell walls is different. Fungi do not contain chloroplasts, as they are heterotrophs, do not have large central vacuoles and do not contain centrioles (as animal cells do). Both fungi and plants are eukaryotic and possess a nucleus.

34. B; Red algae grow in marine environments and are classified along with the green algae.

35. D; Hydrogen bonds are present between nucleic acids in the double helix DNA chain.

36. C; Antibiotic-resistant bacteria possess mechanisms to increase their survival in the environment. They are resistant to many forms of antibiotics humans use to eradicate them.

37. A; Restriction enzymes recognize specific recognition sequences. If a restriction enzyme were to recognize more than one sequence and cut in those locations, the plasmid would not be able to accept a gene of interest during transformation.

38. E; An edge reduces the amount of habitat because their habitat is different. This impacts population size of various populations and can be detrimental because of habitat fragmentation.

39. A; Nitrogen is primarily made available to biotic communities by internal cycling of the element. Without human activities, the amount of nitrogen returned to the atmosphere (denitrification) exceeds withdrawal from the atmosphere (nitrogen fixation and nitrification). Human activities result in an increased amount of nitrates in terrestrial communities.

40. B; Group living has disadvantages and advantages. There is a great investment of time for social behaviors, grooming, and rearing offspring. However, disease can spread easily and disputes can arise over food and habitat.

41. C; Mitochondria have some of their own DNA and ribosomes and can make many of their own proteins. The DNA is circular and lies in the structures called nucleoids.

42. E; Photosynthesis had to come before aerobic respiration because it produced the O_2, and glycolysis is utilized in prokaryotes.

43. D; Without the evolution of the photosystems, the Calvin cycle would not have evolved because without the photosystem I, NADPH would not be produced to feed hydrogens into the formation of carbohydrates.

44. A; The DPIP is used to substitute $NADP^+$. In photosynthesis, electrons are normally transferred to $NADP^+$. However, DPIP will take its place in this experiment. DPIP is normally blue. When it is reduced, or gains electrons, it will turn colorless. DPIP is used to show that photosynthesis is taking place.

45. A; Water has a tendency to move for many reasons. Pressure, solute concentration, water concentration, and the type of solutes all impact its natural tendencies.

46. C; Antibiotic-resistant bacteria will proliferate prolonging infection and making the current antibiotic less useful against the bacteria.

47. B; There are several alternative hypotheses to explain the flow from DNA to RNA to protein in the history of life. The RNA-first hypothesis indicates that RNA would have been the first to evolve, and the first true cell would have had RNA genes.

48. A; Keystone species are integral in preserving ecosystem balance and biodiversity.

49. D; Flagship species are considered treasured for their beauty and motivate the public to preserve biodiversity.

50. B; Exotic species threaten the existence of native species.

51. A; Other keystone species include beavers in wetlands, bison in grasslands, alligators in swamps, and elephants in grasslands.

52. A; This occurs when two species occupy different habitats, even within the same geographic range.

53. D; When animal genitalia or plant floral structures are incompatible, reproduction cannot occur.

54. C; Many animal species have courtship patterns that allow males and females to recognize one another.

55. B; A hybrid zygote may develop into a sterile adult. This is a postzygotic isolating mechanism.

56. E; Acetyl CoA is integral in many biochemical reactions. It also is an important component in the synthesis of acetylcholine.

57. B; The conversion of CO_2 to CH_2O requires NADPH which is converted to $NADP^+$.

58. A; As five molecules of G3P become three molecules of RuBP, three molecules of ATP become three molecules of ADP + P.

59. E; Acetyl CoA metabolism produces ketone bodies.

60. A; Smooth endoplasmic reticulum functions in alcohol and drug detoxification.

61. D; Chloroplasts are the site of photosynthesis, a characteristic of photoautotrophs.

62. B; Protein synthesis (translation) occurs at the ribosome.

63. A; Smooth endoplasmic reticulum functions in alcohol and drug detoxification.

64. B; The retina is rich with photoreceptors (both rods and cones). Rods are sensitive to light, but they do not see color; cones, which require bright light, are sensitive to different wavelengths of light.

65. E; The sclera is a fibrous layer that covers most of the eye for protection and support.

66. D; The iris is the colored portion of the eye and regulates the size of an opening called the pupil.

67. C; The cornea is the window of the eye that refracts light rays.

68. C; Conifers are cone-bearing gymnosperms that include pine, cedar, and spruce trees.

69. A; Angiosperms are flowering land plants.

70. D; Bryophytes include the mosses which do not possess vascular tissue (xylem and phloem) and have neither flowers nor produce seeds, reproducing via spores.

71. A; The number of species of flowering plants is estimated at around 300,000, with most being dicots.

72. E; Synthetic forms of auxin have been used as herbicides to control weeds, but have little effect on grasses.

73. D; Gibberellins induce plant growth and increase the size of flowers. They cause an increase in the space between the grapes, allowing them to grow larger.

74. A; Normally, tomatoes ripen on the vine because the plants produce ethylene. Today, tomato plants can be genetically modified to not produce ethylene and this facilitates shipping. Once they have arrived, they are treated with ethylene to induce ripening.

75. B; ABA closes stomata when a plant is under water stress. It also induces rapid depolymerization of the actin filaments throughout the plant cell, which may be part of the transduction pathways involved in stomatal closure.

76. C. Melatonin. Production of melatonin regulates many circadian rhythms in humans.

77. C; Melatonin is produced primarily at night. When we grow sleepy at night, melatonin levels are high and once daylight returns, they recede.

78. D; Prostaglandins include several hormones produced within cells from arachidonate. While they cause uterine muscles to contract during childbirth, they also are implicated in the pain and discomfort of menstruation.

79. C; It has been shown that melatonin also regulates sexual development by noting that children whose pineal gland has been destroyed due to a brain tumor experience early puberty.

80. B; Testosterone secretions at the time of puberty stimulate the growth of the penis and the testes and also cause secondary sexual characteristics to appear.

81. A; In an autosomal dominance pattern, both homozygote dominant (AA) individuals and heterozygotes (Aa) are afflicted. In this situation, two afflicted individuals (Aa) may have a normal offspring as in the case of 7-8 and offspring 11-12-14. This is not possible in a recessive inheritance pattern.

82. E; In this situation there is a 25% chance for an AA individual, 50% chance for an Aa individual, and 25% chance for an aa individual (normal).

83. D; Individual 15 is heterozygous based upon his parents. Therefore, as in the previous question, the chances of an afflicted child (AA or Aa) is 75%.

84. E; Huntington's disease is autosomal dominant. Sickle cell is autosomal recessive, hemophilia is sex-linked recessive, Down Syndrome is a case of trisomy caused by nondisjunction as is Turner (XO) Syndrome.

85. A; Egg A gained the most water because it was hypertonic.

86. D; Water is moving in order to reach equilibrium through the process of osmosis.

87. A; It is impossible to compare mass changes directly when all the eggs had different masses to begin with. Calculating a % allows the eggs of differing masses to be compared.

88. D; % change is dependent upon time.

89. A; The environment is in equilibrium.

90. A; Counting the number of asci showing crossover is not the same as counting the number of organisms showing crossover. With most species, we count the actual number of offspring showing crossover, so we need to do the same thing here by counting the spores. One ascus contains eight ascospores, which have the potential of becoming eight new happy *Sordaria* organisms. In an ascus that shows crossing over, four of the spores will produce organisms that should be counted as crossovers, and the other four spores would produce organisms that should counted as non-crossovers. Therefore, the one ascus represents four crossover organisms out of eight, for a crossover rate of 50%. Thus, we divide by two.

91. A; A map unit is an arbitrary unit of measure used to describe the relative distances between linked genes. The number of map units between two genes or between a gene and a centromere is equal to the percentage of recombinants.

92. D; 84+28=112 total asci. 84/112=0.75 or 75% asci with crossover events. 0.75/2=0.375 or 37.5% (or map units)

93. B; *Sordaria* are ideal for genetic mapping because we can assess the genotypes of the spores while they are still contained in intact asci.

94. D; Crossing over is only apparent in crosses between two different strains of *Sordaria*.

95. B; Enzymes have optimum temperatures for activity. At this temperature the bonds of the protein may have denatured and the enzyme was rendered unusable at this point.

96. C; Based upon the slope of the line, the maximum rate of enzymatic activity (steepest slope) occurs in the 35°C range.

97. E; This is the point that the enzyme at 35°C flat lines; it would be no different for the catalase at 15°C if allowed to run its course.

98. D; This is a decomposition reaction where hydrogen peroxide decomposes to produce water and oxygen.

99. B; The absence of bacteria indicates that they were unable to proliferate where antibiotics were present.

100. D; Resistant microbes, however, either alter their cell walls so penicillin can't bind or produce enzymes that dismantle the antibiotic.

FRQ 1

MAXIMUM 4 PTS EACH TOPIC, choose 3.

	REGULATORY MECHANISM	HOMEOSTASIS/EQUILIBRIUM
Water movement in vascular plants	**Guard cells** regulate water loss.	Day/night opening; water loss/Abscisic acid/env. changes
	Root hairs increase surface area.	Increase water absorption
	Casparian strip traps water in vascular tissue.	Increase water absorption
	Tonoplast maintains proton pumps →pH gradient	Chemiosmotic potential
	Aquaporins augment water transport.	Affect rate of diffusion but not the water potential gradient
	Abscisic Acid	Affects stomatal opening and closing
Carbohydrate storage in vertebrates	**Insulin** facilitates entry of glucose into tissues; absence of insulin stimulates glycogen breakdown.	Occurs when blood glucose levels are high
	Glucagon decreases the ability of glucose to enter the tissues; stimulates glycogen breakdown.	Occurs when blood glucose levels are low
	Secretin, pancreozymin, enteroglucagon stimulate insulin release.	Occurs when blood glucose levels are high
Proteins and the cell cycle	**Cyclins**	Affect cyclin dependent kinases; cyclins rise and fall with the cell cycle.
	Kinases (Cyclin dependent kinases (CDk))	Levels are stable; activated by bonding with CDks
		Damage to the DNA before entering the S phase
	Anaphase promoting complex (APC)	Degrades mitotic cyclins; triggers sister chromatids to separate
	S-Phase promoting factor (SPF)	Prepares to duplicate DNA
	M-Phase promoting factor	Initiates assembly of the spindle; breakdown of the envelope; condensation of the chromosomes
Temperature in ectotherms	**Behavioral mechanisms include:** climbing to higher ground, entering warm water/air, building nests, lying on hot rocks/sun (or vice versa).	↑ or ↓ of body temperature
	Physiological mechanisms include: folding skin, concealing wings, changing shape, inflation of body, low metabolic rates.	↑ or ↓ of body temperature
	Enzymes that operate at many different temperatures	↑ or ↓ of body temperature
Water balance in freshwater protists	**Contractile vacuole**	Expels excess water; impacted by water potential
		Surrounded by canals; expels water through pores
		Capable of storing extra water

Population size and ecosystem functioning	Limited food supply	Density dependent;
	Predators	Density dependent; increased competition
	Toxic Waste	Density dependent
	Disease	Density dependent; contagious
	Environmental Disasters	Density independent

FRQ 2 (MAX 4 EACH)

	STUCTURE	FUNCTION
Introns in eukaryotic DNA	Interrupted location between exons	Increases crossover frequency; increases gene shuffling
	Long length	Increases crossover frequency →increases genetic diversity
	Precise looping sequence next to the exons	Accurate splicing
	Termination code bypassed	Shuffling
		Sex determination pathway
	Located within introns (nested)	Transcribed in the opposite direction; function as exon
		Reduce or correct mutations (i.e., hemoglobin)
	Polymorphisms within the introns of protease systems	→ metastasis
	Increased variation of introns	Increased gene regulation in disease
	Bonding pattern of introns	Assist in packaging of DNA
	Located near boundaries	Genetic and phenotypic variety
Chloroplasts in autotrophs	Double membrane	Protection
	Stroma: fluid-filled space	Room for starch production
	Thylakoids	Separate ETC
		Increased surface area
	Granum membrane	Chemiosmosis
Angiosperm leaves	Flat/thin	Increased surface area to expose chloroplasts to light
	Deciduous	Reduce water loss; induce dormancy
	Spongy layer (air spaces)	Gas exchange
	Waxy cuticle	Protection against water loss
	Epidermal hairs (trichomes)	Deters feeding by herbivores; protection against frost; break up air flow; reflect solar radiation;
Cellulose in cell walls	Cellulose (alternating alpha- and beta- linkages – with pectin)	Strength; rigidity within crossbridges
	Thicker layer	Increased rigidity

FRQ 3
PART A (MAX 8)
- Identification of independent variable – color of light
- Identification of dependent variable with units, transpiration mL of water lost
- **Testable** hypothesis stated, i.e., If plants are exposed to different colors, then the plants exposed to blue light will show the greatest rates of transpiration. (Actual color of blue not needed, but a specific color should be mentioned)
- Control – plant in day light (NOT plant in no light)
- Constants, at least two variables held constant, amount of light, types of plants, container size/type
- Repeated trials, at least two trials
- Range of IV noted, more than two colors used
- Correct equipment or technique used to measure transpiration, potometer, whole plant, etc.
- Statistical treatment of data, trials averaged

PART B (MAX 4)
- Stomata open when guard cells swell
- Inside walls of guard cells are thickened or cytoskeleton of guard cells radiate out like spokes.
- Physiological mechanism – Calcium ions pumped in, water follows and cell swells.
- Pumping of calcium ions coupled with generation of a membrane potential by active transport of protons out of the guard cell, resulting potential drives calcium ions into cell
- Cues for opening (internal max 2 pts)
  - Circadian rhythms
  - CO2 depletion
  - Blue light receptors
  - Environmental stress, lack of water
- physical mechanisms of stomata for limiting water loss (max 1 pt)
  - placement – underside of leaf
  - hairs on leaf surface
  - pits

FRQ 4
PART A (MAX 6)
Explanation of any of the following as mechanisms:
- Meiosis
- Sexual reproduction
- Random fertilization
- Independent assortment
- Mutation (point, substitution, frame shift, insertion, deletion)
- Chromosomal (translocation, crossing over, nondisjunction, polyploidy)
- Transposons, genetic engineering

PART B (MAX 6)
- Genetically modified organisms (GMOs)
  - Examples: corn, soybeans, glowing tetras and others
- Transformation of bacteria
- Procedures for gene transfer
  - Plasmids as vectors
  - Genomic libraries
  - Complementary DNA
- Transgenic animals
- Ti plasmid, crown gall bacterium for introducing genes into plants
- Genetic "shotgun" – use of gold or tungsten fragments coated with DNA and shot into plants

Correlation for Exam #1

Students and teachers should feel free to highlight the numbers of questions that were skipped and/or answered incorrectly. This will help to determine areas of strength and weakness in preparing for the AP Examination.

I. **Molecules and Cells (25%)**
  A. **Chemistry of Life (7%)**
    1. Water 84, 88, 45
    2. Organic molecules in organisms 35,
    3. Free energy changes
    4. Enzymes 1, 21, 94, 95, 96, 97
  B. **Cells (10%)**
    1. Prokaryotic and eukaryotic cells
    2. Membranes 85, 86, 87
    3. Subcellular organization 5, 10, 9, 60, 62
    4. Cell cycle and its regulation 20, 26,
  C. **Cellular Energetics (8%)**
    1. Coupled reactions
    2. Fermentation and cellular respiration 55, 58
    3. Photosynthesis 42, 43, 44, 56, 57,

II. **Heredity and Evolution (25%)**
  A. **Heredity (8%)**
    1. Meiosis and gametogenesis 8, 15, 89
    2. Eukaryotic chromosomes 5,
    3. Inheritance patterns 2, 18, 80, 81, 82, 83
  B. **Molecular Genetics (9%)**
    1. RNA and DNA structure and function 22, 23
    2. Gene regulation 36, 61, 90, 91
    3. Mutation 93
    4. Viral structure and replication 17
    5. Nucleic acid technology and applications 37
  C. **Evolutionary Biology (8%)**
    1. Early evolution of life 41, 47
    2. Evidence for evolution 4,
    3. Mechanisms of evolution 14, 98, 99, 100

III. **Organisms and Populations (50%)**
  A. **Diversity of Organisms (8%)**
    1. Evolutionary patterns 51, 52, 53, 54
    2. Survey of the diversity of life 13, 25, 34, 92
    3. Phylogenetic classification 30, 31, 69, 70
    4. Evolutionary relationships 33
  B. **Structure and Function of Plants and Animals (32%)**
    1. Reproduction, growth, and development 19, 24, 28, 67, 68,78, 79
    2. Structural, physiological, and behavioral adaptations 6, 9, 12, 16, 27, 29, 63, 64, 66
    3. Response to the environment 3, 65, 71, 72, 73, 74, 75, 76, 77,
  C. **Ecology (10%)**
    1. Population dynamics 11,
    2. Communities and ecosystems 7, 32, 39
    3. Global issues 38, 40, 46

BIOLOGY PRACTICE EXAM II

1. Organic molecules
   A. contain covalent bonds.
   B. do not contain hydrogen.
   C. contain either silica or carbon in long chains.
   D. contain a very small number of atoms.
   E. are ionized.

2. Polysaccharides
   A. serve as the major source of cellular fuel for all living things.
   B. are broken down during cellular respiration.
   C. are transported in the blood of animals.
   D. are the major components of cell membranes.
   E. cannot pass through the plasma membrane.

3. Saturated fats
   A. have double bonds in the carbon chain.
   B. are found in the feet of reindeer to prevent them from freezing.
   C. accumulate in blood vessels and block blood flow.
   D. are polymers of amino acids joined by peptide bonds.
   E. are not preferred by animals to glycogen for long-term energy storage.

4. Which of the following statements is true regarding DNA and RNA?
   A. DNA contains uracil in place of thymine as a nucleic acid.
   B. In DNA, the number of purine bases always equals the number of pyrimidines bases.
   C. RNA is often double stranded and forms a double helix.
   D. RNA is a nucleotide in which adenosine is joined with three phosphate groups.
   E. A specific type of DNA contains codons that are translated during protein synthesis.

5. Which of the following is the smallest?
   A. chloroplast
   B. amino acid
   C. virus
   D. human egg cell
   E. protein

6. An organism that is classified in Domain Archae
   A. has a nucleus with DNA containing introns and exons.
   B. contains ribosomes and endoplasmic reticulum for protein synthesis.
   C. undergoes mitosis during asexual reproduction.
   D. may use several flagella for locomotion.
   E. has a large central vacuole.

7. When a plant cell is placed in a hypertonic environment
    A. the large central vacuole gains water.
    B. turgor pressure increases.
    C. the net movement of water is from the outside of the cell to the inside.
    D. the plasma membrane pushes against the rigid cell wall.
    E. plasmolysis occurs.

8. In *Drosophila,* a cross between a X^rY and X^RX^R yields X^RY and X^RX^r in the F_1 generation where X^R = red eyes and X^r = white eyes. Males with white eyes in the F_2 generation
    A. are homozygous dominant for the trait.
    B. received both dominant alleles from the male parent.
    C. occur due to a mutation.
    D. are heterozygous for the trait.
    E. receive the recessive allele from the female parent.

9. The loosely packed arrangement of cells in the spongy mesophyll of the leaf
    A. increases the amount of surface area for gas exchange.
    B. secretes a waxy cuticle to prevent water loss.
    C. includes stomata to allow CO_2 to enter the leaf.
    D. also bears trichomes, protective hairs that secrete irritating substances.
    E. facilitates the production of ATP from glucose in the mitochondria.

10. Phylogenetic trees do NOT show that
    A. radial symmetry preceded bilaterally symmetry.
    B. deuterostome development occurs in echinoderms and chordates.
    C. sponges are not composed of specialized tissues.
    D. arthropods molt.
    E. cnidarians and sponges consist of two germ layers.

11. Which of the following structures is NOT correctly matched with its function?
    A. carapace – protection
    B. scolex – production of sperm in flatworms
    C. ganglia – nervous integration
    D. water vascular system – locomotion
    E. mantle – secretion of calcium carbonate shell

12. As the body temperature of an athlete increases,
    A. blood vessels constrict.
    B. sweat glands become inactive.
    C. shivering may occur.
    D. blood pools in the vital organs and thoracic cavity.
    E. the hypothalamus is triggered by warm receptors in the skin.

13. A disadvantage of a jointed exoskeleton in arthropods is
    A. an ability to grow with the animal.
    B. that molting makes them vulnerable to predators.
    C. a lack of support of the weight of a large animal.
    D. that it does not allow flexible movements.
    E. that it is not protected by outer tissues.

14. When an organism becomes deceased, a sustained contraction of the muscles called rigor mortis occurs. Rigor mortis is caused by
    A. an inability of the body to produce ATP.
    B. secretion of calcium from the sarcoplasmic reticulum.
    C. secretion of acetylcholine at the neuromuscular junction.
    D. breakdown of lactic acid.
    E. enzymes breakdown the actin-myosin crossbridges.

15. A plant and its pollinators are adapted to one another. Bee-pollinated flowers most likely do NOT
    A. contain nectar guides, highlighting the portion of the flower that contains the reproductive structures.
    B. appear irregular in shape, because they often have a landing platform where the bee can land.
    C. contain a large tube for the bee's feeding apparatus, but too small for other insects to reach the nectar.
    D. open at night and have a strong, sweet perfume that attracts moths and bees.
    E. appear brightly colored and are predominantly blue or yellow with shadings.

16. While walking in the woods you encounter a plant with an extensive fibrous root system. This plant is most suited for
    A. erosion control.
    B. storage of starches.
    C. climbing.
    D. parasitizing host plants.
    E. acquiring oxygen for respiration.

17. The SA node
    A. may be stimulated to reestablish a coordinated heart beat during a heart attack.
    B. signals the ventricles to contract by stimulating the Purkinje fibers throughout the ventricles.
    C. delays the impulse from the AV node to allow the ventricles time to fill with blood.
    D. produces the QRS wave on the ECG.
    E. initiates the heartbeat.

18. An individual who is experiencing a loss of motor coordination, tremors, and involuntary spasms of the eye may have experienced damage to the
    A. frontal lobe.
    B. Wernicke's area.
    C. cerebellum.
    D. brainstem.
    E. corpus callosum.

19. A physiological response to stress in the long term is
    A. increase in blood pressure.
    B. increase in blood glucose levels.
    C. increase in heart rate.
    D. suppression of immune system.
    E. protein and fat metabolism.

20. Autoimmune diseases target lymphatic tissue where
    A. T cells mature.
    B. prevention of the entrance of pathogens occurs.
    C. red blood cells are produced.
    D. drugs are detoxified.
    E. hormones are secreted.

21. When *E. coli* is in the presence of lactose
    A. the repressor will bind to the operator and inactivate enzyme production.
    B. RNA polymerase will bind to the promoter, and the structural genes for lactase will be expressed.
    C. it becomes the preferential sugar in place of glucose.
    D. lactase exons are expressed.
    E. and glucose is present, cAMP levels are high.

22. Prokaryotic and eukaryotic DNA replication differ in that
    A. in prokaryotes, replication can occur in only one direction.
    B. only eukaryotic replication requires DNA polymerase to begin the replication process.
    C. in eukaryotes replication occurs at numerous replication forks.
    D. DNA polymerase is unable to replicate the ends of the linear chromosomes of eukaryotes.
    E. eukaryotic DNA replication occurs at a much faster rate.

23. In sickle cell anemia, an autosomal recessive disorder,
    A. all affected children will have affected parents.
    B. heterozygotes are affected.
    C. males are affected with a higher frequency.
    D. two affected parents will always have affected children.
    E. it is impossible to have carriers.

24. In a dihybrid cross involving F_1 offspring resulting from a homozygous dominant tall, green plant and a short, yellow plant, the F_2 generation would have
    A. 1/16 tall and green plants.
    B. 1/4 short and green plants.
    C. 3/16 tall and yellow plants.
    D. 1/8 short and yellow plants.
    E. all tall and green.

25. Which of the following is NOT true regarding photosynthesis and respiration?
    A. In photosynthesis, water is oxidized and oxygen is released.
    B. In cellular respiration, a carbohydrate is oxidized to CO_2.
    C. Both processes have an electron transport chain within membranes where ATP is produced.
    D. In photosynthesis, the carbon in CO_2 is reduced to a carbohydrate.
    E. In photosynthesis, NADPH is reduced to NADP+ occurs when glucose is metabolized.

26. The most common form of nutrition in bacteria is
    A. photoautotrophic.
    B. heterotrophic.
    C. chemoautotrophic.
    D. saprophytic.
    E. hypertrophic.

27. Asexual reproduction in plants does NOT include
    A. horizontal stems running aboveground.
    B. underground stems that grow laterally and produce buds.
    C. production of suckers to grow new trees.
    D. propagation from stem cuttings.
    E. cross pollination in angiosperms.

28. How many capillary beds does a red blood cell need to travel through if it travels from the right ventricle to the right atrium by way of the kidney?
    A. 1
    B. 3
    C. 4
    D. 2
    E. 0

29. Red blood cells are degraded in the
    A. kidneys.
    B. intestines.
    C. gall bladder.
    D. pancreas.
    E. liver.

30. Countercurrent exchange maximizes which of the following processes in the gills of a fish?
    A. diffusion
    B. osmosis
    C. active transport
    D. phagocytosis
    E. endocytosis

31. The majority of the ATP molecules formed by complete glucose breakdown occurs as a result of
    A. the electron transport chain.
    B. glycolysis.
    C. the Krebs cycle.
    D. the light independent reactions.
    E. carbon fixation.

32. A student attempts to insert a gene for bioluminescence into a bacterial plasmid. Transformation of the bacteria cannot be verified. Which of the following could have occurred?
    A. A restriction enzyme that only cut at one location on the plasmid was used.
    B. Sticky ends were produced on the DNA where the gene is located.
    C. A recognition sequence for the restriction enzyme is present in the gene of interest.
    D. The restriction enzyme used the same sequence on the DNA and on the plasmid.
    E. The restriction enzyme did not make a cut within the DNA of the bioluminescent gene.

33. Fungi were once classified with animals because they
    A. store energy in the form of glycogen.
    B. do not ingest food.
    C. have cell walls containing chitin instead of cellulose.
    D. produce spores and gametes with flagella.
    E. are multicellular parasites.

34. Most of the water taken in by plants
    A. enters through the stomata at night.
    B. travels through the phloem to the leaves.
    C. is lost through stomata by transpiration.
    D. exits the leaves through guttation.
    E. passes through the Casparian strip in the roots.

35. The formation of ATP
    A. requires energy in order to proceed.
    B. releases heat.
    C. is exergonic.
    D. releases energy for protein synthesis and nerve conduction.
    E. is the result of the oxidation of phosphorous.

36. Through coupled reactions, ATP is NOT able to
    A. drive energetically unfavorable processes.
    B. synthesize macromolecules.
    C. form tissues.
    D. maintain the internal composition of the cell.
    E. be produced in large quantities in an anaerobic environment.

37. Microspores develop into
    A. female sporophytes.
    B. male gametophytes.
    C. megaspores.
    D. flowers.
    E. fruits.

38. The immediate inflammatory response to an excruciating paper cut does NOT include
    A. release of histamine.
    B. migration of phagocytes to the affected area.
    C. stimulation of white blood cells.
    D. production of a blood clot.
    E. release of interferons.

39. Which of the following is an example of a positive feedback system?
    A. blood calcium levels
    B. uterine contractions during childbirth
    C. blood glucose levels
    D. temperature
    E. blood pressure regulation and erythropoiesis

40. Prior to metamorphosis, tadpoles are most likely to excrete
    A. uric acid.
    B. urea.
    C. ammonia.
    D. isourea.
    E. carbon dioxide.

41. A population of reindeer grows exponentially for several seasons and then undergoes a sharp decline as a result of overgrazing. The period of exponential growth was characterized by
    A. excessive hunting pressure.
    B. low reproductive rate.
    C. increased number of predators.
    D. unlimited resources.
    E. competition for food.

42. One of the fundamental differences between meiosis and mitosis is
    A. during prophase I of mitosis, crossing over occurs to increase genetic variability of offspring.
    B. meiosis requires one nuclear division.
    C. mitosis produces four haploid daughter cells.
    D. mitosis involves a separation of homologous chromosomes.
    E. meiosis occurs only at certain times in the life cycle of sexually reproducing organisms.

43. When you receive a tetanus shot, you are protecting yourself against
    A. a bacteria that is capable of reproducing aerobically in the lungs.
    B. a virus capable of causing cancer.
    C. a bacterial toxin that prevents relaxation of muscles.
    D. endospores that absorb water in the intestinal tract and begin reproducing as bacterial cells.
    E. a virus that penetrates the colon and causes diarrhea.

44. Which of the following events occurred latest on the geologic timescale?
    A. First chordates appear.
    B. Amphibians diversify.
    C. Bony fish appear.
    D. Reptiles appear.
    E. Placental mammals appear.

45. A single base change accounts for the genetic disorder sickle-cell disease because the incorporation of valine, instead of glutamic acid, causes hemoglobin to be misshapen. This mutation is classified as a
    A. frameshift mutation.
    B. insertion mutation.
    C. deletion mutation.
    D. point mutation.
    E. nonsense mutation.

46. Which of the following types of gene transfer therapy is NOT correctly matched with the condition?
    A. aerosol spray – cystic fibrosis
    B. bone marrow transplant – sickle cell anemia
    C. implantation – diabetes
    D. aerosol spray – Alzheimer's disease
    E. injection – brain tumors

47. Which of the following structures is NOT homologous to the flipper of a whale?
    A. mammalian tail bone
    B. wing of a bat
    C. forelimb of a cat
    D. human arm and hand
    E. wing of a bird

48. Protists are extremely diverse. A protist that is unicellular, colonial, and contains plastids is a
    A. green algae.
    B. water mold.
    C. euglena.
    D. amoeba.
    E. diatom.

49. Due to exposure to two different environments, British land snails have two shell patterns. This is an example of
    A. directional selection.
    B. territoriality.
    C. the founder effect.
    D. disruptive selection.
    E. stabilizing selection.

50. Active genes in eukaryotic cells are NOT associated with
    A. euchromatin.
    B. nucleosomes.
    C. histones.
    D. chromatin remodeling complex.
    E. Barr bodies.

51. The fact that individuals afflicted with polydactyly, the presence of extra digits on hands and feet, can have varying degrees of affliction is due to
    A. incomplete dominance.
    B. penetrance.
    C. pleiotropy.
    D. multiple alleles.
    E. epistasis.

52. Spermatogenesis produces
    A. four haploid sperm.
    B. spermatozoa that differentiate into spermatids.
    C. two haploid spermatozoa through mitosis.
    D. one diploid sperm and three polar bodies.
    E. two haploid sperm cells and two polar bodies.

53. The final electron acceptor of the electron transport chain in photosynthesis is
    A. CO_2.
    B. $C_6H_{12}O_6$.
    C. NADP+.
    D. H_2O.
    E. RuBP.

The groups below consist of five choices followed by a list of numbered phrases. Select the choice that is most closely related. Each choice may be used once, more than once or not at all.

Questions 54–57 refer to the following structures.
    A. fimbriae
    B. plasmid
    C. cilia
    D. flagella
    E. pili

54. One of the primary reasons *E.coli* bacteria can attach to their host and cause disease

55. A primary mechanism for the development of methicillin resistant *Staphlococcus aureus* (MRSA), a specific strain of antibiotic resistant bacteria

56. Spirochetes are specialized with twisting locomotion to cause diseases such as Lyme Disease and syphilis.

57. Lack of this structure in mammalian Fallopian tubes can cause ectopic pregnancy.

Questions 58–61 refer to the following graph.

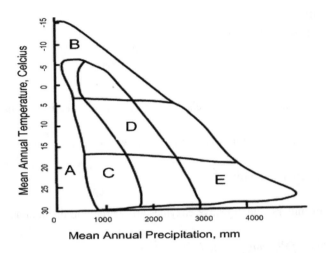

58. This biome is the most productive on Earth, with the highest number of different kinds of species and their abundance.

59. This biome is cold and dark much of the year with extremely long, cold, harsh winters and short summers.

60. Adaptations of plants that live here include thick epidermal layers, water-storing succulent stems and leaves, and the ability to set seeds quickly in the spring.

61. Species with adaptations for living in extreme cold with short growing or feeding seasons live here.

Questions 62–65 refer to the following cellular organelles.

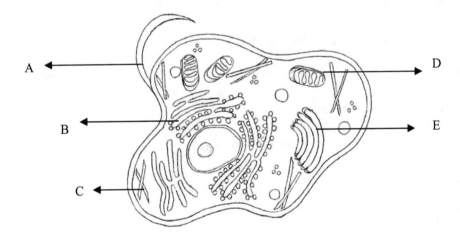

62. Site of ATP synthesis

63. Proteins for export are produced and processed here.

64. Attachment of oligosaccharides to identify self from non-self occurs here.

65. Contains DNA

Questions 66–69 refer to the following processes.

A. exocytosis
B. facilitated diffusion
C. receptor mediated endocytosis
D. osmosis
E. phagocytosis

66. Cholesterol accumulates in the blood vessels due to this type of defective mechanism.

67. Involved in development of immunity to bacterial diseases

68. Maintains turgor pressure in autotrophs

69. Hormones, neurotransmitters, and digestive enzymes are secreted in this manner.

Questions 70–73 refer to the following pyramid.

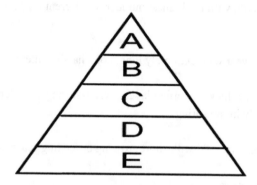

70. The productivity of the ecosystem is dependent upon organisms at this level.

71. This trophic level always contains herbivores.

72. The species at this trophic level is often slow to evolve and easy to drive to extinction.

73. Autotrophs are located at this level.

Questions 74–77 refer to the following classes of molecules.

A. polypeptides
B. lipids
C. monosaccharides
D. polysaccharides
E. nucleotides

74. Form immunoglobulins and hormones

75. Comprise the structural units of DNA and RNA

76. Ideal storage molecules that are insoluble in water and fold into compact shapes

77. Function as enzymes

Questions 78–81 refer to the following classes of vertebrates.

    A. mammals

    B. birds

    C. fish

    D. amphibians

    E. reptiles

78. External fertilization, moist skin with no scales, eggs with no amnion

79. Swim bladder and lungs

80. Lungs and air sacs, but no diaphragm muscle

81. Single loop circulatory system

Questions 82–85 refer to the following experiment.

Students placed three pillbugs on each side of the petri dish setup. Every 30 seconds the students observed how many pillbugs were on each side. They continued data collection for 15 minutes. After five minutes, 5 of the pillbugs were on the wet side. After 10 minutes, 4 pillbugs were on the wet side of the petri dish, and after 15 minutes all 6 pillbugs were on the wet side of the petri dish.

Dry petri dish   Wet petri dish

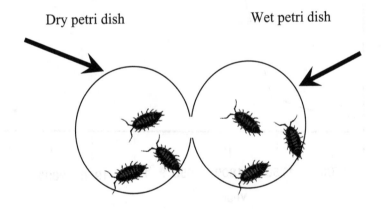

82. What is the independent variable in this experiment?

    A. the different choice chambers

    B. the wet side of the petri dish

    C. the time spent observing the pillbugs

    D. the number of pillbugs on each side of the petri dish

    E. the amount of moisture – wet or dry

83. The movement of the pillbugs can be classified as

    A. kinesis.

    B. territoriality.

    C. imprinting.

    D. taxis.

    E. learned behavior.

84. The first three times the students ran this experiment, they had very random data that did not allow them to make a conclusive statement about pillbug behavior. Which of the following statements is NOT a source of error the students may have committed?
    A. The students observed the pillbugs for 25 minutes each trial.
    B. The students placed a light over the wet side in the first trial, but not the other trials.
    C. The pillbugs were all facing away from the connector site at the start of the experiment.
    D. The filter paper on the wet side was thoroughly saturated, but not dripping wet.
    E. The filter papers were both white in color.

85. An example in nature of a similar behavior is NOT
    A. euglena moving towards light.
    B. fish turning to face the current.
    C. marine bacteria directing movement towards sediment.
    D. larvae of the king crab finding the surface waters.
    E. paramecium moving slowly in random circles to encounter food.

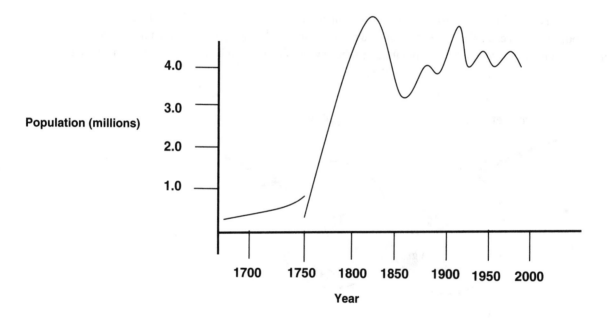

86. The apparent carrying capacity of this population appears to be approximately
    A. 4.5 million.
    B. 2.75 million.
    C. 3. 5 million.
    D. 2.0 million.
    E. 0.5 million.

87. Between 1825 and 1850, which of the following could have occurred?
    A. increased birth rate
    B. increased competition
    C. immigration of prey
    D. decreased predation
    E. emigration of predators

88. Between 1750 and 1800 which of the following may have occurred?
    A. the founder effect
    B. a bottleneck effect
    C. severe drought
    D. outbreak of disease
    E. El Niño event

89. The time period of 1700 to 1750 is often called the lag phase. What accounts for the slow growth during this phase?
    A. low fecundity
    B. increased predation
    C. low population levels
    D. low food levels
    E. high rates of photosynthesis

A group of students set up a potomoter, a device that measures the rate at which a plant is able to draw up water. Transpiration drives this uptake of water; therefore, the students measured the rate of transpiration from the values on the potometer. The students investigated how environmental factors (i.e., wind, light, humidity) affected the rate of transpiration in angiosperms. Results are as follows:

Treatment		Water Loss , mL/m^2 of leaf	
	5 minutes	10 minutes	15 minutes
Room	1.67	4.12	5.87
Fan	3.75	5.98	6.02
Light	3.91	7.01	9.78
Mist	0.95	2.23	7.74

90. When conditions are conducive to stomatal opening
    A. a proton pump drives protons from the guard cells.
    B. K^+ ions diffuse out of the guard cells.
    C. osmotic pressure inside the cell decreases.
    D. water leaves the guard cells via osmosis.
    E. the cell's volume and turgor pressure decrease.

91. Which of the following hormones is responsible for closing of the stomata during times of environmental stress?
    A. auxin
    B. abscisic acid
    C. gibberellin
    D. ethylene
    E. cytokinin

92. Which of the following statements is supported by the data?
    A. The greatest overall rate of transpiration occurred in the plant in the humid condition.
    B. An increase in temperature and decrease in humidity are conditions that will cause the greatest photosynthetic rate.
    C. Transpiration did not occur in the plants in the windy condition.
    D. The rate of photosynthesis is the highest in the plant under light conditions because the leaves are larger than in the other treatments.
    E. At the end of 15 minutes, all plants were actively photosynthesizing.

93. What can account for the changes between 10 and 15 minutes in the plant exposed to the fan?
    A. Photosynthetic rate increased.
    B. The stomata remained open and CO_2 entered the mesophyll.
    C. K^+ rushed out of the cell and the cell lost turgidity, causing the stomata to close.
    D. Photosynthesis was occurring at a higher rate than cellular respiration.
    E. CO_2 was not present in the atmosphere for photosynthesis.

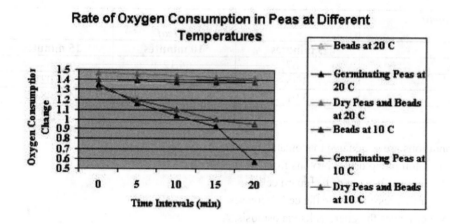

94. The rate of cellular respiration is calculated in the lab by measuring the
    A. consumption of O_2 by the photosynthesis of the peas.
    B. moles of CO_2 produced by chloroplasts.
    C. amount of energy released by respiration.
    D. mL of water produced from oxidation of food.
    E. change in gas pressure due to consumption of O_2.

95. The dependent variable in the above graph is
    A. time.
    B. temperature.
    C. change in oxygen consumption.
    D. the beads.
    E. germinating peas.

96. Which of the following is true regarding the effects of germination and temperature on pea seed respiration?

    A. Germinating seeds have a higher metabolic rate and need more oxygen for survival. *

    B. Nongerminating peas are not living and therefore do not require oxygen.

    C. A decrease in temperature increases oxygen consumption.

    D. Germinating peas have higher rates of photosynthesis than nongerminating peas.

    E. Peas that are in a respirometer are incapable of photosynthesizing.

Population Pyramids from 1990

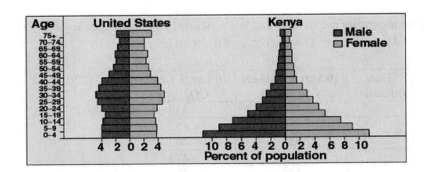

97. Which of the following can account for the broad base of the Kenyan population in comparison to that of the United States?

    A. infant mortality

    B. natural disasters

    C. high birth rate

    D. high death rate

    E. earlier reproductive maturity

98. Which of the following sets of variables best describe the population pyramid of Kenya?

    A. high birth rates followed by high death rates

    B. quality health care and low infant mortality

    C. adequate nutrition and clean sources of drinking water

    D. broadly available education and transportation

    E. adequate economic development and employment

99. Which of the following can specifically account for the narrow base of the population pyramid representing the United States?

    A. low birth rates

    B. technological advances

    C. higher death rates

    D. global warming

    E. adequate health care

100. Which of the following can you conclude about the diagrams?
    A. There was a baby boom in Kenya nearly 40 years ago.
    B. The Kenyan population is aging at a faster rate than that of the United States.
    C. The Kenyan population consists of more males than females.
    D. The birth rate is higher in the United States than in Kenya.
    E. The United States population is growing at a slower rate than that of Kenya.

Free Response Questions

1. Energy transfer occurs throughout ecosystems.
    A. **Explain** how energy flows between trophic levels in a food chain.
    B. **Discuss** how membrane structure affects the synthesis of ATP in photosynthesis.

2. A student measured the respiratory rate of small and large crickets by placing 10 crickets into a vial and measuring the amount of CO_2 released. The data obtained is shown below:

Time (minutes)	10 Small Crickets CO_2 (ppm)	10 Large Crickets CO_2 (ppm)
0	450	450
1	460	455
2	480	460
3	490	470
4	510	475
5	525	490
6	540	510
7	555	520
8	565	535
9	590	540
10	610	550

    A. **Graph** the data, being sure to appropriately label and title your graph.
    B. Determine the respiratory rates for the large and small crickets.
    C. **Explain** the results shown by the data above.
    D. **Discuss** how carbohydrates, fats, and proteins are used as energy sources.

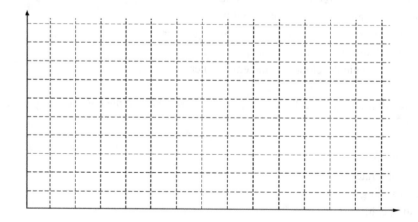

3. Natural selection is the driving force of evolution.
   A. **Explain** the four main components of natural selection.
   B. **Discuss** how the heterozygote advantage assists in the maintenance of genetic diversity.
   C. **Discuss** how male competition and female choice play a role in natural selection.

4. Living organisms must have regulation of chemical and behavioral mechanisms. For FOUR of the following, **explain** how regulation occurs and **discuss** the implications of a lack of regulation.
   - Cell cycle controls
   - Membrane permeability
   - Menstrual cycle
   - Stomatal opening and closing
   - Carrying capacity of a population
   - Osmoregulation in vertebrate fishes

Annotated Answer Key

1. A; Organic molecules, associated with living organisms, contain carbon and hydrogen atoms. They are always joined in covalent bonds and are often quite large molecules, with many atoms.
2. E; Polysaccharides are broken down to release sugar molecules when an organism requires energy. They are not as soluble in water and are much larger than a sugar. Therefore they cannot pass through the lipid bilayer of the cell membrane.
3. C; Saturated fats are solid at room temperature and are of animal origin. Diets high in saturated fats have been associated with circulatory disorders because fat accumulates in the lining of blood vessels.
4. B;

	DNA	**RNA**
Sugar	Deoxyribose	Ribose
Bases	A, T, C, G	A, U, C, G
Strands	Double stranded with base pairing	Single stranded
Double Helix	Yes	No

5. B; Amino acids are very small. The order, from smallest to largest, is amino acid, protein, virus, chloroplast, human egg cell. Viruses are comprised of proteins and nucleic acids.
6. D; Organisms in Domain Archaea are all prokaryotic and therefore do not have a nucleus. They also do not contain membrane-bound organelles such as an endoplasmic reticulum. Prokaryotes reproduce via binary fission, and plants contain large central vacuoles.
7. E; The hypertonic external environment contains a higher concentration of solutes (high solute potential) and a low water potential. Water travels from a high water potential to a low water potential, or a low solute potential to a high solute potential. Water, therefore, will move out of the cell, causing plasmolysis (separation of the cell membrane and cell wall) to occur.
8. E; In a sex-linked trait, males only carry one allele on the single X chromosome that they have. Therefore, they cannot be heterozygous and must receive their X-linked allele from the female parent, as the male parent passes on the Y chromosome to the male offspring.
9. A; The epidermis secretes a waxy cuticle, contains hairs, and includes stomata. The mesophyll is located in the interior of the leaf.
10. E; Sponges do not have germ layers, and cnidarians only have one germ layer.
11. B; The scolex is used for attachment by parasitic flatworms to tissues of the host.

12. E; Warm receptors in the skin trigger the hypothalamus to dilate blood vessels in the skin to dump heat. Shivering and pooling of blood in the vital organs are heat conservation mechanisms.

13. B; During molting, arthropods are vulnerable to predators. However, they must molt in order to grow.

14. A; An inability to produce ATP will prevent the crossbridges from sliding apart to relaxation. Therefore, a sustained contraction occurs, called rigor mortis.

15. D; Bee-pollinated flowers are not open at night with a sweet perfume. Moths are attracted to lightly colored flowers, however bees prefer brightly colored flowers that are open during the day.

16. A; Tap roots function in storage of starches, and aerial roots aid in climbing. The nature of fibrous roots is to spread out and therefore anchor the plant and prevent erosion

17. E; The sinoatrial (SA) node is located at the top of the heart and initiates the heartbeat. It sends an electrical message to the AV node located in the interventricular septum that signals the ventricles to contract.

18. C; The cerebellum is chiefly responsible for coordination and muscle control. Wernicke's area is integral in language production. The brainstem controls involuntary bodily functions, and the frontal lobe (cerebrum) is responsible for cognitive processes.

19. B; A long-term response to stress is protein and fat metabolism; a short-term response involves mobilizing glucose for energy.

20. A; T cells are lymphocytes that are capable of recognizing self from non-self and therefore are the self-pathogens in an autoimmune disease.

21. B; The lac operon is an inducible operon and operates only if lactose is present and glucose levels are low, in which case cAMP levels are low.

22. C; There is only one replication fork in prokaryotes, however, they move in both directions on the linear chromosome at a very efficient rate, much faster than DNA replication in eukaryotes.

23. D; In an autosomal-recessive disorder, if both parents are carriers (Aa x Aa) there is a 25% chance of an affected child being born. However, if both parents are affected (aa x aa) the only possible offspring are affected (aa).

24. C; The predicted outcome from this cross is 9 Tall Green: 3 Tall Yellow: 3 Short Green: 1 Short Yellow.

25. E; NADPH is a component of photosynthesis, not respiration.

26. B; All forms of nutrition are found among the bacteria, but most are heterotrophic. These are beneficial in ecosystems because they are organisms of decay that break down organic remains.

27. E; Some species of plants rely on wind pollination. Much of the plant's energy goes into making pollen to ensure that some pollen grains actually reach a stigma.

28. D; The blood cell will travel through a capillary bed in the lungs and one in the kidney.

29. E; The liver is the site of breakdown of red blood cells, the pancreas produces insulin, the gall bladder stores bile, and the intestines are the site of nutrient breakdown and absorption.

30. A; Countercurrent exchange maximizes diffusion of gases across membranes from higher concentration to lower concentration.

31. A; The electron transport chain contributes substantially more ATP during respiration than any other cycle. The light-independent reactions do not involve glucose breakdown and are part of photosynthesis.

32. C; Restriction enzymes should only cut at one location on the plasmid. Sticky ends facilitate the fusion of the gene of interest and the plasmid. Not making a cut in the gene of interest should facilitate expression.

33. A; Both animals and fungi store energy in the form of glycogen.

34. C; Transpirational pull is the driving force of water as it gets pulled from the roots to the leaves.

35. A; The production of ATP is endergonic (energy requiring) and uses energy from exergonic reactions (e.g., cellular respiration) in order to proceed. The breakdown of ATP to ADP + P releases energy for endergonic reactions (e.g., protein synthesis, nerve conduction, muscle contraction).

36. E; Through coupled reactions, ATP drives forward processes that occur to create the high degree of order essential to life. This includes the production of macromolecules, organization of cells and tissues, and the movement of cellular organelles. B; the lac operon is an inducible operon and operates only if lactose is present and glucose levels are low in which case cAMP levels are low.

37. B; Flowering plants exhibit an alternation of generations life cycle. Flowers borne by the sporophyte produce microspores and megaspores. Microspores develop into a male gametophyte, and megaspores develop into the female gametophyte.

38. E; Interferons are cytokines, soluble proteins that affect the behavior of other cells. Interferons are made by virus-infected cells. They bind to the receptors of noninfected cells, causing them to produce substances that interfere with viral replication.

39. B; Uterine contractions are a positive feedback loop. The production of oxytocin continues to strengthen the contractions until the stimulus ceases.

40. C; Due to the fact that tadpoles are aquatic, they are able to balance their energy expenditure and toxicity by secreting ammonia. Their excretory substance will change as they become terrestrial.

41. D; The period of exponential growth indicates that there are no controls acting on the population (hunting, predation, competition) and that the population is growing at will (unlimited resources).

42. E; Meiosis is a reduction division in which four haploid daughter cells are formed after two divisions. Mitosis produces diploid daughter cells that are identical (no joining of homologous chromosomes or crossing over).

43. C; *Clostridium tetani* bacteria cause tetanus. They secrete a bacterial toxin that causes prolonged contraction of the muscles (lockjaw).

44. C; Insectivores appeared 65.5 MYA, placental mammals 145.5 MYA, reptiles diversified 300 MYA, and the first chordates appeared 500 MYA.

45. D; Point mutations involve a change in a single DNA nucleotide and therefore a possible change in a specific amino acid, as in sickle-cell disease.

46. D; Aerosol spray is used in cavities where there is air flow, as in the lungs. It treats cystic fibrosis and inherited forms of emphysema.

47. A; All of the limbs, whether wings, legs, arms, or flippers are built upon the same basic structure. The mammalian tailbone does not have this same structure.

48. A; Protists in the Archaeplastid super group are distinguished by containing plastids and being unicellular, multicellular, OR colonial. The colonial characteristic is indicative of the algae in Protista.

49. D; There are three types of natural selection. During stabilizing selection, the intermediate phenotype is favored. During directional selection, an extreme phenotype is favored, and during disruptive selection, TWO extreme phenotypes are favored.

50. E; Active genes in eukaryotic cells are associated with more loosely packed chromatin. A nucleosome is a portion of DNA wrapped around a group of histone molecules. A chromatin remodeling complex pushes aside the histone portion of a nucleosome so that access to DNA is not blocked for transcription.

51. B; Many dominant alleles exhibit varying degrees of penetrance. Perhaps the expression of the allele may require additional environmental factors or be influenced by other genes.

52. A; In human males, spermatogenesis occurs within the tests. Primary diploid spermatocytes undergo meiosis I to form two secondary spermatocytes, each with 23 duplicated chromosomes. After meiosis II, four haploid spermatids are produced that differentiate into viable spermatozoa.

53. C; Oxygen is the final electron acceptor in cyclic phosphorylation.

54. A; *E. coli* bacteria can attach via fimbriae—small, bristlelike fibers on the surface of a bacterial cell— which attaches bacteria to a surface.

55. E; Sex pili are capable of transferring plasmids conferring antibiotic resistance to specific bacteria.

56. D; Twisting locomotion requires flagella.

57. C; Cilia line the oviduct and transport the ovum towards the uterus after fertilization in the oviduct.

58. E; E represents the tropical rainforest.

59. B; B represents the tundra.

60. A; A represents deserts.

61. B; The tundra is characterized here.

62. D; The site of ATP synthesis is the mitochondria.

63. B; Proteins for export are produced and processed in the rough endoplasmic reticulum.

64. E; The Golgi apparatus attaches chemical nametags, in the form of oligosaccharides, to molecules here.

65. D; The mitochondria and nucleus contain DNA in eukaryotic animal cells.

66. C; Receptor-mediated endocytosis involves selective uptake of molecules into a cell by vacuole formation after the bind to specific receptor proteins in the plasma membrane.

67. E; Macrophages phagocytize extracellular particles by engulfing them.

68. D; The diffusion of water from areas of high water potential to lower water potential maintains turgor pressure by ensuring that the large central vacuole will remain full of water.

69. A; Large molecules that cannot pass through the phospholipid membrane or protein channels are secreted by exocytosis.

70. E; E represents the autotrophs that determine primary productivity.

71. D; Herbivores consume plants, which are located at level A.

72. A; The top carnivore, where biomass is the lowest, is easily driven to extinction.

73. E; Autotrophs are located at level E.

74. A; Polypeptides are polymers of many amino acids linked by peptide bonds, and they constitute hormones and immunoglobulins.

75. E; DNA and RNA are composed of nucleic acids including adenine, thymine, cytosine, and guanine as well as uracil.

76. D; Polysaccharides (starch in plants and glycogen in animals) are insoluble in water and fold to compact themselves.

77. A; Polypeptides are diverse in function, but certainly do constitute enzymes.

78. D; Amphibians require moist skin for respiration and water for egg laying and fertilization.

79. C; Fish utilize a swim bladder , a gas filled sac whose pressure can be altered to change buoyancy.

80. B; Birds maintain continuous airflow and do not have a diaphragm.

81. C; In fish, blood flows from the heart to the lungs to the body and back to the heart. There is no double loop to the heart as in organisms with 3- or 4-chambered hearts.

82. E; The amount of moisture is the independent variable, and the dependent variable is the number of pillbugs on each side of the choice chamber.

83. D; The pillbugs may first demonstrate kinesis, or random movements, but will over a short time demonstrate taxis, or a nonrandom movement towards or away from a specific stimulus.

84. B; All external conditions must remain the same to determine accurate results. Placing the light over either choice chamber could influence the behaviors of the isopods.

85. E; Random movements, moving more or less in response to a stimulus, occurs when the Paramecium attempts to locate the food in the environment.

86. C; The carrying capacity is located where the line attempts to level off. This is the point at which abiotic resources support the population but are not exploited.

87. B; Increased competition will cause a decrease in population size. The other choices will cause an increase in population size.

88. A; This is an example of the founder effect where an extremely small population, with a unique gene pool, proliferates.

89. C; Low population levels is the primary reason that the growth rate is so slow during the lag phase.

90. A; The opening of the stomata is dependent upon water and K^+ concentrations in the guard cells that surround the stomata. A proton pump drives H+ ions from the guard cells The opening of potassium voltage gated channels causes an uptake of potassium and water into the guard cells, making them more turgid due to the influx of water to the large central vacuole.

91. B; When the roots begin to sense a shortage in the soil, ABA is released. This raises the pH of the cytosol and causes the concentration of Ca^{2+} in the cytosol to increase. This decreases the ability of K^+ to enter the cells, and therefore decreases turgidity and closes the stomata.

92. A; An increased rate of photosynthesis also affects the amount of water than can be lost via the stomata. In the windy condition, the plants initially lost a great deal of water due to transpiration, however, signal transduction pathways in the plant after a short time caused a closing of the stomata to prevent this water loss.

93. B; After a prolonged period under extremely windy conditions, the stomata will close to conserve water, ceasing photosynthesis.

94. E; The carbon dioxide released is precipitated out by the KOH, and the resulting drop in pressure as O2 is consumed is measured by a drop in pressure in the vial.

95. C; The change in oxygen consumption is dependent upon the temperature.

96. A; Germinating peas have a higher water content, more cells, and a higher metabolic rate than nongerminating peas.

97. C; The birth rate is much higher in Kenya than in the United States. However, as indicated by the bars directly above the base, infant mortality is also higher in Kenya.

98. A; A lack of contraceptives increases the birth rate. Many factors, including poor health care, nutrition, and an adequate drinking water supply affect the mortality rate.

99. A; A small percentage of the population is between ages 0-4, therefore showing a decreased birth rate, as compared to Kenya where the base is broad.

100. E; The bars on the right and left in each diagram are similar, indicating that gender is not disparate in either population. The baby boom was in the United States, and the reproductive rate is much higher in Kenya than in the United States as evidenced by the larger bottom of the graph.

FRQ 1

PART A (MAX 6)

- Trophic level definition: total amount of energy present in each trophic level
- ~2% of light energy → chemical energy by photosynthesis
- 10% one trophic level → next trophic level
- Loss of energy from one trophic level to another
- Decrease in energy is due to:
  o waste and conversion of potential energy into kinetic energy
  o heat
  o produce new cells/remains fixed in that organism
- Accompanied by a decrease in biomass
- Number of food chains is limited
- Carnivores fix organic matter more efficiently than herbivores

PART B (MAX 6)

- Integral membrane proteins
- Photosystems: light absorption →
- On thylakoid membrane
- Light harvesting complexes: removes electrons → hydrolysis of water
- Cytochromes: chemiosmosis occurs
  o Requires lipid bilayer, proton pump, protons, and ATPase
  o Chemical energy pumps protons through the pump
  o Into thylakoid space/grana
  o Increases concentration gradient/electrical gradient of H+
  o Facilitated diffusion of protons via →
- ATP synthase: protons exit here
- ATP used in light independent reactions in stroma

FRQ 2

PART A (MAX 3)

- Axis correctly oriented and scaled
- Axis correctly labeled, title for graph
- Points plotted and lines drawn, points connected or line of best fit ok, lines must be labeled or key given to identify the two lines

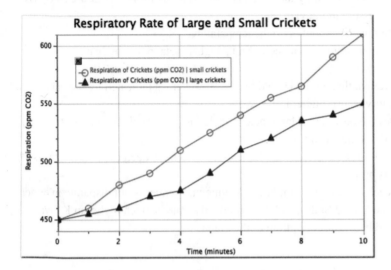

PART B (MAX 2)

Units must be given, work must be shown

- Large crickets = 55 ppm CO_2 / minute for 10 crickets
- Small crickets = 61 ppm CO_2 / minute for 10 crickets
    - Note – could also be expressed as 5.5 and 6.1 ppm CO_2 / minute for 1 cricket

PART C (MAX 2)

- Correct statement of results: Small crickets have a higher rate of respiration than large crickets.
- Explanation: Smaller organisms typically have higher rates of respiration.
- Mass-specific metabolic rate increases with decreasing body size.

PART D (MAX 4)

- glucose →cellular respiration
- fat→ glycerol and three fatty acids → glycolysis
- fats are an efficient form of stored energy
- amino acids → glycolysis or citric acid cycle
- deamination of amino groups → ammonia

FRQ 3

PART A (MAX 6)

- Heritable variation
  - Many traits, many alleles, etc.
- Production of more offspring than the environment can support
  - Limited food/water/physical space → Provides for competition
- Differential reproductive success
  - Acquire more resources and can devote more time to reproduction
- Well adapted population
  - Able to survive and reproduce

PART B (MAX 3)

- Definition: Heterozygote is favored over the two homozygotes.
- Maintains the recessive gene in the population → increases diversity
- Advantageous to maintain the recessive allele → advantage is present

PART C (MAX 4)

- Females invest more energy in producing offspring (gestation/lactation).
- Males are more available then females → competition
- Females choose males with ability to provide resources and advantageous alleles.
- Polygamy may occur.
- Males have a longer reproductive life than females.

FRQ 4

(MAX 3 EACH x 4)

Cell cycle controls

- Checkpoints: internal signal transmission
- Cyclins
- MPF activates CDK
- Apoptosis control
- No regulation → passes checkpoints without environmental conditions being right → cancer/mutation

Membrane permeability

- Phospholipid bilayer is selective → small, noncharged molecules
- Protein channels are specific (channel, carrier, enzymatic associated with permeability)
- No regulation → no ATP production, no nerve conduction, pathogen infection, cell rupture, dehydration

Menstrual cycle

- Follicular phase: FSH → estrogen
- Ovulation: LH
- Corpus luteum: LH → progesterone
- Menstruation: fibrinolysin prevents clotting
- No regulation: infertility, ectopic pregnancy, irregular menstruation, abnormal ovulation, spontaneous abortion, etc.

Stomatal opening and closing

- K^+ control to decrease/increase solute potential → drives osmosis
- Proton pump → establishes gradient
- Flavin pigment → signal transduction → proton pump
- Abscisic acid control → stomata close
- Sunlight → stomata open
- CO_2 concentration influences the stomata opening/closing
- No regulation: dehydration (increased transpiration), decreased photosynthesis

Carrying capacity of a population

- Density-independent controls: abiotic factors (can state any – drought, freeze, hurricane, flood, fire, etc.)
- Density-dependent controls: competition, predation, parasitism, availability of resources
- Reproductive strategies/rates
- Life expectancy
- Innate instability
- No regulation: exponential growth, lag phase, boom and bust cycle, extinction of species

Osmoregulation in vertebrate fishes

- Maintain particular ion concentrations in the blood
- Marine: drink seawater, passive loss of salt through gills, small amount of isotonic urine
- Freshwater: does not drink, large amounts of hypotonic urine with few salts
- No regulation: lack of skeletal, nervous and muscular system homeostasis

Correlation for Exam #2

Students and teachers should feel free to highlight the numbers of questions that were skipped and/or answered incorrectly. This will help to determine areas of strength and weakness in preparing for the AP Examination.

I.   **Molecules and Cells (25%)**
    A.   **Chemistry of Life (7%)**
        1.   Water 7, 34,
        2.   Organic molecules in organisms 1, 2, 3, 74, 76
        3.   Free energy changes
        4.   Enzymes 77
    B.   **Cells (10%)**
        1.   Prokaryotic and eukaryotic cells 6, 54, 57,
        2.   Membranes 66, 68, 69,
        3.   Subcellular organization 5, 56, 62, 63, 64, 65
        4.   Cell cycle and its regulation
    C.   **Cellular Energetics (8%)**
        1.   Coupled reactions 35, 36
        2.   Fermentation and cellular respiration 31, 95, 96, 97
        3.   Photosynthesis 25, 53

II. **Heredity and Evolution (25%)**
    A.   **Heredity (8%)**
        1.   Meiosis and gametogenesis 42, 52
        2.   Eukaryotic chromosomes 50,
        3.   Inheritance patterns 8, 23, 24, 51
    B.   **Molecular Genetics (9%)**
        1.   RNA and DNA structure and function 4, 75
        2.   Gene regulation 21, 22,
        3.   Mutation 45
        4.   Viral structure and replication 43, 46
        5.   Nucleic acid technology and applications 32, 55
    C.   **Evolutionary Biology (8%)**
        1.   Early evolution of life 44, 26, 48
        2.   Evidence for evolution 33, 47
        3.   Mechanisms of evolution 60, 61, 72, 49

III. **Organisms and Populations (50%)**
    A.   **Diversity of Organisms (8%)**
        1.   Evolutionary patterns 10
        2.   Survey of the diversity of life 11, 13, 40
        3.   Phylogenetic classification 78, 79, 80, 81
        4.   Evolutionary relationships
    B.   **Structure and Function of Plants and Animals (32%)**
        1.   Reproduction, growth, and development 9, 27, 37
        2.   Structural, physiological, and behavioral adaptations 12, 14, 16, 17, 18, 20, 39, 67, 38, 28, 29, 30
        3.   Response to the environment 15, 19, 82, 83, 84, 85, 90, 91, 92, 93, 94
    C.   **Ecology (10%)**
        1.   Population dynamics 41, 86, 87, 88, 89
        2.   Communities and ecosystems 58, 59, 70, 71, 73
        3.   Global issues 98, 99, 100